Japan's Empire of Birds

SOAS Studies in Modern and Contemporary Japan

SERIES EDITOR:
Christopher Gerteis (SOAS, University of London, UK)

EDITORIAL BOARD:
Stephen Dodd (SOAS, University of London, UK)
Andrew Gerstle (SOAS, University of London, UK)
Janet Hunter (London School of Economics, UK)
Barak Kushner (University of Cambridge, UK)
Helen Macnaughtan (SOAS, University of London, UK)
Aaron W Moore (University of Edinburgh, UK)
Timon Screech (SOAS, University of London, UK)
Naoko Shimazu (NUS-Yale College, Singapore)

Published in association with the Japan Research Centre at the School of Oriental and African Studies, University of London, UK.

SOAS Studies in Modern and Contemporary Japan features scholarly books on modern and contemporary Japan, showcasing new research monographs as well as translations of scholarship not previously available in English. Its goal is to ensure that current, high quality research on Japan, its history, politics, and culture, is made available to an English-speaking audience.

Published:
Women and Democracy in Cold War Japan, Jan Bardsley
Christianity and Imperialism in Modern Japan, Emily Anderson
The China Problem in Postwar Japan, Robert Hoppens
Media, Propaganda and Politics in 20th Century Japan, The Asahi Shimbun Company
(translated by Barak Kushner)
Contemporary Sino-Japanese Relations on Screen, Griseldis Kirsch
Debating Otaku in Contemporary Japan, edited by Patrick W. Galbraith, Thiam Huat Kam and Björn-Ole Kamm
Politics and Power in 20th-Century Japan, Mikuriya Takashi and Nakamura Takafusa
(translated by Timothy S. George)
Japanese Taiwan, edited by Andrew Morris

Japan's Postwar Military and Civil Society, Tomoyuki Sasaki
The History of Japanese Psychology, Brian J. McVeigh
Postwar Emigration to South America from Japan and the Ryukyu Islands, Pedro Iacobelli
The Uses of Literature in Modern Japan, Sari Kawana
Post-Fascist Japan, Laura Hein
Mass Media, Consumerism and National Identity in Postwar Japan, Martyn David Smith
Japan's Occupation of Java in the Second World War, Ethan Mark
Gathering for Tea in Modern Japan, Taka Oshikiri
Engineering Asia, Hiromi Mizuno, Aaron S. Moore and John DiMoia
Automobility and the City in Japan and Britain, c. 1955–1990, Simon Gunn and Susan Townsend
The Origins of Modern Japanese Bureaucracy, Yuichiro Shimizu (translated by Amin Ghadimi)
Kenkoku University and the Experience of Pan-Asianism, Yuka Hiruma Kishida
Overcoming Empire in Post-Imperial East Asia, Barak Kushner and Sherzod Muminov
Imperial Japan and Defeat in the Second World War, Peter Wetzler
Gender, Culture, and Disaster in Post-3.11 Japan, Mire Koikari
Empire and Constitution in Modern Japan, Junji Banno (translated by Arthur Stockwin)
A History of Economic Thought in Japan, Hiroshi Kawaguchi and Sumiyo Ishii (translated by Ayuko Tanaka and Tadashi Anno)
Haruki Murakami and the Search for Self-Therapy, Jonathan Dil
Japan's Empire of Birds, Annika A. Culver

Japan's Empire of Birds

Aristocrats, Anglo-Americans, and Transwar Ornithology

Annika A. Culver

BLOOMSBURY ACADEMIC
LONDON • NEW YORK • OXFORD • NEW DELHI • SYDNEY

BLOOMSBURY ACADEMIC
Bloomsbury Publishing Plc
50 Bedford Square, London, WC1B 3DP, UK
1385 Broadway, New York, NY 10018, USA
29 Earlsfort Terrace, Dublin 2, Ireland

BLOOMSBURY, BLOOMSBURY ACADEMIC and the Diana logo
are trademarks of Bloomsbury Publishing Plc

First published in Great Britain 2022
Paperback edition first published 2024

Copyright © Annika A. Culver, 2022

Annika A. Culver has asserted their right under the Copyright, Designs and Patents Act, 1988, to be identified as Author of this work.

Cover image: Ornithological Society of Japan, circa 1947. © Oliver L. Austin Photographic Collection, permission granted by curator Annika A. Culver

All rights reserved. No part of this publication may be reproduced or transmitted in any form or by any means, electronic or mechanical, including photocopying, recording, or any information storage or retrieval system, without prior permission in writing from the publishers.

Bloomsbury Publishing Plc does not have any control over, or responsibility for, any third-party websites referred to or in this book. All internet addresses given in this book were correct at the time of going to press. The author and publisher regret any inconvenience caused if addresses have changed or sites have ceased to exist, but can accept no responsibility for any such changes.

A catalogue record for this book is available from the British Library.

A catalog record for this book is available from the Library of Congress.

ISBN: HB: 978-1-3501-8493-0
       PB: 978-1-3501-8611-8
       ePDF: 978-1-3501-8494-7
       eBook: 978-1-3501-8495-4

Typeset by Integra Software Services Pvt. Ltd.

To find out more about our authors and books visit www.bloomsbury.com and sign up for our newsletters.

Dedicated to my parents.

Contents

List of Illustrations — x
Acknowledgments — xi

Introduction: Birds of a Feather Flock Together: *Japanese Aristocrats and the Cosmopolitan Science of Empire* — 1

1. The Practice of Ornithology: *Birds, Hunting, and Social Class in Prewar Japan and the Anglo-American World* — 17
2. Western Villas in Aristocratic Hands: *Spaces of Imperial Mimesis and Informal Scientific Exchange* — 43
3. Cambridge, UK (1925–9)—*From "Scandalous Marquis" to Explorer-Scientist: Japanese in Western Imperial Settings* — 65
4. The Philippines (1929–31)—*A Japanese Ornithologist Encounters the American Empire* — 81
5. Manchukuo and the Japanese Empire (1932–40)—*Deploying Avian Imperialism in the Media, Military, and Scientific Expeditions* — 109
6. Wartime Tokyo and Defeat (1937–45)—*Mobilizing Imperial Japan's Ornithologists and Birds for War* — 135
7. Tokyo under the Allied Occupation (1945–52)—*Yankees with a Mission amongst Threadbare Aristocrats* — 165
8. Tokyo and the United States (1940s–70s)—*Cold War Ornithological Collaborations between Japanese and American Scientists* — 195

Conclusion: Tokyo and Cambridge, UK (1960–Present), *Fledging Global Conservation Policies* — 213

Notes — 223
Works Cited — 284
Index — 295

Illustrations

0.1	Canon 35 mm Camera "Made in Occupied Japan" used by Austin, purchased in Tokyo in 1946 after theft of an American-made Kodak camera at the Pusan, Korea embarkation port	3
0.2	The revived *Chôgakkai* [Bird Society], with Oliver L. Austin at the center, early 1947	4
0.3	Requisitioned House #665, late 1949	6
1.1	March 1936, No. 32 *Shiseidô Gurafu* [Shiseido Graph] cover	26
3.1	Hachisuka Masauji's matriculation record (October 22, 1924)	69
4.1	Map of the "Routes in Mindanao of the Hachisuka Expeditions 1929 & 1930"	98
4.2	"At Lunch," a photo of the Mount Apo Expedition Team	100
4.3	"The Author at Lake Faggamb"	101
4.4	"Inscription on a Rock at the Summit of the Nameless Peak in the Apo Range"	102
5.1	Photograph of the expedition leader	118
5.2	September 22, 1933, photograph of expedition members	122
5.3	Map of Jehol [Rehe] and the route of the Tokunaga Expedition	124
6.1	Front cover of Prince Nobusuke Takatukasa [sic], *Japanese Birds*	145
6.2	Japanese blue flycatcher feeding a Chinese hawk-cuckoo	146
7.1	Oliver L. Austin Jr., circa 1947–1950	184
8.1	Kuroda Nagahisa, circa 1947–1950	199

Acknowledgments

The collective assistance of numerous individuals, organizations, and institutions in Japan, the United States, Germany, and Great Britain aided in this book's completion. Much of this project began fortuitously and percolated through anecdotes, slowly unfolding in a fascinating American expat's journey, which inspired me to investigate the Japanese scientists surrounding him.

In 2008, I first discovered the ornithologist and Allied Occupation official Oliver L. Austin Jr. through colorful stories by his granddaughter Valerie Austin. I thank her father Tony, now deceased, and the immediate Austin family, for generously sharing memories of Austin and his Japan sojourn. Exchanges with Timmy Austin painted a certain portrait of his father, while his older brother Tony expressed another opinion. In November 2013, Tony recounted a full oral history of his late 1940s Tokyo boyhood, conversational English lessons with Crown Prince Akihito, and Austin's role in postwar Japan. This led to donation of his father's collection of nearly 1,000 color slides of postwar Japan to Florida State University's (FSU) archives, now known as the Oliver L. Austin Photographic Collection, which myself and a team of students in the Undergraduate Research Opportunity Program digitized and archived online.[1]

Dr. Yoshi Nakada, Tony's next-door neighbor and childhood friend from Tokyo into his US career as a Bell Labs physicist and inventor, became a frequent correspondent and valuable information source in unpacking the lives and struggles of Japan's transwar ornithologists. He exemplifies how strongly international relations and national pride form integral aspects of Japanese scientists' narratives.

I also thank Professor Satô Yoichi, urban historian at Waseda University, for sharing founts of knowledge on postwar Tokyo's reconstruction during FSU visits in 2016 and late 2018, and during my summer 2018 Tokyo trip, when he introduced me to architectural engineer Yoneno Masahito and building maintenance engineer Komoto Takao at Tokyo Little House. Both Yoneno and Komoto compose a team of Japanese researchers, including Dr. Nakada, who scout exact locations (and names) for the Austin Collection's images to plot them on Google Maps. Along with Dr. Nakada, they continue to enlighten me about immediate postwar Japan as they share rare photographs, films, and music in our Facebook working group.

I am grateful to historian Barak Kushner for inviting me to lecture at the University of Cambridge in October 2014, which also permitted access to the university's library. Japanese Studies Librarian Koyama Noboru kindly shared Japanese aristocrats' personal characteristics and histories, and steered me toward resources on Hachisuka Masauji. Claire Welford-Elkin of the Rare Books Department discovered matriculation records (or lack thereof) and other documents pertaining to Cambridge's Japanese alumni.

I thank the D. Kim Foundation for Science and Technology for a 2015 grant enabling summer research.

I express gratitude to Osamu Honda, Director General, Japan Foundation, New York, and Junichi Chano, Managing Director, Japanese Studies and Intellectual Exchange Department, Tokyo Japan Foundation, for a semester-long research fellowship at Waseda University from August to December 2015. Hiraku Shimoda, historian of modern Japan in Waseda's Law Department, served as my academic host and ensured that I obtained necessary resources. Archivists for Minato Ward and the Tokyo American School also provided valuable research materials. Independent scholar Yoshie Osawa's moral support and source suggestions were invaluable during my Tokyo sojourn and beyond.

At the Yamashina Ornithological Institute, I am grateful for generous assistance from Natural History Curator Dr. Tsurumi Miyako. She also introduced me to Odaya Yoshiya, the Abiko Bird Museum's curator who invited me for band-banding during a late November migratory bird study. The late professor Nakamura Tsukasa kindly recounted postwar memories of his mentor Hachisuka and other ornithologists. I also thank Masako Hachisuka for relating memories of her father and Dr. Hasegawa Michiko for sharing family stories.

In summer 2014, FSU funded early work on this project with a Committee on Faculty Research Support grant. Frederick (Fritz) Davis, now at Purdue University, and Ron Doel offered helpful suggestions as historians of science and the environment. Chuck Upchurch provided useful advice on British masculinity, gender, and sexuality. Robinson A. Herrera read innumerable drafts and suggested theoretical texts. FSU humanities librarian Adam Beauchamp ferreted out illusive sources during the notorious 2020 coronavirus lockdowns, while alumnus Elias Larralde transformed footnotes into a bibliography.

Conversations and presentations with colleagues in the field much enriched this book. I thank Daqing Yang, Frederico Marcon, and Miriam Kingsberg for their comments. The March 2018 workshop "Epistemic Breakdowns and Crises in East Asia" at Princeton University helped me articulate Chapters 1 and 2, while remarks by Benjamin Elman, Frederico, Miriam, Susan Burns, Constance Cook, and Joan Judge urged me to critically examine these scientists' lives and preoccupations.

Any academic study referencing hunting is much enhanced by the author's experience of the sport. Hence, I engaged in experiential ethnography as a participant observer. This included forays into skinning with a pair of doves shot by Tulane University professor Felipe Cruz, which he brought to a party after an FSU lecture. Since specimen collecting by ornithologists involved techniques resembling upland hunting, I discovered resources near Thomasville, Georgia, its southeastern American epicenter. I thank former quail plantation manager George Fuller for teaching me shotgun fundamentals for bird hunting at Horseshoe Plantation on Tallahassee's outskirts in November 2016. I experienced a vintage Browning Auto-5, a shotgun preferred by late 1940s American ornithologists, and progressed to AK74 and SKS rifles as bonuses. Amidst pronounced shifts in US politics at the American Century's eclipse, our encounters generated animated conversations on quail hunting's broader social histories and Southern quail plantation management, which I captured in jaw-dropping oral history interviews aided by my graduate student assistant Zachary Reddick in summer 2017. That fall I obtained my hunter safety license and practiced with rifles and shotguns at Oldfields Plantation in Monticello, Florida, for my National Rifle Association online hunter safety course's experiential portion. These settings allowed me to observe the new American aristocracy's current hunting grounds while encountering politics surrounding gun ownership in contemporary Trump-era America.

In summer 2019, a visit to maternal relatives in northern Germany while writing early chapters revealed that bird-besottedness is a family affliction. My cousin Claudius Böttcher patiently answered questions about hunting and bird-keeping at my late uncle's 1930s-era hunting lodge in Schrum, where he kept racing pigeon aviaries, bred prized roosters, and ate less-spectacular flock members for dinner. My uncle's brother Uwe, former German consul to the Dominican Republic, once owned neighboring land. In 1978, with his partner, he founded Palmitos Park in Maspalomas on Gran Canaria, Spain, which we mockingly referred to as his "bird park."[2] I now respect his sound business acumen in an investment profiting from robust German tourism to the Canary Islands. Like Delacour, Uwe stocked his park with rare animals, protected bird life, and cultivated rare orchids, and like Marquis Takatsukasa, particularly enjoyed *Psittasidae* [Parrots]. My cousin Thomas Böttcher, a former trainer of prize-winning German Shorthaired Pointers, shared family stories, and enlivened my memories as I began writing, while comfortably in exile from Floridian heat near the Danish border.

Lastly, I thank my children for a semester-long move to Japan and numerous travels abroad to archives and universities. My parents and older sister provided valuable childcare and mentoring when I was afield. Amid the book's conception, another child was born. Throughout this project, RH taught me about Florida wildlife while offering proper critique and intellectual stimulation to continue production at a precarious historical time.

Introduction: Birds of a Feather Flock Together: *Japanese Aristocrats and the Cosmopolitan Science of Empire*

In summer 2020, to raise awareness for Black Birders Week and Black Lives Matter (BLM), while prevented from in-person lab research due to pandemic-related precautions, Scott V. Edwards, Alexander Agassiz Professor and Curator of Ornithology at Harvard University's Museum of Comparative Zoology, began riding his bike in a bold solo traversal of the United States. On June 6, Edwards ventured from Newbury Port, Massachusetts, on the Atlantic Ocean, and planned to reach the Pacific Ocean weeks later, after riding 50 to 60 miles daily on a bicycle festooned with political signs, and pitching his tent overnight in local campgrounds. A July 31 *Harvard Gazette* article indicated Edwards' reasons for his meaningful trip, "one he'd longed to make for years":

> as a Black birder and a Black scientist, he decided he also wanted to raise awareness of the movements and give his trip a larger purpose as racial tensions across the country continued to escalate. He added the signs a few days after setting out. "It's important for folks to see that African Americans do enjoy nature. It's important to showcase that we like camping and show it's not just the domain of white people."[1]

The scientist documented his journey on Twitter, describing spaces he navigated, including Black Hills "Indian Country," the Crow Reservation, and Nez Perce lands, while meeting new friends and old colleagues, and garnering thousands of social media followers publicizing his message far beyond the American ornithological world.[2] He explained "[i]t's all about role models" and asserted that "[p]eople need to see people like themselves out there ... they need to be told that they belong in science and that there's a place for them."[3] Only fifty years ago, such an expedition would have endangered a Black man, carried out alone and crossing areas dominated by conceptions of white space and accompanying ideas of political supremacy. In some locations, Edwards changed his BLM signs to more neutral ones publicizing birding or the Cornell Ornithology Lab—highlighting how much ornithology intersected with politics.

As Edwards's remarkable journey, tumultuous 2020 events, and ornithology's earlier histories reveal, the study of birds is deeply political and exposes complex notions of space and who possesses rights to occupy it. Birding thus becomes a metaphor for spatial transgressions and quasi-imperial acts of claiming privilege, which in the United States harbor a deep colonial legacy intertwined within longer histories of settler-colonialism expelling and ethnically cleansing Indigenous peoples while denying Blacks rights to that space. Other European countries and Japan enacted similar imperialist legacies where colonial subalterns made way for white or honorary white settlers (like Japanese colonists) or ceded their lands. Then, as now, non-white ornithologists hunting birds most likely have traversed or will traverse areas dominated by white men—including former and current military bases or reservations often hosting diverse bird life.[4]

Movement through such spaces often includes struggles for who possesses rights to assert their names in intellectual spaces.[5] Intriguingly, these same battles for legitimacy were experienced by transwar Japanese ornithologists, despite their benefiting from suppositions (or carefully curated performances) of honorary whiteness and imperial masculinity. The following narratives describe endeavors of particular Japanese scientists carving out spaces for themselves amidst a broader global history of science where white men dominated. These figures are viewed through an ornithological lens to discover how the politics of class, race, gender, and place filtered through a language of birds expressed in Japanese and English amidst an Anglo-American imperial mindset. Here, imperial masculinities were enacted in drawing rooms, laboratories, and the field, in high-stakes performances fraught with international intrigue.[6] Many aspects remained unarticulated by participants, despite existence of their scientific writings and letters, and thus, must be gleaned as anthropologist Ann Laura Stoler urges historians to humbly "explore the grain with care and read along it first" before one can "read against the grain."[7] Scholar Vicki L. Ruiz once commented upon the historian's craft, where she described creating engaging historical narratives as "creative non-fiction," since one cannot ascertain what research subjects actually thought, felt, or believed beyond diverse documents and records of utterances—often subjected to revision or bias.[8]

Memory is notoriously unreliable, formed from malleable processes of informed-forgetting where narratives are crafted from debris amidst the overwhelming march of existence and time's relentlessness. The memoirs of French ornithologist Jean Théodore Delacour (1890–1985), who interacted closely with the Japanese scientists examined, often misremembered dates and event sequences, while Japanese accounts understood similar persons or experiences in differing ways. Anthropologist Marc Augé calls for historians to acknowledge that memory forms from selective oblivion, where "memories are crafted by oblivion as the outlines of the shore are created by the sea"[9] and "oblivion is the life force of memory and remembrance its product."[10] Rather than experiencing frustration, historians should seek to understand why certain things are forgotten, while others are remembered, in a careful gleaning where Augé posits, "tell me what you forget and I will tell you who you are."[11]

Snapshots of Enduring Imperial Encounters

Held carefully for decades in the hands of family, loved ones, or other trusted individuals, personal collections often match formal and systematized archives. Such was the case for an unassuming black plastic box perched incongruously on my desk

Introduction: Birds of a Feather Flock Together

Figure 0.1 Canon 35 mm Camera "Made in Occupied Japan" used by Austin, purchased in Tokyo in 1946 after theft of an American-made Kodak camera at the Pusan, Korea embarkation port. © Photo taken on September 19, 2016, permission granted by Dr. Yoshi Nakada.

hiding yellowing slides emitting a slightly vinegary smell—a seemingly unlikely treasure trove. When I opened the cover to marvel at its contents, the dusty container revealed nearly a thousand rare Kodachrome slides taken between 1946 and 1950 with a Canon 35 mm camera labeled "Made in Occupied Japan."[12]

Brought to my office by Tony Austin, a retired commercial fisherman, these slides once belonged to his father Oliver L. Austin Jr. (1903–88), a US scientist devoting much of his professional career to studying birds, who last worked as Curator of Ornithology at the University of Florida's Museum of Natural History. In postwar Japan, the Yankee ornithologist engaged in amateur photography while serving the US-led Occupation (1945–52) government in the late 1940s as "scientific consultant" and head of the Wildlife Branch, Fisheries Division. As a scientist, Austin clearly acted as an agent of empire in close friendships with Japanese elites boasting important political connections; notably, a keen passion for birds united them. Snapshots of a rapidly transforming society arising from the war's ashes offer a rare glimpse into US-Japan relations on an intimate yet elite level, from the perspectives of science, nation, class, race, and enduring imperial masculinities.

At first glance, the images resemble a bureaucrat's privileged outsider view of a defeated imperial Japan's scions amidst global postwar American ascendancy.[13] However, two slides particularly caught my attention and led me to wonder about the professional experiences and personal lives of the Japanese men depicted. Framed by the sculpturesque stone ruins of a former prince's burnt-out villa, these proud Japanese scientists in threadbare English suits positioned themselves as Western gentlemen of science in two differing group photographs of the revived *Chôgakkai* [Ornithological Society]. In an informal photo featured on this book's cover, the men, along with one woman, a secretary, are caught unawares while grouping together; some glance downwards or aside while folding or clasping their hands, as others clench hands near their pockets in stances unwittingly expressing discomfort. Only some smile or directly glance toward the camera.

Yet, a second highly staged photo positions Austin at the front and center amidst Japanese ornithologists, of whom many were aristocrats, in an odd resemblance to the famous 1850s shot of Dutch political advisor Guido Verbeck (1830–98) (and his son) amidst restorationist "men of purpose" in the late Tokugawa (1603–1868) era. Austin obviously took the first photo as a candid shot, while in the latter formal one shot by his wife or servant, he placed himself in the middle to denote his importance. The snapshots encapsulate how these aristocrats briefly ceded power in temporary obeisance to the conqueror's representatives to further their own agendas and restart scientific endeavors—sometimes as US Occupation employees. Clearly, the 1947 image framed by ruins and bare trees intended to show the Japanese *Ornithological Society* and dozens of members as a fully resurrected scientific organization.

Such intriguing amateur photographs provide an instant medium for capturing a certain moment with a contemporaneous immediacy and thus are imbued with immediate historicity. Robert Rosenstone, a visual media scholar who examined

Figure 0.2 The revived *Chôgakkai* [Bird Society], with Oliver L. Austin at the center, early 1947. © Oliver L. Austin Photographic Collection, permission granted by Annika A. Culver, curator. <https://austin.as.fsu.edu/files/original/d1ae1af92c600a928e751920cde2b150.jpg>

Americans encountering Meiji Japan (1868–1912),[14] cautions in earlier theoretical essays that historians should treat images as constructed narratives resembling text: "visual narratives will be 'visual fictions'—that is, not mirrors of the past but representations of it."[15] Like film, photographs must be read critically as exemplars of time and place, often embedded within specific historical narratives. Thus, Hayden White's call for historians to view images as reflecting their own discourse holds especially true.[16] For the American scientist, creation of a co-extant affective space with his collaborators amidst ruined or reconstructing surroundings in a postwar environment motivated such representations where space, place, race, and masculinity unwittingly appeared in stark relief.

One can aptly apply Raymond Williams's concept of a "structure of feeling"[17] to affective spaces for these Japanese aristocrats and to Japan's broader vanquished population, as they coped with the American-led occupation as subalterns.[18] Japanese ornithologists, appearing in group photos as *Ornithological Society* members in chastened positions as defeated elites, clearly became subalterns in collaborating with Occupation authorities to recapture narratives of their scientific work. While the Allied Occupation functioned somewhat differently than complete colonization processes, the defeated Japanese were still placed in positions of secondary status; thus, subaltern theories prove useful in decoding photography taken by Occupation officials. Not only can one discern individuals positioned as subaltern (as well as their race, gender, and often, class), but subalterns also redefined dominant colonial narratives in their portrayal.[19]

Such images subtly evince how Austin's aristocratic Japanese associates defined themselves amidst the American "colonizer" while performing whiteness in crossing boundaries of social space.[20] Communications studies scholars Thomas K. Nakayama and Robert L. Krizek assert that whiteness is a rhetorical strategy, one certainly adopted by these scientists in both imperial and postimperial contexts.[21] In postwar Japan under foreign occupation, they revealed how playing "white" meant placing themselves in proximity to American military-led power centers, embracing democratization, and assuming Anglo-Saxon inspired postwar values loosely defined by a victor's sense of moralizing, while adopting his manner and outward appearance through "enclothed cognition," where a person's dress positively informed their behavior in becoming what they wore.[22]

During the Allied Occupation, Japanese elites intently reclaimed their positions by temporarily performing whiteness and inserted themselves into images with Austin through proximity to power in two degrees of separation from Japan's new American Shogun, the Military Governor General MacArthur (1880–1964).[23] Before the war, these former imperialists engaged in similar processes of empire-building by participating in "collecting imperialism" alongside defined political roles.[24] Yet, adopting such physical and social trappings of whiteness took root earlier in the 1920s, when Japanese scientists travelled throughout Japan's empire and Anglo-American world and positioned themselves as quasi-equals to white counterparts—especially in the ornithological field. Since Great Britain and the United States emerged as dominant nations training and supporting ornithologists at the time, it was crucial for Japanese bird researchers to establish transnational connections to scientists and research institutions in these two imperial powers.

The histories surrounding Japanese scientists in the Austin photographs, and more importantly, the transwar encounters that they depict, form the core of my study and resonate with feminist geographer Doreen Massey's sense of space and place.[25] Her theory of "power geometries"[26] indicates a relational approach to these concepts where she urges scholars "to think of the spatial in terms of social relations."[27] This holds true for personal histories of the Japanese ornithologists depicted, many of whom were also aristocrats with political power and cultural cachet. Anthropologist Takie Sugiyama Lebra's study reveals the enduring "status culture" of Japan's aristocracy deep into the postwar.[28] Arising as a separate class distinction amidst Meiji-Era reforms, the relatively short-lived status of *kazoku* ["aristocrat"] was dissolved in 1947 by General Headquarters (GHQ). Nonetheless, these highly Westernized cosmopolitan elites thrived on interactions with Anglo-American foreigners and Westerners, boasting of intergenerational study abroad. Many knew their parents as "Mummy" and "Daddy," and took afternoon black tea with cream and sugar. Numerous Japanese aristocrats lived in Tokyo's Shibuya Ward in Occidentalist surroundings of spacious, well-appointed rooms of plush furniture suffused by the colorful glow of stained glass, reminiscent of church windows, and Tiffany lamps. Oriental rugs lined their shod feet, and servants clad in Western clothing attended to every need—in ministrations resembling Austin's postwar experience surrounded by multiple household personnel. However, with arrival of the American victors, Japan's defeated aristocrats had much to resent in forced requisition of homes and land redistribution.

Figure 0.3 Requisitioned House #665, late 1949. © Oliver L. Austin Photographic Collection, permission granted by Annika A. Culver, curator. <https://austin.as.fsu.edu/files/original/0b57e997e5e28c4777cefb67ecd081f5.jpg>

In exchange, the Occupation overlooked much of imperial Japan's dark wartime past to rebuild an amicable new alliance. Austin welcomed cooperation of zoologists like Cambridge-trained Marquis Hachisuka Masauji (1903–53), Duke Takatsukasa Nobusuke (1890–1959), and former Prince Yamashina Yoshimaro (1900–89), earlier employed by the wartime *Shigen kagaku kenkyû-jo* [Research Institute for Natural Resources], along with Kuroda Nagahisa (1916–2009), who once researched for Japan's military in Tokyo. Though links are unsubstantiated, sources suggest that the Institute's zoological research on rodents likely supported interests of the Manchukuo-based biological weapons facility Unit 731.[29] Obfuscating wartime work as ornithologists and zoologists, of which Allied air raids had burnt many records, Hachisuka, Takatsukasa, Yamashina, Kuroda, and others now focused on helping Austin with postwar scientific contributions.[30] However, private resistance reappears in anecdotes persisting over seventy years later, such as Takatsukasa's criticism of Austin's allegedly imprecise bird skinning techniques[31] and Kuroda's 2004 privately published revision of Austin's 1953 study "Birds of Japan."[32] Despite trepidation and pride, initial cooperation with the American conquerors was key to reviving scholarly contributions emerging from a longer history of Anglo-American transnational connections and exchanges developing in the 1920s.

The Japanese ornithologists' careers traversed the transwar period, a term spanning prewar, wartime, and postwar eras. Historian Nakamura Masanori understands transwar as "persistence," especially regarding the emperor system and the imperial institution's ambiguous postwar reshaping.[33] My use resembles Andrew Gordon's broader perspective, who perceives similarities in 1920s and 1950s economic and political issues within Japanese social and labor history.[34] However, Nakamura's interpretation also relates to the social hierarchies and Anglo-American knowledge networks traversed by aristocratic Japanese scientists. This appears starkly in performances of imperial masculinity, which arose out of the scientists' intricately constructed masculine identities while working as non-white (or honorary white) ornithologists in contested imperial and postimperial spaces controlled by Japanese or Anglo-Americans.

Imperial Masculinity: Gendered Performances and Precarious Subaltern Manhood amongst Japanese Scientists in the Anglo-American World

According to British literature scholar Praseeda Gopinath, within imperial contexts, masculine identities were highly mutable and in continuous need of affirmation: "like imperial-national power, imperial masculinities are labile and in constant need of shoring up."[35] By nature precarious, British manhood was constantly sought and defined to justify imperial rule over subaltern peoples, like Indians under the British Raj (1858–1947), alternately evoking negations of Britishness or (unreachable) exemplars for Indian men. This also pertained to Japanese identity amongst men in imperial Japan's first colony, Taiwan (under Japanese rule from 1894 to 1945), where tensions arose between assimilation (*dôka*) and imperialization (*kôminka*) for both Indigenous peoples and settler Chinese, who never quite became equals to colonizing Japanese.[36] Gopinath reveals how imperial masculinity is integral to imperial formation and composes a dynamic process.[37]

Yet how and where were these imperial masculinities created? British sports historian J. A. Mangan proposes that initial production (and reproduction) of imperial masculinity occurred via inculcation in elite public schools and playing fields—notably on the cricket field—throughout the British Empire, and in colonies for non-white subalterns, like India or the West Indies. Mangan notes that "From approximately 1850, imperial masculinity was methodically 'manufactured' by means of a cultural 'conveyor-belt' ... By this means minds were to be moulded, attitudes were to be constructed and bodies were to be shaped for manhood."[38] In Japanese and American examples, cricket fields were replaced by baseball fields and "private" schools and universities. Yet, these were not the only fields where imperial masculinities were played out, especially for Japanese aristocrats. Fields of study, exploration, and geographical scope all became spaces for making one's name and claiming one's place as an authority.

However, before the following chapters investigate such spaces, it is useful to examine masculinity in a Japanese context as crucial to the ornithologists' imperial (and postimperial) endeavors and, in particular, as constructed in their performance of scientific activities. Here, I rely upon Joan Scott's interpretation that gender is useful as a historical category of analysis,[39] and authenticated through performance, as argued by Judith Butler.[40] In addition, Gopinath's understanding of imperial masculinity supports psychologist Joseph Vandello's theory of "precarious masculinity," where manhood is constantly at risk and requires re-validation.[41] Imperial masculinity is also more convincingly performed when coupled with "enclothed cognition," where a person's clothing informs behavior in becoming what they wear.[42] These concepts serve as integral aspects of these scientists' expressions of imperial masculinity while engaging in ornithological endeavors through exploration in colonial areas, proactive seeking out of Anglo-American networks, and strong conservation advocacy.

Though Butler's studies avoid singling out femininity or masculinity, and concentrate on how the body can (or is made to) perform gender, especially for those gendered as women, her assertions of gender as performative and socially constructed are supported by contemporary psychological studies on masculinity as a gendered construct. In 2008, psychologists Joseph A. Vandello, Jennifer K. Bosson, Dov Cohen, Rochelle M. Burnaford, and Jonathan R. Weaver conducted studies with American undergraduates to examine the common cross-cultural phenomenon of "precarious manhood," where they discovered masculinity is so socially constructed that it even surpasses physical markers deemed male.[43] They concluded that American society views "manhood as elusive and tenuous" which "requires social proof,"[44] and initially posited that "actions that should most effectively prove manhood are those that are public and exclusive .... actions that involve a degree of danger or risk, and those that display physical toughness, should be appealing demonstrations of manhood."[45] Their studies relied on impressions of young adult American students, so these conclusions may also relate to generational perceptions. Nevertheless, Vandello and his colleagues' findings are valuable in discovering scientifically measurable bases for ideas on gender.⁴⁶ In conclusion, Vandello and his team elucidated "precarious manhood" where "men experience their gender as a tenuous status that they may at any time lose and about which they readily experience anxiety and threat."⁴⁷ In a recent 2019 *The Atlantic* interview, Vandello refined this term as "precarious masculinity," defined as "the sense of insecurity that a lot of men feel about their status as masculine."⁴⁸ Precarious masculinity closely aligns with humanities gender theorists like Butler and resonates with my work on Japanese scientists embodying imperial masculinity.

For much of Japanese history, gender has seemingly been performative, with a particular set of behaviors, along with specific language and speech patterns expected for individuals embodying roles of men or women of certain ages, status ranks, or professions.⁴⁹ This includes conventions of dress, and what one might consider enclothed cognition, where certain social roles or professions required proper clothing to acquire authenticity. An important aspect of performances of gender, and masculinity in general, in Japan and the Anglo-American world, was adoption of proper clothing, which reinforced embodiment of a particular role, like the specific Western adage, "Clothes make the man," or in Japanese, *mago ni mo ishô* ["Even a (lowly) pack driver can look fine with proper comportment and dress"]. For premodern Japanese, gender was not necessarily a biological given, but could be assumed with clothing, actions, and speech.⁵⁰

However, these more fluid gender norms began to change with state imposition of outside legal structures in the late nineteenth century, which attempted to codify gender roles. In imperial Japan, Meiji reformers intended for the 1898 Meiji Civil Code to bring the empire in line with heteronormative Anglo-American social norms and enacted compulsory heterosexuality for Japanese subjects where households were now legally determined as under a married man's patriarchal authority over his wife and children, even though many households were still multi-generational, with concubinage and sexual relationships between men as common.⁵¹ Historians Barbara Maloney and Kathleen Uno assert that, in Japan's modern era, tensions existed between actual personal practice and state dictates as "individuals have constructed and experienced complex, fluid identities, while social institutions and the state have attempted to craft and enforce unitary constructions of gender."⁵² Expressions of imperial masculinity often intersected with political identities. Tensions between personal desires and broader expectations for Japanese aristocrats, also political elites, are especially revealed in Hachisuka's 1929 explorations in the Philippines after his failed engagement to an imperial princess, Yamashina's 1920s quitting of an expected military career to focus on ornithology, and Kuroda's postwar work for a former American enemy after serving the imperial Japanese Army during wartime.

As I asserted earlier, photographs perhaps best reveal historical narratives, including those surrounding imperial masculinity, through the visual narratives they evince. Two 1920s-era images of Marquis Hachisuka evocatively capture these characteristics. One depicts him in Western-style khaki dress and jodhpurs with twin ammo cartridges crossing his chest while holding a scoped rifle, possibly a Holland and Holland deluxe from a manufacturer of highly regarded firearms commonly used to shoot big game in Africa, while another displays him in his plane wearing full aviator's gear complete with leather gloves. These two images clearly reveal the effects of enclothed cognition upon Hachisuka's embodiment of the twin tropes of cosmopolitan imperial scientist and explorer-hunter as key components of prewar imperial masculinities for Japanese ornithologists.

Additionally, the iconic photograph with Austin in the mid-forefront, evoking earlier images of "foreign hired hands" [*oyatoi-gaikokujin*] surrounded by pupils during Meiji times, exposes further aspects of imperial masculinity following defeat by a wartime rival. The Japanese men, excepting Hachisuka's secretary and alleged mistress Usui Mitsue (unknown) as the sole woman in a dark skirt and red sweater, all wear threadbare yet formal Western suits in subdued, dark colors, amidst ruins of Yamashina's Western-style estate, surrounding an American official from the war's victorious side. Besides in photographs from a historical period and place, how can one *see* gender as a historian? And, how might this fit into social configurations, especially amongst men of a certain class, race, and nation?

In the following chapters, I argue that in the prewar period, imperial masculinity was a crucial trope for the Japanese scientists that I examine and strongly motivated the majority of their endeavors as non-white actors in an Anglo-American world privileging graduates of certain institutions and members of select organizations. Some, like Hachisuka and Takatsukasa, successfully penetrated this world, which during the Pacific War (1941–5) distanced itself from and even rejected them, but several years afterwards they rejoined the fold amidst new power configurations, notably the US-Japan Alliance denoted by the 1951 Treaty of San Francisco. Their forms of manhood expressed as explorers, hunters, collectors, military scientists, or conservationists throughout the transwar period were incredibly performative and involved careful calibrations of behavior while moving through diverse spaces ranging from imperial colonies and scientific research institutes to Western-furnished drawing rooms. Clearly, imperial and postimperial masculinities are not the only lenses through which to examine these men. Yet, they provide an important window into their private and public lives as scientists attempting to position themselves in a world marked by shifting centers of power where non-white actors were still relegated to subaltern status even amidst the postwar detritus of empire.

The Book's Focus

Japan's Empire of Birds examines the public scientific endeavors and private lives of a group of mid-century Japanese scientific elites circulating in the Japanese empire and Anglo-American world. The book interrogates questions of social space illuminating national identity, gender, race, and social status during the transwar period from the 1920s until 1950s, including the interwar, wartime, Allied Occupation, and early Cold War periods. The book follows a peripatetic cast of characters, mostly aristocratic Japanese, who migrate within imperial Japan's colonies and domestic drawing rooms to satiate a common fascination for birds in the name of science, and in the process, they redefine the politics of engaging in scientific work as non-white male actors. After Japan's defeat and transcending the Allied Occupation, they again collaborate with Anglo-American counterparts in new political and social contexts. In this cultural history of science, the book explores how ornithology becomes an obsession for these zoologists who created knowledge networks throughout the empire and beyond. It explores the language of science and ornithology in a defeated empire's rebuilding of relations with a former enemy, while revealing the enduring transwar social hierarchies of interactions persisting into the postwar period.

In the Japanese prewar scientific community, I focus on zoologists, broadly defined, because of strong connections to the imperial family and ruling elites, with proximity to political power centers. Associations like the *Ornithological Society* (est. 1912) and *Yachô no kai* [Wild Bird Society (est. 1934)] formalized their circles, while they published research findings in English and Japanese. As a formal ornithological organization based in Tokyo, imperial Japan's capital, the *Ornithological Society* published its own internationally recognized journal, *Tori* [Birds], and enabled members to exchange scholarly work. In contrast, the *Wild Bird Society* enjoyed a more informal membership, with many literary figures and amateur bird lovers attending meetings and bird-watching hikes in the Tokyo area. Outside of these organizations, these scientists exerted lasting political impact on imperial Japan and postwar democratization due to their House of Peers memberships. Fashioning themselves into gentlemen of science or explorer scientists with laboratories on lavish estates, they initially served the state as imperial agents, and after the war, furthered democratic internationalism through conservation. By the 1920s, the formerly dilettantish study of ornithology by aristocrats became an important epistemological field with its own linguistic paradigms. Until the Pacific War, Hachisuka, Takatsukasa, Kuroda Nagamichi (1889–1978), Kuroda Nagahisa, and Yamashina moved freely between Tokyo and England, the Philippines, Manchuria, and China for study, scientific exchange, and exploration. During the war, some scientists worked for the *Research Institute for Natural Resources* on projects like finding new sources of protein for the Japanese population, while others, like Uchida Seinosuke (1884–1975), studied parasites in mammals and birds, possibly intersecting with research on biological warfare.

In the postwar, to distance these scientists from controversy, ornithology served to symbolize the nation's new democratic ideals and allowed close access to American military elites to broker influence and decide the fallen empire's fate. Imperial Japan's August 1945 defeat in the Asia-Pacific War and corresponding dissolution of its empire led to a stark epistemological breakdown for these Japanese scientists whose aims were deeply enmeshed within the state's imperialistic goals. Their subsequent encounter with a conquering force occupying their nation in surrender's wake forced them to rethink their scientific roles by now directly dealing with the American overlords. Fraught by radically different conceptions of social space superimposed upon occupied landscapes in Tokyo and beyond as the Allies requisitioned estates and redistributed lands of the former Japanese aristocracy and elites, these men attempted to both charm and hold at bay American counterparts by interacting with them as uneasy "benefactors" rather than wartime enemies.

After the 1950s, renewed connections with the British also led to Japanese becoming active forces in the Cambridge-based International Council for Bird Preservation (ICBP), like Yamashina and Hachisuka's disciple Nakamura Tsukasa (1926–2018). Billed as "the world's oldest international conservation organisation," the ICBP was founded in London in 1922 by an international group including French ornithologist Jean Théodore Delacour (1890–1985), president of the *Ligue Pour la Protection des Oiseaux* [League for the Protection of Birds] (LPO); Frank E. Lemon, honorary secretary of the Royal Society for the Protection of Birds (RSPB); American conservationist Thomas Gilbert Pearson (1873–1943), president of the National Audubon Societies; and Dutch financier and conservationist P.G. Van Tienhoven (1875–1953), and Dr. A. Burdet of the Netherlands.[53] This organization, now known as Bird Life International since 1993, boasts Princess Takamado (b. 1953) as honorary president since 2004 and remains important through worldwide bird conservation and protection efforts.

Japan's Empire of Birds seeks to answer several key questions. How does a language of science develop amidst an atmosphere where zoological and biological endeavors supported a militaristic state's aims during wartime, but with defeat, were rebranded as "peaceful"? How did these scientists reintegrate themselves into Western international communities, where they had once circulated freely until the late 1930s? I argue that transwar imperial continuities, where Japanese and Anglo-American scientists enjoyed embedded networks of knowledge collaboration and competition preexisting the wartime conflict, allowed for a rapid reintegration of the Japanese after defeat when they adopted a new language of scientific research decoupled from empire to reestablish earlier paradigms of peaceful internationalism through global wildlife conservation, genetic research, and bird migration studies. This book examines the 1920s beginnings of such interactions and their evolution from wartime into the postwar period.

Scholarly Contributions

Japan's Empire of Birds fosters deeper understandings of imperial and postwar Japanese contributions to the transnational history of science, while shedding light on the social and cultural histories of an elite group of cosmopolitan Japanese scientists concentrating on ornithology and operating in arenas alongside their Western peers, whether from the United States or Europe. Centered on social class and expressions of masculinity in endeavors ranging from specimen hunting expeditions to intellectual sparring in aristocratic drawing rooms, my book also investigates their public and private performances of gender and nationhood.[54] I therefore examine the men's lives and hybrid milieu amidst what Barbara Fuchs calls "cultural mimesis,"[55] where they fashion themselves after Anglo-American gentlemen of science and "honorary whites,"[56] while focusing their scientific passions upon birds. The study of zoology, and particularly ornithology, fostered transnational encounters transcending borders and political barriers. Thus, individual chapters focus on a particular scientist or group and the geographical areas where they lived and worked, arranged roughly chronologically.

This practice of transnationality was also crucial to the Japanese ornithologists' imperial masculinity; Marquis Hachisuka flew his plane across the Sahara to hunt lions with his Holland and Holland rifle, while Prince Yamashina cultivated his Shibuya-based specimen laboratory, exchanging bird skins and knowledge with American and British counterparts. These chapters therefore critically examine the role of gender within Japanese scientists' performances of masculinity in Japanese and Western contexts, while I also investigate social class, race, and place for these politically influential scientists in their work-related peregrinations—to Tokyo and England, the Philippines, Manchuria, and China for study, scientific exchange, and exploration. *Japan's Empire of Birds* especially delves into the understudied interwar and wartime periods from the 1920s into 1940s, and continues into the postwar period to highlight trans-imperial continuities for interconnecting Anglo-American knowledge networks absorbing these Japanese ornithologists.

Now classic historical studies on postwar Japan, such John Dower's *Embracing Defeat* (1999)[57] and Naoko Shibusawa's *America's Geisha Ally* (2010),[58] focused on the Allied Occupation's cultural aspects and its effects upon a defeated Japanese population that briefly courted its conquerors. Yet, more recently, historians Morris Low and Hiromi Mizuno extensively investigated the role of Japanese scientists in sustaining imperial endeavors within postwar reconfigurations of national identity through science. Like the scientists I examine here, Low (2005) asserts that Japanese physicists, including Nishina Yoshio (1890–1951) and 1949 Nobel Prize winner Hideki Yukawa (1907–81), welcomed earlier international exchanges and embraced a hybrid cultural identity in their careers, while also involving themselves in postwar policy-making.[59] Mizuno (2008) also reveals how prewar and wartime technocrats, Marxists, and popular science writers engaged with science as a discourse, which she terms "scientific nationalism" or "a kind of nationalism that believes that science and technology are the most urgent and important assets for the integrity, survival, and progress of the nation."[60]

Such a discourse persisted into the postwar period, but was repackaged as sustaining a democratic country. More recently, Mizuno, Aaron S. Moore, and John DiMoia (2018) compiled a volume arguing that the postwar success of both Japanese reconstruction and Southeast Asia's development resulted from Japanese engineers whose knowledge networks spanned the colonial era and prewar period, after which they invested energies into Cold War-era international overseas development projects.[61] Integrating themes previously explored by these scholars, Miriam L. Kingsberg Kadia's *Into the Field*[62] explores a generational cohort of Japanese social scientists, focusing particularly on Izumi Sei'ichi (1915–70), to investigate knowledge production amongst anthropologists, archeologists, and other fieldworkers in transwar Japan amidst political transformations from empire to democracy.

The aforementioned works to some extent examine knowledge production by Japanese researchers in science, technology, environmental, and mathematically related fields, and assess science's importance in establishing, and maintaining, a particularly Japanese national identity from prewar times into the postwar period, when science transformed radically in its meanings for its practitioners and their nation. In addition, these scholars clearly connect science with national political consequences. Yet, only Kingsberg Kadia examines the scientists' private lives and personal struggles as men negotiating their place as representatives from a non-white empire and postwar democratic regional power.

Two best-selling books, *The Invention of Nature* (2016)[63] and *Founding Gardeners* (2012), by journalist Andrea Wulf writing as a cultural historian, reveal general readers' appeal for understanding how broader historical trends in politics and foreign policy, albeit during the eighteenth and nineteenth centuries in a Euro-American context, interwove in scientific endeavors and interactions between individuals of disparate backgrounds.[64] Wulf's remarkable scope, while bringing readers into the jungles and gardens of Enlightenment-era Anglo-European gentleman-scientists, unfolds a narrative of these men epistemologically imposing order upon the natural world and their own political spaces in attempts to gain empirical knowledge. This is strikingly similar to what Izumi and his cohort later accomplished in the Japanese empire and beyond, as they ventured afield into Manchuria, Mongolia, and Papua New Guinea in the prewar and wartime periods, and rural Japan, Brazil, and Peru in postwar times. Both Kingsberg Kadia and Wulf inspired me to imagine similar pathways to unpacking the lives and roles of ornithologists in the following chapters.

Besides Kingsberg Kadia's excellent book, few studies examine issues in the history of science during Japan's transwar period (1920s–50s). My book fills an important gap as a cultural historian's perspective of knowledge networks and cultures of scientific inquiry during a critical time. Many English-language works emphasize the transnationality of Japanese scientific networks, and form the basis for what I call "transwar imperial continuities"—the persistence of prewar engagements with science and earlier networks within imperial frameworks (marked by imperial Japan later defeated by a rising American and waning British empire) and continuing into the postwar period. Building upon such contributions, my book investigates the transwar careers, Anglo-American knowledge networks, and personal lives of a select group of highly influential scientists to emphasize their enduring political clout and influence amidst an ornithological backdrop.

Sources and Methodology

Amongst Japan's prewar scientific community, ornithology was viewed as an aristocratic pastime, while the zoologists engaging in it enjoyed intimate connections with political elites and the imperial family. Consequently, Japanese zoologists enjoyed unique proximity to political power centers and interacted in cosmopolitan arenas within global networks. Thus, I interrogate how the practice of ornithology could be viewed as a political act. Subsequent chapters explain my methodology and use of theory in interpreting materials related to the Japanese "gentlemen of science" and cosmopolitan imperial scientists examined in this book. I arrange the book's chapters around specific geographical loci or spaces marked by class status and imperial designs, a form of organization inspired by geographer Doreen Massey's theory of "power geometries"[65] and advocacy of an approach where scholars attempt "to think of the spatial in terms of social relations."[66] Accordingly, I investigate these scientists in various geographical foci and how gendered identities transformed to garner social capital and scientific credibility in transnational associations and travel.

What began with slides shot by Austin expanded into an international investigation and painstaking process of piecing together oftentimes seemingly unconnected fragments of information. Rooted in my expertise as a historian of modern Japan, *Japan's Empire of Birds* interrogates interdisciplinary and multidisciplinary sources, including personal letters, oral histories, memoirs, scientific writings, photographs, maps, architectural blueprints, and even bird specimens. These are supplemented with analyses of articles, essays, and obituaries in leading Anglo-American ornithological journals like *The Auk*, *Ibis*, and *Nature*, along with interpretations of pieces in their Japanese equivalent, *Tori*. I also investigate modes of "seeing" space or social status and utilize photographic interpretation as well as other document types, like personal letters, architectural floor plans, travel narratives, oral histories, government reports, and so on. I consulted university archives in Japan (Waseda University and Tokyo University), the United States (Harvard University, University of Florida, and Cornell University), and England (University of Cambridge)—where these scientists studied, researched, and corresponded. I also include personal correspondence and oral history interviews to provide an intimate glimpse into their world.

In addition, I incorporate interdisciplinary and multidisciplinary research methods beyond the field to examine the complex social relations arising out of scientific encounters between Japanese and Western counterparts in the transwar period. I visited and investigated the locations where the scientists lived and performed their activities, and also engaged in fieldwork and examined ornithology as a practice. Traditional document-based research and oral history interviews were enhanced by what Ann L. Stoler calls "historical ethnography." Here, she urges scholars to view archives as dynamic spaces suffused by the personalities of the materials' creators.[67]

Methods from cultural anthropology were also used. To effectively understand cultural others, anthropologist Neil Whitehead proposes adoption of "performative ethnography," where "in order to understand desire we must become desiring subjects ourselves," and a state of humble immersion allows "auto-ethnographic description and overtly positioned observation."[68] Thus, I engaged in extensive practical research to permit me to briefly experience the scientists' particular social spaces and daily life activities, including their passion for birds. Besides interviewing Japanese ornithologists and the retired manager of a local north Florida quail plantation, this included learning about upland hunting and shooting skeet, activities similar to targeting bird specimens for ornithologists. In addition, I viewed specimens collected by Austin, Hachisuka, and others currently housed at the Harvard Museum of Comparative Zoology, the Florida Museum of Natural History at University of Florida, and the Yamashina Institute for Ornithology (YIO). I also participated in bird banding with YIO researchers in Chiba prefecture, Japan, and examined how volunteers banded songbirds in Florida's Wakulla Springs State Park. In Schrum, Germany, my cousin Claudius Böttcher, a former taxidermist and carrier pigeon racer who now breeds prized racers and extravagantly plumed chickens, taught me about shooting, skinning, and mounting birds during forest walks and teas at his hunting lodge. As peacocks casually strolled near his fishponds hosting edible carp, he mentioned that weaker flock members often ended up in the saucepan—similar to Yamashina's postwar breeding experiments. Lastly, I also skinned doves to directly experience the technical difficulties of collecting and preparing bird specimens. These investigations permitted me to experientially comprehend the scientists that I examined, while I even shot their favored guns, which provided deeper understandings of the worlds that ornithologists navigated and aided me in writing more vivid, personalized descriptions. Such "collages of memory"[69] afforded by oral histories and non-traditional sources, including material objects and photographs, allow for a layered view of these scientists' lives and politics while immersed in the early-to-mid twentieth century's monumental events.

Conclusion

Within the broader narratives of these scientists, one can examine the confluence of Japanese and Anglo-American imperial ambitions, and how male scientific explorers, and especially ornithologists—as birds brook no borders—often transgressed into spaces of shifting national boundaries, social class, race, and place while engaged in scientific study, expressing national endeavors (like surveying of land or species) in a language of scientific inquiry and imperial masculinity. The following chapters investigate the book's central premise—what it means to migrate as a Japanese imperial (and postimperial) agent into multiple geographical loci and encounter contemporary Anglo-American imperialisms through the expedient lens of ornithology.

In the process, these scientists' careers and connections strikingly reveal intimate political aspirations couched in research on birds. The following is a book about ornithology and empire, and the Japanese scientists whose passion for birds mobilized imperial and national ambitions throughout the transwar period.

1

The Practice of Ornithology: *Birds, Hunting, and Social Class in Prewar Japan and the Anglo-American World*

Introduction

Prewar Japanese scientists engaging in the study of birds were usually men from aristocratic backgrounds with disproportionate political influence in the National Diet's House of Peers; thus, any book examining their lives begs a penetrating look into ornithology and its political aspects. Yet, one must first understand how the field began as an offshoot of zoology supported by existing Japanese traditions of popular enjoyment of birds, including in the arts, bird keeping, and various hunting methods. Associations like the *Ornithological Society* (est. 1912) and *Wild Bird Society* (est. 1934) later formalized circles of ornithologists and enthusiasts, who also published research findings in English and Japanese in publications like *Birds* to make their work readily accessible to the broader Anglo-American world. The formation of this world, and its initial intersections with Japanese ornithological endeavors, will be examined to underline why these scientists believed in close interactions with it to legitimate their contributions.

In the zoology field, and specifically, ornithology, "gentlemanly" scientific exchange included observing birds and hunting within vast lands and lavish estates in areas near Tokyo. It also meant engaging in collecting forays or expeditions within Japan and its imperial borders, topics explored in Chapters 4 and 5, and to some extent, Chapter 6. Such pursuits related to capturing birds were important means for Japanese scientists to perform imperial masculinity through the lens of what I call "avian imperialism."[1] In prewar Japan, Westernized elites enjoyed hunting into the wartime and early postwar periods. British and American upper-class hunting practices informed hunting in Japan, where Japanese outfitted themselves with foreign firearms and garments. Aspects of such imperial mimicry appear in the early twentieth-century magazine *Ryôyû* [Hunter's Companion], which detailed hunts hosted by Japanese and also attended by Anglo-Americans. Here, two traditions encountered each other and became opportunities for informal diplomacy, which revived during the postwar Allied Occupation.

Numerous Japanese aristocrats engaged in scientific study, primarily at the state-funded Tokyo Imperial University with some at the private Waseda University, while others like Marquis Hachisuka Masauji informally studied at University of Cambridge (detailed in Chapter 3). Here, several key issues offer investigation in relation to power, social hierarchies, and historical understandings of birds. What did science and scientific research mean for imperial Japan's elites and what attracted them toward fields of study within zoology like ornithology? Ornithology also benefited from a long history of bird-keeping and Japanese falconry, or *takagari* [hunting with falcons] practiced by pre-modern samurai elites, the nobility, and Imperial Household-supported falconers until the immediate postwar period. Moreover, a history of Tokugawa-era investigations into zoology also influenced later twentieth-century ornithological studies. Lastly, what aspects of social class infused interpersonal relationships, and how could wealth and status ensure the privilege (and vast resources) to freely engage in scientific research in private and public spaces?

Training Birds for Popular Entertainment and Hunting in the Premodern Era: Warblers, *Yamagara*, and Falcons

Such questions can initially be addressed by understanding what factors influenced ornithology's early foundations in Japan. A longstanding indigenous tradition of trained birds as performers and hunters beginning in premodern eras and continuing into prewar times provided an important base for more formalized study that began in the late nineteenth century at Tokyo Imperial University. On an informal level, before zoology's professionalization as an academic discipline, and ornithology's concomitant codification as a natural science, premodern Japan boasted a long history of birds in art, along with popular and practical knowledge of birds through training of avian species for public spectacle or hunting game, as in falconry.

Neurobiologist and ornithologist Sachiko Koyama highlights Japan's extensive history of bird-keeping and training birds to perform songs and tricks in pre-modernity, with possible Chinese origins; aristocrats initiated these practices in the Nara Period (710–794), which expanded during the Heian (794–1185) and Kamakura (1185–1333) eras, but reached a heyday during Edo (1603–1868) times when they spread amongst commoners.[2] According to Koyama, three wild bird species were most suited to keeping for birdsong: "the Japanese bush warbler, blue-and-white flycatcher (*Cyanoptila cyanomelana*), and Japanese robin (*Erithacus akahige*) are together called 'sandaimeichô' which means the three birds with the most beautiful voices."[3] She notes how adult male bush warblers became "song-tutors" for hand-raised nestlings, while owners manipulated day-length using artificial light to coax birds into singing earlier.[4] Thus, bucolic birdsong even suffused living quarters of premodern Japan's densest cities, where eighteenth-century Tokyo with one million inhabitants was one of the world's largest urban centers.[5]

From the seventeenth-century onwards, trained songbirds also assisted in luring small birds to nearly invisible silk thread *kasumi-ami* [mist-nets] at a *toyaba* [literally, "bird hut place"], a mountainside area hosting vertically assembled nets with pockets, live bird decoys, and a flagging platform where a person jerked down a flag attached by string to a ten-foot bamboo pole, ostensibly mimicking a falcon's motions to agitate birds into waiting net pockets.[6] Before the netting harvest, practitioners chose a bird decoy from numerous candidates in a small hut filled with cages, or *torisha*.[7] Oliver L. Austin Jr, in his late 1940s report on Japanese mist-netting for the NRS wildlife branch, noted how, in Ishikawa prefecture since the mid-Tokugawa era, mist-netters utilized songbirds trained to sing in earlier seasons to attract passerines into downwind nets to harvest wild bird meat.[8] Once caught, sparrows and other songbirds were plucked and broiled for *yakitori* [roasted bird], or for thrushes, skinned and pickled in *koji* [rice bran lees] packed in barrels for transport into towns and cities as seasonal dishes in fall and early winter.[9] The Occupation soon banned songbird mist-netting, but Japan's Ministry of the Environment still uses nets to collect migratory birds for bird-banding, though now, taped birdsong replaces live decoys.

Trained birds not only aided Japanese netters and hunters, but also performed entertaining antics on command. Koyama's work also covers the cultural history of *yamagara* (varied tit or *Sittiparus varius*) tricks through a behavioral science lens.[10] In her observations, she believes that Japanese custom, rather than the bird's skill or natural qualities, and whose aptitude for complex tricks indeed implies "high cognitive powers in this species," led to the *yamagara*'s singling out as the song-training species of choice.[11] During his early Meiji-era Japan sojourn, zoologist Edward S. Morse (1838–1925) listed in his recollections eight different tricks *yamagara* often performed, including "horse-riding," donating to a shrine or temple for fortune-telling, and water-drawing, among others.[12] Fortune telling by birds, especially with *yamagara*, delighted children and adults at shrine festivals and other occasions from the Tokugawa period into mid-twentieth century. After imperial Japan's Second World War defeat, in cities like Tokyo, demobilized veterans often performed such street-side entertainments, but by the 1970s, practitioners nearly all died out or retired.[13]

Falcon hunting, involving captured raptors and flourishing in premodern times, also diminished after imperial Japan's defeat. According to Japan's earliest written records from the eighth-century *Nihon shoki* [Chronicles of Japan], state-sponsored falconry began in 355 A.D. with the Korean hawk trainer Sakenokimi coaching a bird received by Emperor Nintoku (r. 319–399).[14] Pleased by a gift that skillfully caught him pheasants, the Emperor created the Hawking Guild, renamed Hawk-Keeping Officers by Emperor Monmu (r. 697–707) in the eighth century.[15] Falconry continued under imperial patronage despite Heian and Kamakura Era political vicissitudes, and reached its zenith in the Tokugawa Era, where the pastime extended to *daimyô* [feudal lord] families. Historian Morgan Pitelka illustrates falconry's deeply political aspects, where feudal lords used hunting for prized birds to gather intelligence on topography, terrain, and natural features of neighboring lands, a practice beginning in the medieval era, accelerating in the Kamakura period, and continuing into the late Tokugawa period.[16] Sport falconer Naoko Otsuka notes that, as favored species, "Traditionally used types are the Northern Goshawk, the European Sparrow Hawk, the Japanese Lesser Sparrow Hawk, the Peregrine Falcon and the Merlin; more recently, there are areas that have employed Hodgson's Hawk Eagle."[17] Falcons, hawks, and other raptors also served as high-status gifts to cement relationships, where "the exchange of birds as well as the exchange of prey became opportunities for sociability, with varying degrees of formality and institutionalization."[18] Later, for Japanese aristocrats involved in ornithology, exchange of bird specimens and skins held the same function.

Raptors were originally trained for hunting birds of prey or game birds, but became useful for flushing ducks in carefully tended aristocratic preserves. The imperial household hired hereditary Tokugawa family retainers to serve as falconers [*takajo*, or "raptor masters"] for pheasant hunting or duck catching at fabulously costly game preserves like their Tokyo Hamarikyû estate, where Hanami Kaoru was last in line of sixteen generations of masters.[19] This long-standing historical practice of maintaining duck decoy ponds was an invented tradition from the mid-Tokugawa era when the Dutch established one in Nagasaki.[20] Now a park in central Tokyo near the famed Tsukiji fish market, the Hamarikyû gardens were opened to the public by Japan's imperial family in April 1946. This location originally featured two *motodamari*, or duck ponds, built in 1771 and 1791,[21] where hunting guests waited hidden behind dugouts with peepholes facing a pond where trained "decoy" ducks attracted their wild fellows; after a signal, hunters rushed out and climbed the moat's levee and caught flying ducks with a hand-held *sadeami* [scoop net].[22] The falcons or hawks, tethered on long strings by the falconer, caught stragglers and were presented with the birds' hearts and lungs as rewards.[23] After the hunt, guests enjoyed a *sukiyaki* [meat simmered in iron pots with sweet soy sauce broth] or *kuwayaki* [a cooking style utilizing hot-steel plates, popularized in postwar Osaka] reception prepared by cooks with pre-dressed duck meat.[24]

In the immediate postwar period, this genteel eighteenth-century hunting practice merged with politically laden socializing between the imperial family and high-ranking Occupation officials; though Emperor Hirohito (1900–89) disliked hunting, Austin noted how "the imperial household subsidizes the preserves and continues the netting for its social and political benefits."[25] Just like in prewar times, the invitation to an imperial duck-hunt in the Occupation era denoted entrance into a selective group with proximity to Japan's former ruling elites, which often entailed diplomats and politicians. The Imperial Family still maintains two wild duck preserves mainly for conservation: the Saitama *kamoba* in Saitama prefecture's Koshigaya City and Chiba prefecture's Shinhama *kamoba* in Ichikawa city.[26]

Such elite hunting practices patronized by Japan's aristocracy were already rare by the early twentieth century, while other traditional bird-harvesting forms were viewed as old-fashioned customs in decline. From 1932 until 1939, during imperial Japan's militarily embroilment in Manchuria and China, Horiuchi Sanmi, while employed by the popular pictorial *Asahi gurafu* [Morning News Graph], documented traditional Japanese hunting practices, including those of the Imperial Household Agency's trained raptors. He shot three thousand photographs of "hunting methods including the use of air-nets, water-nets, blinds, hooks and bait, baskets, box-traps, cage-traps, rope-snares, hawking/falconry, cormorants, decoys and more standard methods such as shooting,"[27] where 229 appeared in his book.[28] Horiuchi's efforts resulted in a 95-page grass-green textile-covered book published in 1939 on avian hunting traditions, featuring a stylized cover with a woodblock print of birds flying into a vertical net,[29] and a Spartan plain-covered 66-page second edition issued in 1942.[30] However, his comprehensive volume on Japanese hunting practices was not published until 1984.[31]

Notably, Horiuchi's first book *Nippon chôrui shuryô-hô* [*Bird Hunting Methods in Japan*] included two prefaces by two well-known ornithologists, Uchida Seinosuke and Yamashina Yoshimaro, while the preface notes that Uchida and Matsuyama Shiro, a Ministry of Agriculture and Forestry technician, assisted Horiuchi throughout its creation. Why had Horiuchi found his topic so important when imperial Japan mired itself in prosecuting war upon the Asian continent? His preface indicated that his aim was "to record hunting methods which are about to die out."[32] Amidst rising militarism with men drafted into military service to hunt Chinese enemies, use of Western guns and pistols since the late nineteenth century soon supplanted a traditional culture of nets and other devices used to peacefully harvest game noiselessly and without apparent violence in Japan's woodlands, countryside, and imperial reserves. By the 1930s, a new era appeared on the horizon, with traditional methods rapidly replaced by more modern and efficient varieties, including gun hunting.

Historically close interactions with birds in premodern Japan as artistic subjects, colorful caged companions, clever entertainers, and hunters of other birds provided important foundations for ornithology's flourishing amongst Japanese elites. Yet, continuities between the premodern and modern persisted. Though falconry continued into the modern era amongst aristocrats, sport hunting with firearms also became extremely popular amongst wider ranges of participants. However, Horiuchi's documentary work surrounding older practices of training and hunting birds without firearms offers the question of why strong connections between sport hunting and ornithological science arose in prewar Japan during the early twentieth century. Notably, ornithological pursuits worldwide involved shooting one's research subjects to amass collections.

"Give the Pheasants a Chance": Enacting Imperial Masculinity in Hunting and Ornithology

Historian Greg Gillespie's concept of "hunting for empire,"[33] where sport hunting with firearms was intrinsically connected to imperialism, and what I call "avian imperialism,"[34] were essential aspects of ornithologists' personal identities developed amidst their imperial migrations throughout the Anglo-American world and Japan's empire. To understand these notions explored in the following chapters, one must examine how a gun hunting culture emerged in Japan, and how this intersected with ornithology and aristocratic political privilege. Hunting for birds amidst a collecting discourse enabled the traversal and even transgressing of multiple national, cultural, and racial-ethnic spaces as natural environments were penetrated by the great (honorary) white hunter's guise. For prewar Japanese zoologists and ornithologists, shooting shotguns and rifles allowed specimen collecting, a practice requiring proper clothing, gear, and, most importantly, the right guns. To collect smaller songbirds, a Japanese mist-net worked efficiently, along with raptors trained to catch falcons until the early twentieth century. However, the most thrilling quarry was brought down on the wing.

Science writer Carol G. Gould asserts that, in Europe and Asia, "some of the bird world's greatest benefactors were themselves sportsmen."[35] However, Japanese evidence reveals that late nineteenth-century gun hunting spurred interest in birds and ornithology, and not vice versa. Setoguchi Akihisa asserts that ornithology was a zoological subfield emerging from Japan's gun-hunting culture, which he calls *yûryô* [sport hunting], established amongst Japan's nobility in the mid-Meiji era.[36] In imperial Japan, rifle hunting developed its reputation as an aristocratic pastime after Meiji Emperor Mutsuhito (1852–1912) embraced it in 1881, and established *goryôba* [private-access royal hunting grounds], which lasted until their late 1920s absorption into national forests.[37] By the turn-of-the-century, sport hunting also rapidly expanded amongst a wealthy bourgeoisie, with the leisure and cash to enjoy it.

Amidst the late Meiji explosion of print materials, sport hunting's popularity was highlighted in its own periodical, and given blessing in its inaugural issue by Iijima Isao (1861–1921), Tokyo Imperial University's most distinguished zoologist and the *Ornithological Society*'s founder. Published from 1900 until 1923, *Ryôyû* [Hunter's Companion] provides a valuable window into Japan's sport-hunting world spanning the late Meiji and Taishô (1912–26) eras, when Western-style gun hunting became trendy for Japanese elites. It features photographs of prominent scientists like Iijima and other officials in front portions of monthly editions, with images of hunting bounties appearing throughout. In *Hunter's Companion*'s first edition, issued on February 1, 1900, Iijima's image graces an entire page following the table of contents, with his short article congratulating its inaugural printing.[38] Iijima enthusiastically claims that the magazine resembles a favored young child, transmitting like father to son his encouraging words for the newborn publication: "you are truly like a beloved child that the hunting world cannot miss, and (I suppose that) you will be just like a beacon in the hunting world."[39] In front matter photographs in *Hunter's Companion*, the zoologist appears in a small, square image, dwarfed by an oval shot below revealing his abundant bounty of foxes, pheasants, and other game birds hanging profusely in a faunal bouquet from a bamboo fence, topped by his hat and ammunition bandolier above a shotgun bisecting his quarry to intimate the absent hunter-scientist.[40] Unbelievable amounts of game like Iijima's amassed in elite hunting forays became centerpieces for carefully staged photographs commemorating hunts, with aristocrats, government officials, professional hunters, industrialists, store owners, and even foreigners participating.

Yet, many hunts acquired little quarry and were purely symbolic, instead instilling sociability amongst Japanese elites and overseas Anglo-American guests to foster international relations. A foreigner's report, published in English and Japanese translation in a special April 1900 *Hunter's Companion* edition, betrays the writer's exoticism-imbued descriptions of Japan's countryside. The American Samuel D. I. Emerson (1855–1931), a temporary expatriate in Japan and US company owner, observed the *Makigari* [a traditional hunt with hunters surrounding game from four sides] in a managed hunt progressing up Mount Fuji's slopes, beginning March 31, 1900, with the actual hunt occurring in early April. He describes the anticipatory enchantment:

> Armed to the teeth we gathered in front of the house. Lanterns were visible here and there as by twos and threes our force was increased. When all was ready we silently began the march, stealthily threading our way in single file along the bye

[sic] paths of the sleeping village. To me it seemed as if we had dropped back a few centuries and constituted a band of Iyeyasu's [sic] soldiery bent on some night attack, or marauding expedition.[41]

However, as the damp cold soaked Emerson's overcoat, and two cans of corned beef (to supplement his unaccustomedly light rice-based meals) became heavy in his pockets during the ascent, his outlook became discouraging. Although the *makigari* resembled the British Raj's famed tiger hunts, complete with "beaters," it failed to flush game for Emerson's group.[42] The group's huge size—including the Tokyo *Hunter's Association*'s sixty members and totaling almost a thousand sportsmen—plus certain noise, undoubtedly contributed to the area's animal exodus. Despite uncomfortably dreary weather, Emerson ultimately enjoyed himself: "The rain still falling, and no conveyance being procurable, we decided to remain at the farmhouse, where we were indeed quite comfortable and jolly. We received calls from various members of the Club and my friends discussed with them till late in the evening the adventures of the chase."[43] Such hunting outings with Japanese elites were clearly coveted invitations for Anglo-Americans.

Almost a year later, *Hunter's Companion* published an opinionated missive from British hunter Gordon Smith, who lamented his failure to attend meetings of the "Tokio Gun Club" during his Japan trip due to illness. On December 23, 1900, he also expressed regrets that hunting in Japan had become too plebeian.[44] Though his letter appears arrogant, his allegations rang true. In *Hunter's Companion* one month before, a New Year's advertisement for an American-made Crescent Firearms Company Model 0 (Armory Steel) double-barreled shotgun highlights this phenomenon, indicating the popularity of private gun ownership amongst aspiring classes in Japan, since its quality and price were lower than firearms preferred by elites like costlier, superior Browning or Parker models.[45] The prevalence of firearm ownership is revealed in numbers of individuals with hunting permits, still quite low during Gordon's time. After Meiji Japan instituted restrictive new rules for gun ownership with the *Daijōkan*'s [Central Government Administration] Proclamation on Gun Control Regulation,[46] only 166 individuals held a sport-hunting permit in 1877;[47] yet, gun-hunting permit holders accelerated in the Taishô period, with 86,007 individuals in 1912, and 206,146 in 1921.[48] In the late Meiji period, such hunting social occasions still held an exclusive cachet.

Nevertheless, in 1900, Gordon grumbled that even the bourgeoisie could afford licenses, which ranged between one to ten *yen* (a day to a week's salary for white-collar workers—a considerable sum for ordinary people), and believed that good game opportunities were diminishing.[49] As remedy, Gordon, in an early form of Western *gaiatsu* [outside pressure] upon Japan's government, whose Diet members likely hunted and read *Hunter's Companion*, proposed raising licensing costs to twenty *yen* to keep hunting amidst wealthy ruling classes better positioned to steward natural assets.[50] His letter, sent from Kobe portside before a planned Burmese elephant hunting expedition, is republished in entirety; its colonial tone of measured praise and admonishment toward Japan as an imperial upstart presages Austin's postwar tirades against mist-netting and indiscriminate game bird hunting irrespective of season:

For a conservative country like Japan where shooting and such sports might be called social rights, it is a marvel to me that such a scale as from 1 *yen* to 10 *yen* could have become law. You must admit it is [sic] hardly sense. At present, you meet youths of eighteen years of age by the dozen, with guns, and professional hunters and a host of others all paying [sic] 1 *yen* license ... It [net hunting] is interesting, clever, and sporting, but it should be confined to migrating birds such as *Ducks, woodcocks, snipe, quail,* and *plover*. Give the pheasants a chance.[51]

Despite over-predation of birds with older netting practices and alleged democratization of licensing costs, Gordon rightly indicated that forays into woods and fields with Western-style shotguns and rifles provided imperial Japan's elites and aspiring bourgeoisie an excuse for sociability first and harvesting game next.

A successful hunt necessitated firearms knowledge and proper guns. Guns connoted imperial masculinity while permitting fruitful hunting forays; yet, they harbored varying reputations for safety and reliability. Access to guns was also carefully controlled by Japan's government for centuries. Following decades of warfare during the Sengoku period (1467–1603), firearms fell out of favor in Japan, with only the nobility and warrior classes permitted to carry them after Hideyoshi's 1588 "Sword Hunt."[52] Nevertheless, historian David Howell asserts that, amidst the hard-won early Tokugawa peace, despite the Shogunate's next round of edicts confiscating "illicit weapons" in 1657,[53] firearms still circulated in peasant households, where they more likely repelled roving boar and deer potentially harming crops than were used to hunt in the mountains.[54] After over two centuries of disuse and scarcity, guns again became readily available after 1858 with international trade. However, those manufactured in Anglo-American countries were expensive, and 1870s regulations restricted their use in Japan.[55]

In 1918, hunting laws also limited access to the wealthy,[56] while legal statutes maintained that, to obtain a gun license, "the applicant must be neither 'a minor; a fool nor an idiot' and ... should a licensed gunner become 'foolish or crazy' the local officials must cancel his licence."[57] Availability of guns was further restricted since few places of sale existed. In the early twentieth century, Japan's government only permitted five gun shops in Tokyo as the capital or in major ports like Kobe, while other prefectures were allowed only three: moreover, all gun sales were registered.[58] According to statistics from the Ministry of Agriculture and Forestry, which issued hunting licenses both without (A-class) and with firearms (B-class), from 1924 to 1947, numbers of licensed hunters with firearms ranged from 8,000 (1924) to around 19,000 (1947), with a median of 10,000 (1936), while those without firearms, like mist-netters, ranged from 110,000 (1924) to 147,000 (1947), with a median of 78,000 (1936).[59] In comparison to total population numbers, imperial Japan held relatively few gun hunters, who were likely wealthy elites or a rising bourgeoisie, since access to firearms was limited in both scope and high cost, despite cheaper varieties of American shotguns becoming relatively more available in Japan.

Notably, Hachisuka, as an expert hunter whose beautifully made custom firearms validated his elite identity, favored a British Holland and Holland deluxe bolt-action magazine 375-caliber rifle for big game,⁶⁰ and held a J. Purdey and Sons 16-gauge shotgun for birds in some photographs.⁶¹ Some Japanese and Americans like Austin vouched for the Browning Auto-5 (A-5), a semi-automatic shotgun made by Fabrique Nationale de Herstal (FN) for the American company in Belgium from 1903 until 1939, and subsequently, by Remington Arms from 1940 until 1946.⁶² Designed in 1898 by John Browning (1855–1926) and produced until 1998, it was the world's first successfully mass-produced semi-automatic shotgun where each fired round caused the mechanism to load another round; previously, shotguns were either side-by-side (still preferred for upland hunting), over-and-under, or pump action (desired by waterfowl hunters).⁶³ Notably, the A-5 could be used in any environment as a highly reliable firearm, including during Japan's characteristically damp or rainy late spring or early autumn weather. With a distinctive "hump-backed" design, it feels comfortably positioned against the shoulder while standing up and taking aim with successive shots for flying game.⁶⁴ These firearms and others became favorites of ornithological and game hunters, who were often one and the same.

Other valuable accompaniments to hunting excursions for winged game were trained bird dogs. Retrieving quarry in impenetrable thickets and muddy meadows usually involved canines, preferably purebred with superb bloodlines from northern Europe—traits valued in both the Anglo-American world and Japan since the nineteenth century. In Japan and other semi-colonial or colonial territories, "dogs were placed within an imperial hierarchy that esteemed Western dogs more than those of colonial areas."⁶⁵ Notably, the November 1, 1900, edition of *Hunter's Companion* depicted popular breeds for hunting, including American, short-haired German, and English pointers.⁶⁶ The following December issue depicted an Irish setter, captioned "The Thistle."⁶⁷ Remarkably, no recommended breed is Japanese; the magazine's photographs of hunters showcase European pointers, though a *Shiba-inu* [traditional Japanese hunting dog] occasionally appears. Aside from guns and gear, a superb imported bird dog provided credibility in performing a convincing imperial masculinity through sporting. No women appeared in the magazine's early issues when hunting remained a purely homosocial activity until the 1920s. By the mid-1930s, though rare, women hunters, complete with the requisite pointer, appeared in high-end commercial publications like *Shiseidô Gurafu* [Shiseido Graph], a magazine for women cosmetics customers. Such images connoted the brand with an elite, exclusive, and cosmopolitan leisure atmosphere emphasizing Western modernity and its trappings—including foreign dogs.

Once a hunter was properly outfitted with guns, gear, and a suitable canine working companion, he also required shooting expertise aside from observing friends in the field. Prewar hunting practices in Japan much resembled those in Anglo-American countries, with parallels to US upland hunting. A 1939 Japanese hunting manual written and published by Ikuta Shigeru, a textile wholesaler from Tokyo's Nihonbashi area, features sketches displaying safe and unsafe ways of carrying, transporting, and laying down a shotgun for a rest.⁶⁸ Its message differs little from cartoon images in my autumn 2017 online National Rifle Association (NRA) Hunter's Safety Course.⁶⁹ Ikuta, representing a growing bourgeois presence in a former elite pursuit, noted that he wrote the manual as an entrepreneur who became interested in hunting

Figure 1.1 March 1936, No. 32 *Shiseidô Gurafu* [Shiseido Graph] cover, MIT Visualizing Cultures website, "Selling Shiseido—II: Cosmetics Advertising, & Design in Early 20th-Century Japan," Visual Narratives, "Leisure & the Smart Set: Shiseido Graph (*monthly*), 1933–1937," https://visualizingcultures.mit.edu/shiseido_02/sh_visnav05.html, [sho2_GR_1936_03_3211]. © Permission through Creative Commons fair use.[70]

with business colleagues as a hobby.[71] Establishment artist Takama Sôshichi (1903–86), who judged the state-supported Ministry of Education's yearly art exhibition, provided sketches from a hunter's standpoint with thirty years of experience—showing how deeply hunting had penetrated bourgeois Japanese society so that business owners and artists now engaged in the practice from youth.[72]

Notable aspects in Ikuta's manual include depictions of masculine-gendered stick figures portraying general hunters employing unsafe firearm methods, contrasting with more-developed sketches of a gentlemanly, middle-aged Japanese in characteristic upland hunting garb demonstrating proper shooting posture ["safe shooting position"], transporting methods, and other safety highlights involving shotguns.[73] His deep-pocketed vest for storing pheasant and quail would be familiar to genteel American hunters visiting quail plantations near Thomasville, Georgia, where I first ventured into the field at Horseshoe Plantation with its former manager George Fuller to understand upland hunting,[74] a process somewhat resembling ornithologists' collecting of birds during expeditions. However, the tweed knickerbocker pants with knee socks, sensible leather shoes, and a smart beaver-felt fedora worn by Ikuta's sketched gentleman assumes a style adopted from upper-class Britain. Even if little game were caught, with the manual's proper use, one could look the part and enjoy an outdoor activity safely with one's fellows.

Nevertheless, hunts might garner little game or even none. Prewar (and contemporary) upland hunting in Japan and the Anglo-American world proffered intimate sociability to participants, where, by engaging in a somewhat life-threatening activity together, elite men who could afford the hunt's proper trappings widened social ties and tightened bonds. Subsequent chapters investigate deeper aspects of such sociability and imperial masculinities, along with the domestic spaces where aristocratic ornithologists circulated after the hunt to showcase their collections to fellow enthusiasts.

Mutual comparisons of quarry and close encounters with birds on the wing often led hunters to wonder about the different varieties of bird life they encountered. Notably, Duke Takatsukasa Nobusuke, a respected ornithologist, also headed the pre- and postwar *Hunter's Association*. Indeed, competition to amass and find rarer varieties of birds, involving shooting and collecting their skins, belonged to a global nineteenth-century desire to describe the natural world as it was harnessed and slowly disappeared. Many thinkers also believed in Auguste Comte's (1798–1857) positivism, where every phenomenon, including human behavior, could be rationally explained through scientific laws. After American zoologist Edward S. Morse invited Ernest Fenellosa (1853–1908) to teach political philosophy at Tokyo Imperial University in 1878, Comte's ideas and Herbert Spencer's (1820–1903) notions of Social Darwinism were imported into a curriculum that many students, including Iijima and Takatsukasa, absorbed during their studies.[75] As members of the House of Peers, they saw themselves as naturally pre-determined to govern and assert their power over surrounding environments, which included naming and claiming new species throughout the empire. Their understandings of nature blended indigenous and imported knowledge to develop ornithology's foundations in Japan amidst the professionalization of zoological studies.

The Rise of Zoology as an Academic Discipline in Imperial Japan

Prior to the arrival of large-scale Westernization, a native tradition of scientific inquiry into bird behaviors, varieties, and morphologies had enjoyed an extensive history in Japan, as had a tradition of bird-watching and ritualized enjoyment of nature (covered in Chapter 2). In addition, long-term quotidian close contact with birds prompted intimate knowledge of their habits and behaviors, and possibly led Japanese to engage in scientific studies like biology and zoology during Meiji period modernizations and beyond, which naturally converged toward an interest in ornithology. Arguably, aside from a concurrently growing sport-hunting trend, a historical close proximity to birds for Japan's imperial elites may have also fostered interest in ornithological studies.

The professionalization of ornithology, beyond employing birds for entertainment or hunting, began during the late Meiji period, but maintained earlier antecedents in an indigenous tradition of scientific inquiry attached to native and imported forms of categorization and observation. During the Tokugawa period, trade with the Dutch and Chinese brought rare objects and animals including birds into Japan, while traders also brought the Shogun exotic gifts. Western precision instruments allowed new ways of perceiving and observing the world, like magnifying glasses, microscopes, and portable telescopes, which enabled close scrutiny of very small or far objects.[76] Painters of *kachôga* ["bird and flower paintings"], a popular form of Chinese-style artistic representation for literati, were energized by these new observational forms along with exposure to realism and empiricism in European art brought by the Dutch.[77]

Additionally, perusal of late Ming dynasty (1368–1644) Chinese texts and imported Dutch medical manuals composing *Rangaku*, or "Dutch Learning," prompted Japanese scholars and intellectuals to believe that nature could be quantified and observed as a collection of objects, or *banbutsu* [myriad things].[78] As plant or animal specimens, these objects "became concrete bearers of abstract characteristics."[79] The relationship between Japanese painters and scientific study was exceptionally close,[80] and arguably, contributed to a passion for birds translating into codified study by the nineteenth century. Here, aesthetics directly contributed to avian science's evolution into ornithology in Japan. Often, *Rangaku-sha*, or adherents of the new Dutch Learning, were also artists, and documented detailed observations of living things, including birds, which permitted careful development of taxonomies, and built upon emerging scientific consciousness in eighteenth-century Japan.[81]

Science historian Frederico Marcon reveals how new epistemologies influenced early modern scientific views in Japan distanced from indigenous spiritual practices: "The ensuing secularization of nature sprang from a parceling of nature in myriads of discrete objects to be described, analyzed, consumed, or accumulated in the form of standardized and quantifiable units as products, natural species, or collectibles."[82] Scholars and collectors now began to view "plants and animals as intellectual commodities."[83] He proposes that these earlier conceptions merged with imported Western ideas of the scientific method and systematics for categorizing species during the early Meiji period.

Nonetheless, Marcon argues that "Many of the practices, institutions, and knowledges of *honzôgaku* were not lost or abandoned when the Western sciences were introduced in the Meiji period to sustain the modernization of Japan but would rather be translated, adapted, and incorporated in the language and forms of the new disciplines of botany, zoology, and botany."[84] He defines *honzôgaku* [the "study of fundamental herbs"] as "a scholarly field that encompassed subjects ranging from *materia medica* and agronomy to natural history."[85] These trends in Japan's early modern scientific world were first solidified within the 1596 *Honzô kômoku* [Systematic *Materia Medica*], written by Ming Dynasty physician and naturalist Li Shizhen (1518–93), but published posthumously in Nanjing.[86] This text arrived in Japan in the early 1600s and joined an indigenous intellectual tradition occupying a parallel space to European Renaissance-era scientific inquiry separate from religious conceptions of the natural world. This syncretic parallelism, whereby distinct epistemologies coexisted, also reflected the hybrid, cosmopolitan educational backgrounds and personal lives of aristocratic Japanese scientists focusing on ornithology.

A pure separation of religion and science never occurred in Japan, and instead, coexisted in parallel, where practitioners were often one and the same. This particularly applies to the early twentieth century. Emperor Hirohito, while serving as the Ise Shrine's prewar and wartime head priest of State Shintô, was also a marine biologist. Politician and Peer, Duke Takatsukasa, who presided as head priest of the Meiji Shrine (where Emperor Mutsuhito's soul was enshrined), was also a published ornithologist. Nakanishi Gotô (1895–1984), the *Wild Bird Society*'s director, engaged in Zen ascetic practices in his youth, while his later anti-authoritarian ideals resonated with discovering "Buddha-nature" in quotidian encounters with living things. In a prewar Japanese context, a religious background undoubtedly *furthered* investigation of the natural world, as did proximity to imperial power centers and membership in aristocratic social circles.[87] To understand how these paradigms arose, it is useful to investigate how zoology became a distinct form of scientific inquiry in late nineteenth-century Japan, merging with the domestic knowledge base of *hôzongaku* posited by Marcon.

The beginnings of the professionalization of Meiji-era zoology, of which ornithology was an outgrowth, began amidst "hired foreigners" employed by the precursors of imperial Japan's key educational institutions. The first state-hired foreign zoologist was German scientist and paleobiologist Franz Martin Hilgendorf (1839–1904), who taught natural history at the Tokyo Medical School (later, Tokyo Imperial University's medical department) from 1873 to 1876, and introduced Charles Darwin's (1809–82) theory of evolution[88]; he nonetheless imparted little influence upon zoological studies in Japan because he focused on a preparatory course without "special students," the equivalent of modern-day graduate students who further reproduced academic lineages.[89]

This task fell to an influential American, the aforementioned naturalist, autodidact artist, and bird-lover Edward S. Morse. After Tokyo Imperial University's 1877 founding, his hiring composed a cohort of fifteen professors (twelve foreign) within the new Department of Science.[90] After attracting attention for his technical drawing and engraving skills, Morse received endorsements from malacologist Philip Pearsall Carpenter (1819–77), who recommended him to Swiss geobiologist Louis Agassiz (1807–73). From 1859 until 1861, he served as Agassiz's assistant at Harvard's MCZ, where he helped conserve and draw collections of brachiopods and other molluscs while receiving training under Agassiz's personal direction.[91] This institution, along with the Smithsonian in Washington, DC, and the AMNH in New York City, formed a trifecta of key American centers of scientific study for zoology—and ornithology in particular. Later, at Bowdoin College, Morse worked as an anatomy and zoology professor from 1871 until 1874, when he became a lecturer at Harvard University to succeed Agassiz, and in 1876, received a fellowship in the prestigious American Academy of Science.[92] Morse had reached a critical stage in his career as a top authority in his field, a fact well-recognized by admirers in Japan.

In June 1877, Morse first traveled to Japan with intentions to study coastal brachiopods, and while searching for shelled sea creatures on a beach near Tokyo, a well-connected student discovered the foreigner's useful talents and mobilized his networks to provide venues for Morse to share his knowledge. Soon after, he was invited to lecture at Tokyo Imperial University, whereupon Meiji *genrô* ["principle elder" or oligarch] Katô Hiroyuki (1836–1913), then superintendent of the university's Departments of Law, Science, and Literature, offered him a position as its first professor of zoology from July 12, 1877, until August 31, 1879.[93] At this moment, Katô's thought evolved into what historian Julia Adeney Thomas calls "a nuanced defence of oligarchic power that combines elements of Social Darwinism and organic social theory with the promotion of an inventive elite transcending evolution and capable of directing it."[94] With Katô's blessing, Morse's scientific knowledge and understandings of evolution would shape the young men he accepted as students to form new imperial elite applying scientific values to the world around them.

At Tokyo Imperial University, Morse remained a total of three years, during which he created the university's Zoological Institute, set up a seaside laboratory at Enoshima in Kanagawa Prefecture, and accepted "special students." Betraying an Orientalist gaze, Morse initially wrote "it seemed as if I were lecturing before a class of girls"[95] with students in skirt-like *hakama*, but quickly gained respect for them, despite (to him) their "feminized" appearance: "These young men are the sons of samurai, some of great wealth, others poor, but all modest and polite to one another and very quiet and attentive. Each one has jet-black hair and dark eyes and all are dressed in bluish-colored clothes, the hakama so like a divided skirt that it seems as if I had a class of girls."[96] Morse soon comprehended their elite provenance and proficient English earned at the Peer's School [*Gakushûin*], which also provided scholarships for talented commoners. Such young men initially came to Tokyo Imperial University pre-steeped in the School's educational mission grooming aristocrats for rule, which formed a basis for imperial Japan's next generation of scientific elites upon university graduation.

Of Morse's five "special students," four later became influential zoologists, including Matsura Sayonhiko (who died of typhus in 1878), Sasaki Chûjirô, Iijima Isao (1861–1921), Iwakawa Tomotarô, and Ishikawa Tomomatsu.[97] In particular, Ijima was Morse's most successful disciple who later acceded to his position at Tokyo Imperial University. He also worked with Morse and Sasaki on *Shell Mounds of Omori*, published in 1879 with drawings and lithographs created by Japanese collaborators.[98] Morse "discovered" this historically rich shell mound while riding a train leaving from Omori Station in Tokyo.[99] In 1882, as an appendix to the 1879 book, Iijima and Sasaki added their own findings in another location, the *Okadaira Shell Mound at Hitachi*.[100] These prehistoric middens served as important caches of anthropological evidence from early Japanese peoples from thousands of years ago, including remains of shells, human bones with curiously flattened tibia, and Jômon era (14,000–300 B.C.E.) pottery shards, discarded after ancient feasts allegedly involving cannibalism.[101] Morse thus popularized archeology and marine biology amongst Japan's elites, including imperial family scions, and in extensive lectures on Darwin's theories of evolution, also laid groundwork for interest in ornithology and systematics.

Morse later wrote about his experiences in the detailed 1917 memoir *Japan Day by Day*,[102] which describes his time as an academic, plus deep appreciation for the country and its people with keen anthropological sketches noting minutia of his surroundings. Historian Kerim Yasar calls his book "practically an encyclopaedia of the soundscapes of early Meiji Japan" which captured the sounds of working people where "vocal utterance serves as a guide and accompaniment to work."[103] In addition, Morse wrote *Japanese Homes and Their Surroundings* (1885)[104] and compiled the

Catalogue of the Morse Collection of Japanese Pottery (1901)[105] to list and classify works in his collection, which he donated to Boston's Museum of Fine Arts in a new position as "Keeper of Japanese Pottery." During his years in Japan, he collected thousands of pieces of ceramic and pottery, and back in the United States, his stories captivated elite Bostonians like astronomer Percival Lowell (1855-1916) and inspired him and others to travel to Japan. Toward Morse's death, upon hearing of Tokyo's destruction amidst the 1923 Kantô Earthquake's catastrophic fires, he bequeathed his collection of nearly 12,000 thousand books and other natural science materials to Tokyo Imperial University's library.[106] The same disaster burnt all bird specimens from Tokyo's National Science Museum, which included ornithologist Momiyama Tokutarô's (1895-1962) collection.[107]

After Morse's academic tenure in Japan, he returned for subsequent visits to purchase pottery and exchange knowledge with former students, some who worked with his successor, zoologist Charles Otis Whitman (1842-1910), who until his 1881 departure, taught them microscopy techniques gleaned from Germany and Italy. Following tutelage under Morse, Ijima trained in Germany at the University of Leipzig, and later succeeded Whitman as a celebrated professor of zoology immortalized as Japan's "father of parasitology," also remembered as a noted ornithologist who wrote scientific reports in English, German, and Japanese.[108] In Germany, he studied parasitology and morphology under Rudolf Leuckart (1822-98), and acquired knowledge of German medical and bacteriological models, then dominating the field.[109]

Historian Setoguchi Akihisa notes that "In those days, most parasitologists in Japan were students of Iijima, and therefore, Japanese parasitology started as a part of zoology rather than as a medical discipline."[110] Unsurprisingly, birds with keratinaceous feathers and thin skins, much like fur-bearing rodents with tendencies to live in close quarters in colonies, are notably infested with parasites. Some species of bird lice are host-specific and prefer certain species of birds, which aids zoologists in solving classification conundrums, especially for systematists.[111] However, birds are thought more pleasant to catch, study, and handle than rodents due to largely positive cultural attitudes—which partially explain why scientists studying parasitology often become interested in ornithology.[112] These connections continued into the early twentieth century and beyond, where ornithologists trained in zoology were also proficient in parasitology like Uchida Seinosuke, a specialist on *mallophaga*, or "biting lice," a suborder of lice infesting birds.[113]

These academic lineages with interconnections between marine biology, parasitology, and ornithology within zoology's broader field persisted until the present, with influential professors training successors through the university or within professional organizations like the *Ornithological Society*. Ijima ultimately graduated under supervision of Itô Keisuke (1803-1901), who began working at Tokyo Imperial University as an adjunct professor in 1881.[114] Combining native *honzôgaku* with new principles adopted from German medicine and natural sciences, Itô had studied with German doctor and naturalist Philipp Franz von Siebold (1796-1866) in Nagasaki while serving as his specimen collector in the 1820s. Thus, Ijima's academic lineage bridged old and new scientific practices from both Europe and Japan in spanning the Tokugawa and Meiji periods. In 1893, Ijima described his first bird, marking the beginnings of a strong Japanese presence in the global field of ornithology.[115]

On May 3, 1912, Ijima initiated professionalization of ornithology in Japan by forming and heading the *Ornithological Society* as its first president (1912–22); his students Takatsukasa and Uchida served as the organization's second (1922–46) and third (1946–7) heads spanning its wartime dissolution, while Kuroda Nagamichi, another student, served as its fourth head (1947–63) after its postwar revival.[116] In 1915, the *Ornithological Society* also began publishing its representative journal *Birds*, subtitled "Bulletin of the Ornithological Society of Japan," which appeared semi-annually from 1915 to 1944, but was suspended during the war and its aftermath from 1945 to 1946; it was renamed *Japanese Journal of Ornithology* in 1986.[117] Although the *Ornithological Society* was a relative latecomer to global ornithological endeavors, members sought international connections and venues for their work, especially in the 1920s and 1930s. During those years, the worlds of the British Ornithologists' Union (BOU), American Ornithological Union (AOU), and to some extent, the Cooper Ornithological Club (COC), began to intersect with *Ornithological Society* members like Hachisuka, Takatsukasa, and Yamashina. However, how did these organizations initially develop in Great Britain and the United States, and what did these connections mean for broader scientific exchange and transnational knowledge networks? Most importantly, what similarities did the shaping of ornithology in the Anglo-American world share with Japan?

Avian Imperialism: The Scientific Study of Ornithology in the Anglo-American World

With their dizzying varieties, and at times, fascinatingly human-like behavior, birds understandably are appealing subjects of scientific study, even for lay people, although ornithologists take this passion to a wholly unprecedented level. As an offshoot of the late eighteenth-century Enlightenment, the discipline developed in the Anglo-American world amidst zoologists, collectors, and popularizers of ornithology, and flourished in the nineteenth century amidst general scientific endeavors connected with growing informal understandings of local and regional bird varieties. In the late nineteenth century, ornithology further expanded with exploration and categorization of new territories and species during the imperial era.

The mid-nineteenth century in Great Britain and the United States was marked by great interest in amateur science, of which bird-watching was one component. Impacted by railroad expansion and urbanization, residents of cities yearned to return to nature and recapture an allegedly "lost" romanticized rural idyll with rivers and valleys whose auditory landscapes were punctuated by birdsong. Historian of "Animal Studies" Harriet Ritvo explains this as a process of successful human control over nature: "Once nature ceased to be a constant antagonist, it could be viewed with affection and even, as the scales tipped to the human side, with nostalgia."[118] Broader technological and social transformations like trains, electric lighting, and expanding educational opportunities associated with control over nature translated into collective fascination for natural history, while moralists promoted collecting as recreation and as an edifying activity for the masses.[119] With the rise of railway networks linking cities to the countryside and seashore, shells, rocks, insects, birds, small animals, and plants like ferns were widely accessible to anyone who desired to observe and collect them.[120] Amateur scientists enthusiastically read books and pamphlets on the natural world to gain more specialized knowledge of a multifarious environment.[121] What education scholar Steven Cowen calls "public literacy," or acquisition of literacy independent from schooling, also greatly expanded amongst all classes from the eighteenth into nineteenth centuries with astounding amounts of circulating printed materials.[122] This potential explosion of knowledge via a thriving Anglo-American print culture allowed laypeople to gather information of their choice in a purportedly empirical means to find out more about the world around them.

Notably, an iconic British manual of ornithology dating from 1840 written by Scottish ornithologist William MacGillivray (1796–1852) describes itself as "available to students of every class" and "of so small a size as to be conveniently portable."[123] In England, MacGillivray befriended French-American illustrator and specimen collector John James Audubon (1785–1851) during his overseas trip and collaborated with him to complete *Ornithological Biographies* (1831–9) in five companion volumes to Audubon's four volumes of *The Birds of America* (1827–38). Unfortunately, with its gorgeous 435 colorized plates, *The Birds of America* cost a princely sum to publish, thus making it accessible only to a select group, but MacGillivray's manual allowed a popularization of ornithology from casual bird-watchers to the most serious naturalists. Dedicated to Dutch naturalist M. J. C. (Jacob Coenraad) Temminck (1778–1858), author of the 1815 *Manuel d'ornithologie, ou Tableau systématique des oiseaux qui se trouvent en Europe* [Manual of Ornithology, or Systematic Classification of Birds Found in Europe], MacGillivray's volume was clearly meant to supplant it as the most current authoritative text.[124] Temminck was also familiar to Japanese practitioners of zoology, since he contributed to portions on mammals in German physician and botanist Philipp Franz von Siebold's (1796–1866) *Fauna Japonica* [Animal Life of Japan] (1833–50).[125] Introducing his manual, MacGillivray writes an enthusiastic paean to birds and ornithology's associated pleasures available to all:

> No branch of Natural History has been more cultivated than Ornithology. The great beauty and liveliness of Birds, the diversity exhibited in their actions and mode of life, their wonderful migrations, the variety of modulated sounds which they emit, the facility with which many of them may be domesticated, the degree in which they are subservient to our wants, and various other considerations, render them objects of attraction to persons of every age and condition in society. At no former period has this study been more zealously and successfully prosecuted than at the present day.[126]

He even intimates that the Victorian Era (1837–1901) might be viewed as the "age of birds," where birds and means to study them finally became widely accessible, and appealed to human observers in multifaceted ways.

In a contemporary American context, MacGillivray's passage from nearly two centuries ago remarkably presages that of Irby J. Lovette and John W. Fitzpatrick, writers of a popular bird biology textbook:

> Birds have a special place in human science and culture: they capture our hearts, arouse our curiosities, and inspire a sense of wonder. We may revel in the diversity and simple beauty of their forms, but birds also fuel fascinations that drive us towards deeper scientific inquiries into their varied ways of life. Every student of ornithology—from recreational birders to career scientists—will find much to learn and appreciate in the extraordinary physical and behavioral adaptations of birds, their rich evolutionary history, and their commanding global presence.[127]

Clearly, the long persistence of informal (and formal university-based) study of birds in Great Britain and the United States was supported by institutionalization of knowledge acquisition beyond manuals and books. Rather, birds became deeply embedded into the Anglo-American cultural world, and in caged and stuffed form, represented humankind's triumph over nature. When enjoyed outdoors, they assumed a presence as a symbol of unlimited freedom, and in traversing boundaries, served as a clear metaphor for political aspects that both the British and Americans held dear. Their appeal was highlighted in personal and public collections, like Tring Park, which housed the Walter Rothschild Zoological Museum (now known as the Natural History Museum at Tring), built in 1889 as a twenty-first birthday present for Lord Lionel Walter Rothschild (1868–1937) from his father, which opened to the public in 1892, with a Bird Wing, Library, and Lepidoptera Hall added between 1906 and 1912.[128]

In addition to an explosion of materials on ornithology and natural histories, the mid-to-late nineteenth century saw the founding of numerous national institutions open to a scientifically curious public. In the United States, the Smithsonian Institution located in the American capital of Washington, DC, initially known as the United States National Museum, was established on August 10, 1846, with funds from British scientist James Smithson's estate (1765–1829), with its first American history museum opening in 1855 in a building known as "The Castle," a National Zoological Park added in 1889, and the National Museum of Natural History founded in 1911.[129] In Great Britain, the British Museum, billed as "the first public national museum in the world," was founded in 1753, and opened in 1759 in the renovated seventeenth-century mansion Montagu House, supported with funds from physician Sir Hans Sloane's (1660–1753) estate, whose collection of 71,000 artifacts provided the basis for initial exhibits. The original building was demolished in 1823, while its still extant iconic Greek revival buildings were constructed in the 1850s, with collections expanding amidst British colonialism's growing reach.[130]

Both of these representative institutions were indelibly linked with imperial expansion and the collecting of a vast array of artifacts and specimens gathered and wrested from colonial territories and native peoples from throughout the world. US historian and former Smithsonian administrator Steven Lubar points out the political aspects of objects contained in a collection's space within museums:

Objects properly displayed could reveal the hierarchies of nature and culture. Objects in museums also served social purposes. The formation of museums with permanent collections, places where experts vouched for the authenticity of art and artefact, and arranged it to tell useful stories, reinforced the authority of political, economic, cultural, and scholarly elites. Museum displays made–and still make–powerful arguments for the stability of culture.[131]

Museums, in displays and collections, reflected current narratives about how the world was structured around certain cultural frameworks and supported allegedly inherent hierarchies. Nowhere was this more obvious than in ornithological collections, where taxonomists carefully categorized birds in a painstaking process often highly debated and endlessly revised with new discoveries from the field. A bird's point of collection, often from an "exotic" place or imperial colony, and its general provenance, lent an image of global bounty and heroic exploration to the museum's space.

Cultural historian Aaron Skabelund, in his 2011 book *Empire of Dogs*, makes a strong case for "the metaphorical manipulation of beasts in imperializing and colonizing projects," and focuses in particular on "canine imperialism," which he defines as a "dynamic" that historically links imperialism and animals, through "a configuration of relationships between colonizers and colonized, between dogs who accompanied the colonizer from the metropole (their home countries) and canines who lived in colonized and colonizable regions, and between different cultural modes of human interaction with dogs."[132] The same pertains to birds in belonging to a discourse that I term "avian imperialism." Lord Rothschild's obsessive devotion to his occasionally belligerent herd of exotic Cassowaries, a flightless bird found in Australia and New Guinea in the Commonwealth, also represents a kind of avian imperialism, that in the British context, highlighted an animal's subservient status, which in undomesticated (or undomesticatible) form represented hierarchies of civilization and subaltern status. Despite fascination for these dinosaur-like birds, he eventually padlocked them due to their bellicosity towards visitors. Rothschild enjoyed greater success with zebras, which he hitched to his carriage, contrary to other zoologists' negative predictions that they were untrainable. Similarly, relationships of imperial actors to birds within a so-called avian imperialism reveal much about how ornithologists categorized the natural world and even understood political hierarchies.

Inspired by Gayatri C. Spivak's now-iconic essay, Skabelund wonders: "Can the Subaltern bark?"[133] He concludes that, although humans cannot decode barking as "language," one can write a history of canines through animal-human interactions, and more importantly, understand their histories through "cultural artifacts," or bodies that they left behind as stuffed specimens (which he calls "taxidermic objects") and their photographic images.[134] Certainly, the same could apply to birds, collected throughout imperial possessions, and returned to the metropole either dead or alive to be housed in zoos, stuffed on mounts, and cached as skins in specimen drawers. Incidentally, ornithological curators in both the Smithsonian and the British Museum would correspond with Duke Takatsukasa, Marquis Hachisuka, and other aristocratic Japanese ornithologists, who donated rare specimens like the Okinawan woodpecker to their collections. As I assert in this book, such exchanges were a means for these

Japanese scientists to integrate (and ingratiate) themselves into Anglo-American networks. These networks included the British Ornithologists' Union and the American Ornithological Union. However, as will be examined, these relationships surrounding birds were strangely mutable, and also directly related to broader political situations encompassing Japan and the Anglo-American world.

The origins of these associations often began around privileged coteries connected to either academic or research institutions. In 1858, a circle of University of Cambridge-based naturalists, including the famed ornithologist Alfred Newton (1829–1907), founded the BOU and first met in his rooms in Magdalene College at University of Cambridge.[135] A year later, the BOU initiated its key journal *Ibis*, where Newton served as editor from 1865 to 1870.[136] From 1866 until his death, still associated with Magdalene, he worked as the University's first Professor of Zoology and Comparative Anatomy,[137] where he was an early proponent of Darwin's theories of evolution. However, despite a coveted position at one of England's elite universities, Newton sorely disapproved of the cliquish British Ornithologists' Club (BOC), which arose out of an 1892 BOU meeting in London as an exclusive monthly social club where BOU members could present papers and exchange specimens, while dining afterwards.[138] The BOC, whose first chairman was Philip L. Sclater (1829–1913), the Zoological Society of London's Secretary from 1859 to 1902, also published a seasonal journal, *Bulletin of the British Ornithologists' Club*, which was initially edited by Richard Bowdler Sharpe (1847–1909), the British Museum's Curator of Birds in its Natural History Section.[139] The interconnectedness of male naturalists associated with universities, museums, and research societies is clearly evident in the early histories of these ornithological associations, a pattern later echoed in the United States and Japan.

The BOU's American equivalent, the AOU, was founded in 1883, and intended to follow the British model, with its first convention meeting that year in the library of the New York-based American Museum of Natural History (AMNH). The AOU's successes included the 1889 founding of the Biological Survey Unit associated with the Smithsonian Institution's National Museum of Natural History (now part of the US Fish and Wildlife Service)[140] and the 1905 establishment of the National Audubon Society, active in conservation.[141] (The Audubon Society's Japanese equivalent was the *Wild Bird Society*, established in 1934 and likewise concerned with bird preservation, discussed in Chapter 2.) The AOU's founding members included William Brewster (1851–1919), curator of mammals and birds at Harvard University's MCZ; ornithologist and Army surgeon Elliott Ladd Coues (1842–99), who wrote key surveys of Western American birds; and Joel Asaph Allen (1838–1921), the AMNH's first Curator of Birds and Mammals and a member of the Nuttall Ornithological Club (founded in 1873) in Cambridge, Massachusetts near Harvard University. *Bulletin of the Nuttall Ornithological Club*, in publication from 1876 to 1883, was the predecessor to *The Auk*, the AOU's key journal. Among these founders was also the brilliant taxonomist and systematics specialist Robert Ridgeway (1850–1929), the US National Museum's (later known as the Smithsonian) first Curator of Birds, who served as *The Auk*'s initial editor. Between 1901 and 1929, on behalf of the Smithsonian, Ridgeway worked on an eleven-volume, 6,000-page *The Birds of North and Middle America* to categorize what he believed as a definitive systematic study of birds ranging "from the Arctic Islands to the Isthmus of Panama, the West Indies and Other Islands of the Caribbean Sea, and the Galapagos Archipelago."[142]

Nevertheless, despite ambitious activities seemingly devoted to categorizing and protecting American bird life, historian Mark Barrow describes the AOU as an elitist East Coast institution serving its own agenda in promoting its members' activities. Barrow cites an 1891 limerick initially published in *Ornithologist and Oologist* that poked fun at exclusive membership policies favoring attractive, arrogant men: "He was an aesthetic young fellow/ Who laughed in a tone very mellow,/ He joined the A.O.U,/ With a tremendous ado,/ And now he is swell, swelled and sweller."[143] In 1893, West Coast ornithologists, who chafed at the AOU's alleged Yankee elitism which allowed only fifty active members (those with power to make the organization's decisions) at any given time, formed the Cooper Ornithological Club (COC), later known as the Cooper Ornithological Society (COS), and began publishing their representative journal *The Condor*. In 2016, COS merged with the AOU to become the Chicago-based American Ornithological Society (AOS), which now publishes both former organizations' journals, along with the AOS Checklist of American Birds.[144]

In 1900, to address issues related to exclusivity, COS member ornithologist Richard MacGregor (1871–1936), later known for surveys of Philippine birds, suggested a division of AOU associate members into two categories, "amateur ornithologists and the bird protectionists," while AOU member Frank Chapman (1864–1945), and assistant to Allen at the AMNH who initiated the Christmas Bird Count that same year, recommended adding a new class of one hundred "senior associates."[145] By the early twentieth century, the AOU began to expand membership toward less professional practitioners of ornithology to improve its revenue base, but few women had joined (less than 5 percent of elected membership) and, like many such organizations then in the United States and Great Britain, had overwhelmingly attracted white Anglo-Saxon Protestant (WASP) men.[146] By the twenties and thirties, with the AUO leadership's efforts to expand, active and associate members successfully grew to nearly 2,000 with a rise of popular interest in both bird watching and ornithology.[147]

Around this time, Hachisuka also became a corresponding member (which he calls "corresponding fellow") in the AOU.[148] The mid-1920s and 1930s would mark a fruitful time of exchange when Japanese ornithologists in the *Ornithological Society* like Hachisuka, Takatsukasa, and Kuroda made names for themselves amongst Western counterparts in the BOU and AOU. These transnational connections flourished and led to strong Japanese representation in top Western ornithological circles where science was strongly connected to empire.

Zoology, Aristocrats, and Ornithologists as Imperial Agents

In the early twentieth century, Japanese understandings of science became domesticated, while a period of study abroad still remained customary for those seeking international networks for collaboration like Hachisuka. Historian of science Hiromi Mizuno asserts that, after the 1910s, "the conception of science as specifically Western disappeared in Japan," where "the latest in science and technology … was integral to the everyday modernity of Japan."[149] Also, she notes how technology became a quotidian part of the nation's landscape with the First World War's ushering in heavy industries supported by growth of conglomerates like Mitsubishi and the environment's technological refashioning through large-scale projects like hydroelectric dams.[150] By this time, a generation of scientists mentored by foreign experts began to teach their own students, like Iijima, who at Tokyo Imperial University fostered the studies of Takatsukasa, Kuroda Nagamichi, and Uchida. Imperial Japan now began to develop its own scientific "tradition" which merged with an earlier base issuing from indigenous practices of *honzôgaku*.

The indomitable Marquis Hachisuka, who appears as a key figure in Chapters 3 and 4, aptly represents this new generation trained locally through informal networks via Takatsukasa, Iijima's former student, and who spread their wings into Anglo-American networks.[151] Thus, in his personage, the increasingly common prewar Japanese trope of what I term "cosmopolitan imperial scientist" intersected in multiple social and political categories: aristocrat, scientist, explorer, and imperialist. My designation is inspired by sociologist Barney G. Glaser's mid-twentieth-century research on organizational behavior and "scientists as organizational men,"[152] where he found that highly motivated scientists worked *both* for the benefit of their organizations and for their professions in a hybrid focus oriented to each of these loci equally, though with greater allocations of time than most.[153] In Hachisuka's case, while known in Japanese ornithology circles in particular (local), he was concurrently working to further his reputation as a *Japanese* zoologist amongst internationally known ornithological associations in general (cosmopolitan).[154] From the very beginnings of his scientific career, not only had he joined Japan's *Ornithological Society* in 1919, but he also entered England's equivalent organization the BOU in the early 1920s, during which he became a Fellow of the Royal Zoological Society (FZS)—or masqueraded as one.[155] This local and cosmopolitan focus also pertained to the prolific domestic and transnational activities and writings of Takatsukasa, Yamashina, and Kuroda Nagahisa at certain points in their careers, when they interacted with Anglo-American scientific circles to varying degrees in the transwar period.

Japanese aristocrats, especially, patterned themselves after their British counterparts as agents of empire, and attempted to be viewed as equals on the scientific playing field by finding new species or updating previous Anglo-American research.[156] Most were also influential members of families intermarrying with Japan's formerly Kyoto-based imperial family and regional *daimyô* elites stemming from feudal times. In addition, they held disproportionate political power due to membership in the House of Peers, Japan's bicameral legislature in the imperial Diet based on German and British models of representative legislative power, and many had enjoyed study abroad in England. Here, well-born Japanese like Hachisuka benefited from privileges of honorary whiteness as a member of a nation once belonging to the Anglo-Japanese Alliance who wrote and communicated in fluent King's English while performing the role of an Anglo explorer-scientist. In 1925, while at University of Cambridge, Hachisuka published *A Comparative Hand List of the Birds of Japan and the British Isles*[157] and *Birds of Egypt* a year later.[158] Preceding his 1929 Philippine foray inspired by his mentor Lord Rothschild that resulted in a four-volume ornithological taxonomic study, Hachisuka even flew his plane to engage in big game hunting in Africa: his intimate knowledge of North Africa by air, car, and on foot during hunting safaris culminated in a 1932 book on the Sahara desert.[159] Yet, he is best known for his comprehensive 1953 study on the extinct Dodo, conceived at Cambridge and the Natural History Museum at Tring in England, but written while in Pasadena, California recovering from an illness in the mid-1930s.[160]

Hachisuka's 1925 comparison of avian species from the two island empires of Japan and Great Britain—indeed, two of the most powerful nations and naval powers on their respective continents at the time—evinced his positing of himself as an equal to British counterparts. Toward this end, Hachisuka not only improved upon findings of British naturalists like Thomas Wright Blakiston (1932–1891), who first described Japanese birds in the 1880s, but also one-upped Blakiston with his 1925 *Hand List*.[161] From 1862 until 1884, the English explorer Blakiston lived mainly in Hokkaidô, and published the first "formal" hand-list of Japanese birds in 1880,[162] which he revised in 1882.[163] After spending over two decades in Japan, he moved to the United States, while much of his specimen collection remained in Hokkaidô. Besides his initial hand-list, Blakiston first proposed a connection between avian species in Hokkaidô and northeast Asian ones, with species types from Honshû resembling more those of southern Asia. This division between Hokkaidô and the rest of Japan at the Tsugaru Straits was later termed "Blakiston's line," or the zoographical boundary denoted by the remote Tsugaru Strait. In the Philippines, based on findings of his expeditions from January to April 1929, Hachisuka noted a similar phenomenon for bird life in the tropics where there also appeared a dividing line between Pacific and East Asian continental varieties.[164]

In subsequent years, the geographic focus of Hachisuka's publication record resembled imperial Japan's expansion to connote avian imperialism with several books in English (with some appearing simultaneously in Japanese, and later, French) featuring the birds of the Philippines (1931–5),[165] Jehol [Manchukuo] (1935),[166] Hainan Island in South China (1939),[167] South and West China (1940),[168] Indochina (1941),[169] Thailand (1941),[170] and Micronesia (1942).[171] In two of these studies, for *Birds of Jehol* (1935), described in Chapter 5, Hachisuka collaborated with Takatsukasa, Kuroda Nagamichi, Uchida, and Yamashina, and for *A List of the Birds of Micronesia under Japanese Mandatory Rule* (1942), he worked with Kuroda (Nagamichi) and Yamashina— who were all active in the House of Peers and involved in domestic legislation while codifying the Japanese empire's avifauna. These endeavors set Hachisuka and his politically influential scholarly colleagues firmly in a pantheon of like-oriented Anglo-American imperial explorer scientists.

Hachisuka's counterparts, including Kuroda (Nagamichi) and Yamashina, revealed similar publishing trends or engaged in revisions of existing Anglo-American systematics work. European and American scientists within their respective empires followed their governments' geopolitical preoccupations, but for Japanese, ornithology was a field where they enjoyed a particular edge. As a former imperial prince and Emperor Hirohito's cousin, Yamashina quit the military in his late twenties and distinguished himself in ornithology, as did many Japanese aristocrats.[172] From a Kyoto aristocratic family intermarrying with the imperial line, he was a more refined and less colorful figure than Hachisuka, who nonetheless interacted closely with

Anglo-American counterparts, but from afar. At the *Peers School*, he learned English, the contemporary language of science by the early twentieth century, and in 1926, even published "How to Breed Fancy Birds" in English.[173] Like Hachisuka, he aimed to outdo Anglo-Americans; yet, he also helped commemorate their work, when in 1932, he, along with Inukai Tetsuo (1897–1989) and Natori Bûko (1905–87), published a new handlist of British explorer and naturalist Blakiston's survey of Hokkaidô's birds, which also covered the influential foreigner's time in Japan.[174]

Nevertheless, by the mid-1930s, a rigorously scientific method of field observation for bird behavior superseded the primacy of collecting reminiscent of Victorian natural history.[175] Systematics purely pertaining to morphology to determine taxonomy was rapidly surpassed by newer ideas forming in the early 1930s later encapsulated in influential monographs by Anglo-Americans. British evolutionary biologist Sir Julian Huxley (1887–1975), whose *The New Systematics* (1940) and *Evolution: The Modern Synthesis* (1942), and the US-based German scientist Ernst Mayr's (1904–2005) *Systematics and the Origin of Species from the Point of View of a Zoologist* (1942), combined Darwin's concept of evolution through natural selection with Mendellian genetics working amidst systematics to explain population genetics in an evolutionary synthesis. During the Second World War's (1939–45) early years, in the UK and United States, Huxley and Mayr devised a basis for future genetic research, and later used cytology in the postwar period, a method enthusiastically embraced by Yamashina, whose research favored water birds, and especially ducks. Members of the *Anatidae* (duck) family are notoriously difficult to categorize, even using mDNA analysis, because of fertile hybridization and parallel evolution for different species.[176]

However, until the Pacific War cut off extensive networks with Anglo-American counterparts, Japanese ornithologists continued to focus on the systematics model based on taxonomies determined by morphology. This form of investigation slowly becoming outmoded in the Anglo-American world also predicated extensive time in the field collecting specimens, either personally, or outsourced to a knowledgeable collector. As seen in Hachisuka's list of publications where he often collaborated with other Japanese ornithologists, such scientific inquiry neatly intersected with Japan's imperial designs—which accelerated in the mid-1930s and whose focus in Manchukuo (1932–45) is explored in Chapter 5. Similar to how ornithology supported early American imperial endeavors in the Philippines in the late nineteenth and early twentieth century, as examined in Chapter 4, Japanese ornithologists spent a considerable amount of time exploring and categorizing areas absorbed into Japan's empire in the early Meiji period, like Okinawa and Taiwan, with Micronesia added as a Mandate following the First World War (1914–18), and then later, Singapore, southeast China, and Indochina invaded and occupied during the Asia-Pacific War (1937–45).

Conclusion

Like the Anglo-American world, Japan underwent vast change in the nineteenth century, but experienced an apparently ubiquitous efflorescence of modernity in a more rapid and compressed fashion where nature seemed controlled and domesticated by the early Taishô period. Indigenous traditions of *honzôgaku* now merged with imported scientific knowledge, and became codified in institutions like Tokyo Imperial University, which initially hired foreign instructors in zoology and other disciplines to cultivate students who soon became authorities in their respective fields. Initially trained by foreigners in the late nineteenth century, Japanese zoologists specializing in ornithology also emerged on the Anglo-American stage through study abroad or by imbedding themselves in organizations that became transnationalized beyond the European world through their participation. In Japan, ornithology was a direct outgrowth of zoology, and also intersected with parasitology, a specialty of Iijima Isao, Tokyo Imperial University's first Japanese professor of zoology and an avid researcher of birds who also enthusiastically hunted winged game. His students became world-renowned ornithologists competing globally on an unequal playing field as scientists of color in a realm largely dominated by Anglo-Americans. Yet, under their auspices, amidst larger imperial projects of naming and claiming, flora and fauna from Japan's domestic and imperial areas became subsumed under the empire's known parameters in a process of organization mirroring Japanese bureaucratic preoccupations. These scientists often belonged to the Japanese government apparatus as lawmakers in the House of Peers and thus held disproportionate political power in their society.

More generally, since the late nineteenth century in Japan, a growing amateur interest in birds, along with native traditions of bird-keeping and falconry, combined with a rising hunting culture to mirror trends in the Anglo-American world. In Britain and the United States, organized groups like the BUO, AOU, and COC specializing in ornithology developed in countries similarly boasting strong traditions of upland hunting practiced by the upper-classes. These foreign associations also inspired Ijima's 1912 founding of Japan's *Ornithological Society*, whose members sought legitimacy in broader global circles by expanding overseas connections. In prewar Japan, informal and formal ornithological networks merged to support the scientific study of birds under the umbrellas of zoology and various associations that codified their observation and classification.

Many *Ornithological Society* members were hunters just like its founder Iijima. As seen in Japanese publications like *Hunter's Companion* and *Birds*, hunting for birds allowed these scientists to enact imperial masculinity through collecting myriad varieties throughout Japan's empire, and therein carried out an avian imperialism whereby ornithologists served as imperial agents. Their stories unfold in successive chapters focusing on their personal collections, research expeditions, scientific connections with the Anglo-American world, and activities supporting the state.

2

Western Villas in Aristocratic Hands: *Spaces of Imperial Mimesis and Informal Scientific Exchange*

Introduction

As agents of empire, Japanese zoologists and ornithologists were public figures, with domestic spaces influenced by their public personas while entertaining guests and enacting social practices representative of their class. Their passion for birds greatly marked their estates and private living quarters—often occupying considerable room in collections, private museums, and aviaries. To understand how they lived and worked, it is useful to examine spaces where these scientists circulated within Tokyo and its environs, as imperial and postwar Japan's political capital. Japanese aristocrats like Yamashina Yoshimaro, Kuroda Nagamichi, and Hachisuka Masauji built their own private museums, laboratories, and aviaries on large, landed urban estates and at seaside homes. These museums mirrored British counterparts like Tring Park, owned by amateur zoologist Lord Rothschild, Hachisuka's friend.

These scientists' hybrid Western and Japanese professional and domestic milieux, and performances of "honorary whiteness"[1] within, expressed a distinct "cultural mimesis," or a "deliberate performance of sameness that necessarily threatens, or at least modifies, the original."[2] Displaying specimens in home collections also allowed them pleasure in revealing and unveiling an exclusive commodity as a curiosity, with private consumption of these treasures involving the ritualistic opening of specimen drawers and unearthing of carefully warehoused boxes. The rarity and fragility of avifaunal specimens, coupled with narratives attached to the complexities surrounding their acquisition and care, allowed for the creation of an intense ornithological homosociality for these aristocratic men. Specimen exchange functioned like ritualized gift-giving, in a process of political transfer couched in reciprocity. In the postwar period, this practice revived with the reestablishment of peace between Japan and the Anglo-American world.

A unique form of socializing evolved amongst aristocratic Japanese involving the viewing of birds and specimen collections in very distinctive hybrid domestic spaces. Ornithological colleagues were entertained within Western parlors or formal Japanese *tatami*-matted rooms, with hospitality that progressed toward showing specimens in personal laboratories and specimen-viewing rooms. Specimen exchange as a practice, and with whom it was conducted, revealed illuminative hierarchies. These endeavors were primarily masculine, and involved homosocial rituals surrounding collecting;

women marginally engaged in bird-watching or assisted spouses, lovers, and male relatives. Prewar Japanese ornithology served as a highly gendered endeavor where men asserted imperial masculinity through scientific inquiry and exploration.

Spaces of Cultural Mimesis and Scientific Exchange

Comparative literature scholar Barbara Fuchs refers to the reproduction of imperial cultures in early modern empires, like New Spain, as "cultural mimesis," defined as "the fun-house mirror, the reflection that dazzles, the impersonator, the sneaky copy, the double-agent—mimesis, that is, as a deliberate performance of sameness that necessarily threatens, or at least modifies, the original."[3] Such imitation essentially led to "incomplete" copies or hybridization. In the late nineteenth century, imperial Japan similarly underwent a process whereby many social, political, and economic forms were adopted from Western Europe and the United States. In early Meiji Japan during overseas empire building, Emperor Mutsuhito and Empress Shôken (1849–1914) appropriated Western dress at government functions, government officials sported beards and satin coat-tails, ladies-in-waiting stopped blackening teeth, and Japanese aristocrats wholeheartedly adopted Euro-American social customs to appear modern and accrue respect on the global stage. This extended to the household's private arenas, including in language exchanges between family members. Anthropologist Takie Sugiyama Lebra notes that Japan's prewar cosmopolitan elites became enamored of British customs as status markers and encouraged their children to call them "Papa" and "daddy" or "mama" and "mommy" instead of otôsan (father) or okâsan; during postwar democratization, this practice even spread amongst middle classes.[4] Living spaces and leisure activities also acquired hybridization of Japanese and Western styles, with Japanese upper-class dwellings showcasing a new cosmopolitan urbanity.

Historian Jordan Sand underlines the home's crucial nature for a late Meiji national project constructing a particular public-oriented bourgeois domesticity in urban Japan.[5] Japanese reformers pinpointed domestic space and worried how Westerners accordingly gauged Japan's level of civilization, a trend continuing into the early twentieth century.[6] Western-style business areas and estates built by Japan's business elites and aristocratic classes, often members of the House of Peers, showcased the nation's imperial modernity. Particular magnificent examples of Meiji-era architecture remain today, serving as guesthouses for foreign dignitaries and corporations, or as museums sheltering Japan's artistic patrimony.

In 1877, the Japanese government hired University of London-educated British architect Josiah Condor (1852–1920) as a foreign hired hand to teach at the Imperial College of Engineering, later absorbed into Tokyo Imperial University. As Meiji reformers and students recognized Condor's talents, he was commissioned to construct the Ueno-based Imperial Museum (1881), and the famed Rokumeikan ["Deer Cry Pavilion"] (1883), a two-story guesthouse for foreign dignitaries housing a Western-style banquet hall and dancing pavilion. In 1890, he advised Baron Iwasaki Yanosuke (1851–1908), the Mitsubishi Corporation's second president, and enlisted his disciple Sone Tatsuzô (1853–1937) to help design and construct the Marunouchi business district's (1894) brick buildings to resemble London's.⁷ In 1912, Sone designed the neo-gothic Old University Library Building for Keio University's Mita area campus.⁸ The Marunouchi district and Sone's library both survived the September 1, 1923, Great Kantô Earthquake and Allied air raids—revealing their high-quality craftsmanship and the brick and stone buildings' resiliency.

Because of high acclaim, Condor also designed Yanosuke's Western-style estate in Takanawa of Tokyo's Minato ward, modeled after British palace façades like Blenheim Castle's newer portions, and known as *Kaitôkaku* ["Opening Up the East Pavilion"], now the Mitsubishi Corporation's guesthouse.⁹ Sadly, in 1907 after moving in, Yanosuke was hospitalized with cancer and died in spring 1908.¹⁰ A mausoleum built in 1910 lies nearby, housing the ashes of Yanosuke and his wife, plus his son Iwasaki Koyata (1879–1945) and heirs. Like many aristocrats, Yanosuke and Koyata collected priceless Chinese and Japanese antiquities in their *Seikadô Bunko* ["Hall of Tranquil Praise" Library] built in Tokyo's Tamagawa area in 1924, with portions in 1930, caching 200,000 books and 6,500 artworks in the contemporaneous Seikadô Bunko Art Museum.¹¹ This two-story library in reddish-yellow brick resembles neo-Tudor residences common to early twentieth-century affluent Anglo-American suburbs.

By the 1920s, Americanization in popular cultural trends influenced by Hollywood films introduced Japanese to US cityscapes and interiors, plus middle-to-upper-class American living spaces. Wealthy Japanese often sojourned in London, Paris, and New York, and later hired foreign architects to construct Western-inspired mansions on land harboring Japanese *yashiki* [residences, or estates]. The French-inspired Art Deco villa, or *Teien* [Estate Park], of Prince Asaka Yasuhiko (1887–1981), Emperor Hirohito's uncle by marriage, was built in 1933 by architect Gondô Yôkichi (1895–1970), who following a European and American study tour, constructed imperial residences for the Imperial Household Agency's Works Bureau. In 1922, Prince Asaka studied military affairs at St. Cyr in France, but an automobile accident caused a leg limp, which prompted his wife to join him overseas, where they remained until 1925.¹² During the couple's French sojourn, the September 1, 1923, Kantô Earthquake damaged their palace and necessitated new lodgings. The 1925 Paris International Exposition for Decorative Arts inspired their project and helped them attract French contractors.¹³

The residence's interiors feature decorations by iconic French designers, including René Lalique (1860–1945), who created the entrance hall's glass-relief doors and dining hall's fruit-shaped chandeliers; and Henri Rapin (1873–1939), who decorated the mansion's seven publicly oriented rooms, and outfitted them with Art Deco furniture, wall paintings, and the front hall's unique white marble *Kôsuitô* [Perfume Fountain].¹⁴ The villa became an Art Deco marvel transplanted upon Japanese soil to showcase early twentieth-century aristocratic sociability; Japanese scholars today still reference it as a sublime overseas example of the original French movement.¹⁵

Though contemporary French design elements predominate, two bronze lions guard the *Teien*'s front portal; these Chinese symbols of good fortune represent the building's few East Asian features. Despite talismans evoking a broader Chinese cultural heritage, Prince Asaka failed to enjoy long-term happiness in his Art Deco residence upon his wife's death months after completion. Four years later, military duties compelled Prince Asaka into the Second Sino-Japanese War (1937–45): he was the military official whose order allegedly led to the brutal December 1937 attack on Nanking.[16] During the Allied Occupation, the Japanese government requisitioned his home until 1954 as Prime Minister Yoshida Shigeru's (1878–1967) official residence. From 1955 until 1974, the estate evolved into a state guesthouse; in 1983, it opened its doors to the public as an art museum.[17] Despite the *Teien*'s continuous years of service as a national showcase, its paneled and muraled rooms still dazzle with grand spaces of muted elegance transcending their age with faded velvet curtains, crushed plush Art Deco chairs, and yellowed oak furnishings.

Like the *Teien*, numerous estates in Japan's imperial capital survived the war to later serve as requisitioned lodgings for high-level Allied Occupation personnel or embassy residences. The Shibuya villa of Ariyoshi Chûichi (1873–1947), a colonial administrator in Korea and Yokohama's former mayor, became requisitioned "US House 665" to lodge SCAP official and ornithologist Oliver L. Austin and family.[18] The Australian Embassy requisitioned the Mita area Hachisuka mansion, and purchased it from the Marquis in 1951. Later, these villas succumbed to developers' wrecking balls in the heady 1980s, when Tokyo land prices soared to astronomical heights. Numerous records for prewar estates, including photographs and floor plans, still exist for homes surviving Allied fire bombings, while those succumbing to the flames require extensive sleuthing in municipal archives. Yet, the careful study of blueprints and images reveals important clues about how wealthy Japanese aristocrats lived and exchanged scientific knowledge amidst collections of books, antiquities, and specimens.

Hachisuka's British-Style Estate in Tokyo's Mita Ward and Spanish Colonial Revival-Influenced Atami *Bessô* [Second Home]

In 1927, on 13,000 square meters of land in Tokyo's fashionable Mita district, Hachisuka Masa'aki (1871–1932) and his son Masauji built a British-inspired mansion reflecting their studies at University of Cambridge.[19] In 1884, the Hachisuka family had acquired this parcel of land and built a sprawling Japanese-style estate.[20] The new Western-style villa was thus constructed upon land hosting an older traditionally Japanese house, a common practice for upper-class Japanese elites bifurcating their lives between hybrid Western and Japanese spaces. The new construction was meant to lodge Masauji and his fiancé, Princess Kitashirakawa Mineko (1910–70). Yet, as Chapters 3 and 4 illustrate, this engagement dissolved, while Masauji left Japan for an expedition in the Philippines to return only twice to Tokyo in a decade, including for his father's 1932 death. As head of the National Diet, Masa'aki entertained hereditary politicians in a villa mimicking British aristocratic homes.

The Mita Ward Museum still houses features of this historic property, including two walnut wooden doors topped with stained-glass, and a hexagonal bronze light-fixture, salvaged for archiving as markers of the area's Western atmosphere when the Australian Embassy razed the mansion in 1988 for its ultramodern mini-skyscraper. The ancient garden with portions from two Tokugawa Era *daimyô* was spared to retain a traditional Japanese ambiance mirrored against its glassy steel façade.[21]

Prior to the 1927-built villa's removal, embassy staff took numerous color photographs, now cached in a tan photo album resembling those collected by middle-class Japanese and Anglo-American families from the 1980s to 1990s. When I examined them in November 2015, the nostalgic faded snapshots captured walnut-paneled interiors with crown moldings, wooden baseboards, and Anglo-American furnishings, evoking design elements from early twentieth-century Ivy League universities and Seven Sisters colleges.[22] Exquisite Japanese screens featuring diverse birds, and superlative collector's objects in glass display cases, including ancient Chinese or Japanese tea bowls and religious statuary, revealed a location beyond wealthy British or American suburbs. Some formal rooms included tatami-matting, but featured Art-Deco touches in round windows, and geometric Mondrian-like shapes for built-in *tansu* [traditional drawers] housing antiques and collectibles, including specimens. The photographs' mid-century modern furnishings were likely imported from the UK via Australia in the 1950s, while brass chandeliers, stained-glass windows over a wooden staircase, leaded windowpanes, and walnut paneled rooms were the Hachisuka family's requested 1920s-era design features.

As highlighted by Sand, elite domestic space in Japan featured highly cosmopolitan Western elements mixed with traditional Japanese architectural characteristics, and provided a stage for the performance of blended social customs.[23] In the 1920s, Japanese recognized this house as a *bunka jûtaku* ["culture house"], represented by locked doors, two or more stories, and compartmentalization of private space with rooms for entertaining separated from bedrooms.[24] These dwellings represented a radical reconfiguration of space, where inhabitants wore shoes (or slippers) inside and oriented themselves toward tables and chairs arranged in floor plans emphasizing empty space, free circulation of movement, and high ceilings, while traditional Japanese homes were geared toward floors, favoring crouching, sitting, and scooting in smaller, yet more economical, spaces. The generous proportions of individual seating in such residences' primarily Western-style rooms better suited guests in Western attire, though most prewar Japanese women retained kimonos: a 1925 study by ethnologist Kon Wajirô (1888–1973) of 1,180 pedestrians in the Ginza shopping area found that 1 percent of women wore Western-style clothing while 67 percent of men adopted such dress.[25] Western-style residences also allowed Japanese elites to showcase their bodies' modernity amidst domestic spaces in a spectacular ritual of cultural mimesis for public consumption.

Like Baron Iwasaki and Prince Asaka, Hachisuka never enjoyed his lodgings due to the souring of relations with his father Masa'aki and the broken engagement, which propelled a long overseas absence. After the 1937 death of Hachisuka's sister Fueko (1898–1937), he made a final return to Tokyo, and designed a separate property after his 1938 engagement to Chiyeko Nagamine (1909–1997), a wealthy California agricultural developer's Japanese-American daughter. Intended for seclusion from Tokyo's bustle and hectic social scene following his 1939 marriage to Nagamine, Hachisuka built a Spanish-style villa in Atami, possibly modeled upon homes he visited in Mexico City during a mid-1930s expedition to collect birds or on those in Los Angeles where he recovered afterwards from a devastating illness. The Atami villa was also a space where Hachisuka's disciple Nakamura Tsukasa remembered that, in the late 1940s, his mentor insisted on speaking English to prepare him for graduate studies at University of California at Berkeley.[26] This exquisite three-story home clad in limestone three hours from Tokyo by train served as a hybrid cultural setting for Hachisuka's work and personal affairs until his sudden 1953 death.

Designed by Matsunoi Kanji (1895–1982), head of the Tokyo branch of American architect and Christian lay missionary William Merrell Vories' (1880–1964) architectural firm,[27] the house sported a distinctive tower, whose setting reminded Nakamura of a Spanish-style villa set on the Mediterranean like in a foreign film.[28] Beginning in 1919, Matsunoi studied at Columbia University in New York, where he learned architectural design for three years, and remained in the United States until 1932.[29] There, he observed popular American home design styles like Mission Revival (1890–1915) or Spanish Colonial Revival (1915–31), propagated by the 1915 Panama-California Exhibition in San Diego, and featured in Hollywood films displaying exotic suburbs like Beverly Hills and Bel Air.[30] Spanish Colonial Revival architecture also enjoyed global adoption, including in Shanghai and the Philippines by the 1920s.[31] In Tokyo, these design elements appeared in the still-extant Ogasawara-Tei [Ogasawara Estate],[32] built in 1927 in Shinjuku by Count Ogasawara Nagayoshi (1885–1935), a House of Peers member also studying at University of Cambridge who later served as an Imperial Household Agency official[33]; the estate now hosts a banquet hall and wedding venue.

For the W. M. Vories Company, Matsunoi also designed Toyô Eiwa Girls' Academy, completed in 1933.[34] This school, with its distinctive white mission-style buildings featuring Art Deco and Spanish-style elements in its orange roof tiles and white-washed stucco walls, still exists in Tokyo's swanky Minato Ward Azabu area up a steep hill along Toriizaka Road. Its buildings line the street bordering the International House (hosting the 1960 meeting of the International Council for Bird Protection), built in the early 1950s on Mitsubishi conglomerate chairman Baron Iwasaki Koyata's former estate. In its hilltop location pre-dating Tokyo's distinctive postwar Roppongi skyscrapers, the Academy was likely visible from Hachisuka's British-style mansion's top floor. Undoubtedly, he admired the school's architectural style referencing his southern California sojourn or Mexico trip. After completion, the newer Atami house served as Hachisuka's work retreat amidst collections of live exotic animals, books, and specimens distanced from his wife, with whom his relationship worsened as wartime progressed.

The Atami Spanish Colonial Revival villa also served to entertain ornithological visitors like the poet Nakanishi Gotô who praised its Western qualities in his prose poem *Hakua no yakata* [Limestone Mansion, or "White House"].[35] Near the home's entrance was a parlor, enhanced by Hachisuka's trophies from African expeditions, including black rhinoceros heads, and Japanese serows gracing its walls, along with a tiger skin rug. In his poem, Nakanishi called this room a "private space," with a bedroom next door located far from a staircase to Hachisuka's wife's room upstairs. In this study, Lord Rothschild's photograph was perched atop a fireplace mantle clad in Moorish tiles, peering at numerous books and periodicals. Western visitors remained shod like Hachisuka, who provided Japanese guests with slippers as he remained barefoot upon heated floors.[36] This near-simulacrum of Westernized space provided a frame for Hachisuka's ornithological activities, and served as an arena with generous spatial proportions and doors facing a tropical garden where he finally felt relaxed at home.

Decoding meanings of entertaining guests in such mimetic spaces, and understanding how ornithological knowledge was transmitted there, reveals much about Japanese scientists' social relationships and how national identity was expressed within "private space." Specimen exchanges additionally played important social roles for these highly politically connected collectors to craft alliances and express rivalries.

Specimen Collecting and Exchanges as Politicized Sociality

Ornithologists' grand explorations—later codified into published texts solidifying scientific research and placing it upon imperial maps—unfolded in a dilettantish fashion at home, where specimen collection evolved into a personal endeavor bolstering social status and highlighting broader national roles and self-identity in domestic Japan. Showing off rare exotic bird skins collected in far-flung parts of the empire and beyond by the scientist or his collectors became a social art. Systematists, or specialists in taxonomy like Hachisuka, Kuroda Nagamichi, his son Nagahisa, and Austin all reveled in collecting a perfect male-female pair or completing a "full set," or all avian species within a particular region or country, while naming new species. In prewar times, Japanese scientists formed a collecting culture allowing them to privately show expensive or rare specimens to friends and colleagues.

In examining such collecting practices, it is tempting to find similarities with contemporary *otaku*,[37] a term first used in 1983 by Japanese essayist Nakamori Akio to mean "your honor"—or, literally, "honorable dwelling," to mock excessive politeness by nerdy male *manga* [comics or graphic novels] and *anime* [animated films] enthusiasts in closed online communities[38] who proudly collected these media at home in an obsessive hobby.[39] These (usually male) fans, who collect diverse mass-produced pop-culture products or consumer items like watches and Pokemon toys, devote a considerable time to their hobby to quasi-scientifically assess an item's total characteristics. How might contemporary collecting of consumer products intersect with past and present collections of bird specimens, a natural and often hard-to-acquire commodity?

Walter Benjamin, in his 1930s sketch on "The Collector" within his unfinished *The Arcades Project* (1927–40), evocatively describes early twentieth-century collectors and their collecting habits.[40] Here, Benjamin asserts that collected objects become micro-histories of owners' preoccupations set in a particular moment, but are constantly added to (and pruned through gifting) in efforts to finally attain completeness or perfection. For early twentieth-century Japanese ornithologists, acquiring all representatives of certain studied species, and discovering new ones, required scouring the empire and other colonial environs for elusive, new additions in enacting avian imperialism. Ex-Smithsonian curator Steven Lubar posits that objects can inform political and social hierarchies,[41] so Japanese ornithologists' *natural history* specimen collections also form a key *national history* of Japan's imperial elites. As agents of empire, the domestic spaces of Japanese scientists represented their public personas in entertaining and socializing.

In a similar vein, British science historian Samuel J. M. M. Alberti describes how wealthy collectors and the British bourgeoisie understood cabinets or displays of curios and specimens "as emblems of their cultural and social status."[42] A similar dynamic spread amongst aristocrats in Meiji Japan, where collections of rare objects and natural history specimens conferred cultural and social capital to owners in a politicized form of sociality when shown and revealed to guests and visitors. Displaying of specimens from collections additionally allowed the pleasure of revealing and unveiling an exclusive commodity as curiosity.

This practice resembles "spectacular consumption" where Tokugawa-era *daimyô* at seasonal moments or on special occasions revealed rare ancient tea utensils and bowls from China, sometimes from the Song dynasty (960–1279) boasting artfully cracked glazes prized for simplicity, in a process emphasizing "acquisition, exchange, and display" amongst feudal elites.[43] Spectacular consumption extended to acquisition of falcons, enabling intelligence activities conducted across large territories in surveillance expeditions seeking an elusive quarry later used to hunt; these predatory birds then served as gifts to create and solidify social and political alliances. Tea ceremony objects and trained raptors associated with men in authority created value and agency through their usage in socialization and exchange amongst masculine elites.[44] This deeply political process contributed to establishment of an ever-fluctuating social hierarchy whereby spectacular accumulation and exchange of rare items strengthened ties, while atrophying the practice could erode such ties.

Rarities could include quasi-legendary items like ultra-rare Song dynasty tea bowls, of which mere hundreds exist, in a fascination for collecting similarly extended to rare bird specimens. Of many priceless heirlooms passed down from Matsudaira ancestors to Marquis Kuroda Nagashige (1867–1939) and his son Nagamichi, their most spectacular item was a deceptively simple black *jian* tea bowl from the Southern Song period (1127–1279), registered as an "Important Art Object" on December 18, 1935, and featuring a lavishly sublime *yûteki tenmoku* [oil spot] glaze upon the blackish-brown ceramic.[45] Nagamichi also owned an exceptional pair of stuffed and mounted crested sheldrakes (*Tadorna cristata*), believed extinct, and known as the Korean Mandarin Duck that he described as *Pseudotadorna cristata* in 1917; this pair poised on wooden mounts featured a male collected in Korea in 1913 and a female in 1916.[46] Amidst a convivial atmosphere of conversation and drink provided by servants, the Kurodas displayed these precious objects to guests invited to their estate in a leisurely process of unveiling prized treasures.

In late nineteenth-century and early twentieth-century imperial Japan, with the rise of the public museum inextricably tied to national aims,[47] zoologists' and ornithologists' private displays also deeply referenced imperial aims, since these politically connected men represented Japan's empire by displaying scientific prowess and intrepid exploration of hard-to-access corners of the empire and its borders. Acquisition of dead specimens and living animals also translated into territorial possession, similar to imperialization of Indigenous peoples by pushing them off native lands and onto contained reservations (like for northern Japan's Ainu). According to historian Ian Miller, animals from far-flung territories served as "imperial trophies."[48] The rarity and fragility of such avifaunal specimens, coupled with narratives attached to complex acquisition and care, allowed for the creation of intense sociality amongst aristocratic men who also enjoyed showing off birds within their estates' aviaries.

In nineteenth-century Great Britain, Alberti notes that "in their private residences ... collectors—like other proud owners of large eighteenth-century houses—welcomed esteemed visitors, who in turn conferred status upon the museum or mansion. The connoisseur therefore had to make his residence accessible to those persons deemed expert enough to judge the collections, and preferably to account for them."[49] Conferral of status by validating guests also pertained to prewar Japanese ornithologists, where socialization came through exclusive access via extensive entertaining couched as scientific exchange. Contrary to the open, static public display of museum collections, consumption of private treasures involved ritualistic openings of specimen drawers and cached boxes in personalized unveilings. Japanese aristocrats built private museums on their estates, including Yamashina's Shibuya laboratory, which survived Allied air raids and became the Yamashina Institute for Ornithology from 1942 until its 1984 move to Abiko in Chiba Prefecture. Such museums mirrored British counterparts like Rothschild's Tring Park, opened to the public in 1892 and owned by Hachisuka's friend and mentor; both men carefully tended living, and taxidermied, collections at their country or seaside estates.

In prewar times, Japanese and Anglo-American individuals and institutions also exchanged rare bird specimens. Hachisuka's contributions initially appear in a 1935 Smithsonian Institution "Report on the Progress and Condition of the United States National Museum."[50] In fiscal year 1939 to 1940, for the Smithsonian's *The Report of the National Museum*, Division of History Curator Theodore T. Belote (d.1953) noted 67 accessions, "including 2,628 specimens," and remarked this was "somewhat less than those recorded during the previous year,"[51] possibly due to European wartime conditions. From 1906 until 1917, Belote served as Assistant Curator, after which he worked as Curator from 1917 until his June 30, 1949, retirement.[52] In notes on accessions received in 1940, contributions of "Marquess Masauji Hachisuka" appear as "1 skin and 154 alcoholic [sic] of Noguchi's woodpecker (154697, exchange)."[53] Hachisuka likely exchanged these for other specimens in the ample Smithsonian natural history collections. Today, the Noguchi Woodpecker (*Dendrocopus noguchii*), or Okinawa woodpecker, is critically endangered, with 50–249 mature adults remaining.[54] Even then, when American travel to tropical Japanese imperial possessions like Okinawa was difficult and uncomfortable for a Westerner, receipt of a skin and multiple jarred specimens of this rare species was a coup for the museum and represented Hachisuka's largess. However, this gift and exchange moreso represented his desire to gain prominence amongst American scientific circles.

Such exchanges of specimens and scientific knowledge with Anglo-American ornithological contacts continued deep into the Pacific War. In 1942, Hachisuka entrusted the *Ornithological Society*'s latest *Handlist of Japanese Birds* to two soon-to-be deported Australian diplomats; it then traveled nearly two years on a nebulous wartime sea route from Japan which Austin supposed was on the repurposed Swedish cruise ship MS Gripsholm, chartered by the US Department of State from 1942 to 1946 for repatriation duties,[55] after which it apparently journeyed from the Philippines to San Francisco, from whence it travelled to Curator of Ornithology "Jimmy" (James

Lee) Peters (1889–1952) at Harvard University's Museum of Comparative Zoology (MCZ).[56] Sent from the other direction on the American side, right before the December 1941 Japanese Pearl Harbor attack, Peters likely mailed Yamashina a hawk specimen.[57]

As a top American ornithological authority, and the AOU's vice president from 1938 to 1942, and president from 1942 until 1945, Peters maintained a keen interest in Asian birds.[58] With the simultaneous Japanese invasion of the Philippines, a US colony, the wooden box stalled on its journey at a Manila warehouse, but with peace in the late 1940s, it arrived at Yamashina's Shibuya laboratory, with a miraculously unharmed specimen.

This practice of reciprocal exchanges continued into the postwar, when Japanese aristocrats (re-)incorporated their American overlords into these exchanges. During the Allied Occupation, fascination for raptors embedded in domestic Japanese aristocratic social practices melded with popular American admiration for symbolism of predatory birds, like eagles, amongst high-ranking US military brass. Yamashina's autobiography mentions a fateful unveiling of specimens (and rare eagle egg gift) to his former enemy, whose weekly visits helped him survive the harsh early postwar period by accessing extra food and medicine, while engendering the conquerors' amity with his ornithological expertise:

> Major General Wolfe[59] from the Yokohama military base headquarters came every week on Saturday afternoon by driving his own car to look at eagle and hawk specimens and investigate my books, and then returned. Since these American military personnel at such occasions left items during their visit, this really helped at times without resources. Major General Wolfe was transferred back to return to the United States after about a year, but at this time, when I gave him the egg of a Japanese mountain-hawk eagle (*Nisaetus nipalensis orientalis*), he was extremely glad, and after he returned back to his country, he also sent me letters.[60]

This measure of reciprocity also benefited the Americans, who needed Yamashina and his colleague Duke Takatsukasa to push through legislation amenable to US interests as members of the National Diet's House of Peers. Hachisuka also held influence there through his deceased father's connections, and saw work with Austin for the Occupation as crucial to economic survival while restarting his scientific endeavors. By 1950, he donated numerous collections to American ornithologist S. Dillon Ripley (1913–2001), who incorporated them into Yale University's Peabody Museum of Natural History, where he served as head curator.[61]

Such exchanges of spectacular objects from personal collections derived from a relatively long practice of ritualized gift-giving tied to political ramifications. Sociologist Marcel Mauss in his iconic 1925 work *The Gift* discusses reciprocity based on social exchanges where gifts could imply continuing relationships.[62] Mauss named reciprocal exchanges in premodern societies prior to the existence of markets as a "system of total services," where the "gift" presupposes an item in return in embodying the giver's spiritual essence, and thus is not "free."[63] For Mauss, any gift presumes reciprocity: "The most important feature among these spiritual mechanisms is clearly one that obliges a person to reciprocate the present that has been received."⁶⁴ Reciprocal gift-giving and politicized sociability between a French aristocrat and a Japanese Duke in the mid-1920s reveal how a European scientist viewed and interacted with Japanese counterparts, thus embedding himself in a particularly Japanese form of homosociality surrounding birds.

Japanese Ornithologists' Interactions with a Leading Doyen of the European Bird World

Much like his British counterpart Lord Rothschild, French ornithologist Jean Théodore Delacour first became renowned in Western Europe and the Anglo-American bird world due to his extensive aviaries and well-stocked personal zoo. Filled with rare avifauna located at Clères, his Normandy chateau was purchased in 1920 after the First World War's worst tank battles completely destroyed his Villiers family home in northern France's Picardy region and Somme department.⁶⁵ Delacour also maintained extensive ornithological connections in England and the United States through travel and correspondence, and became an important nodal point for Japanese ornithologists. In the mid-1920s, following travels in French colonial Indochina, he met and befriended imperial Japan's key bird experts. These relationships transcended the wartime period, momentarily arresting their exchanges, but revived in the postwar, when Delacour and Japanese counterparts like Yamashina became global advocates for bird conservation. In 1922, the International Council for Bird Preservation (ICBP), the precursor to Bird Life International, was formed by Delacour and other Anglo-Americans and Dutchmen.⁶⁶ Besides Yamashina, Takatsukasa, Hachisuka, and Nakamura would play significant roles in supporting the activities of this Cambridge-based British organization partially sponsored by Delacour.

In 1914, Hachisuka's first mentor Takatsukasa graduated from Tokyo Imperial University with a degree in zoology, which he earned after studying with parasitologist and ornithologist Iijima Isao. In 1919, as an *Ornithological Society* member whose first president was Iijima, Takatsukasa invited a sixteen-year-old Hachisuka to join, which marked Takatsukasa's accession to a status where he could accept disciples, or *odeshi*, in transmitting scientific knowledge from senior to younger scientists. A year later, Hachisuka traveled to Great Britain and soon decided to study ornithology at University of Cambridge. In 1922, after Iijima's death, Takatsukasa became the *Ornithological Society*'s president, and thus garnered the credentials to extend his reach into Anglo-American networks.

Between 1924 and 1925, Takatsukasa traveled extensively throughout the United States and Europe to pursue ornithological endeavors and build international connections, even posting messages in key Anglo-American ornithological journals, while popular Japanese newspapers like *Asahi Shimbun* noted his travels. On April 20, 1925, right before he embarked on travels to Europe, *Asahi Shimbun* reported a send-off ceremony at its headquarters to wish him success.⁶⁷ Takatsukasa's European journey for ornithological research thus garnered national publicity and a media social event. In July 1924 in the AOU's journal *The Auk*, the "Notes and News" section indicates that "Prince N. Taka-Tsukasa [sic], President of the Ornithological Society of Japan, has recently been making a brief tour of the United States and examining the leading ornithological collections. The prince is especially interested in the Psittasidae [Parrots]."[68] Though Takatsukasa rightfully was a Duke, it describes him as "Prince" to augment his social status, while his leadership of Japan's *Ornithological Society* highlights his authority to gain potential entrée into exclusive American ornithological circles like the AOU. The article nudged likeminded readers to extend invitations to display their parrot collections to him.

Much of Takatsukasa's mid-1920s international ornithological activities and outreach to globally influential ornithologists can be gleaned from Delacour's writings in *The Avicultural Magazine*, a British periodical published by the UK-based Avicultural Society, whose history arose from "a small group of British and foreign bird-keeping enthusiasts [that] met at Brighton in 1894 with a view to forming a society devoted to their interests" and who coined the word "aviculture" to reference breeding and raising birds.[69] From 1920 until 1940 when the war exiled him to the United States, Delacour spent two decades ensconced at his Edenic Normandy chateau Clères, which boasted aviaries and a zoo; there, he wrote periodic reports for this magazine for the Avicultural Society to which he belonged. In the February, March, and April 1922 issues of *The Avicultural Magazine*, Takatsukasa introduced European readers to Japan's more than 1,700-year history of aviculture in a series of articles.[70] These essays keenly captured Delacour's attention due to his own bird-keeping at Clères of myriad species collected in the field or gifted by others.

In his 1966 autobiography *The Living Air*, Delacour complained about the Normandy countryside's dark barren northern winters, which drew him to tropical climes beginning in autumn, after which he returned in spring "to enjoy the first flowers and the first nests."[71] Accustomed to winter trips to North Africa and Indochina, but intrigued by Takatsukasa's descriptions of Japanese bird-keeping before finally meeting the ornithologist in Paris in 1925, he decided to travel thrice to Japan in the mid-to-late 1920s. Also, his budding friendship with a young Hachisuka, who he first met in Paris through Takatsukasa's introduction, helped inform this decision. In *The Living Air*, Delacour even describes him as Takatsukasa's "young cousin."[72] Three successive articles appearing in the July, August, and September 1926 editions of *The Avicultural Magazine* detailed Delacour's first trip to Japan, where he extensively described personal impressions of Japanese bird-keeping and the remarkable birds he received as gifts from his new aristocratic ornithological friends.

In a July 1926 article, Delacour lists a remarkable forty-three birds shipped home from Japan amidst winter travels in Asia:

3 pairs of Copper Pheasants, 1 male Ijima Copper Pheasant, 3 pairs of Green Japanese Pheasants, 1 Chinese Spot-billed Duck, 4 Japanese Blue Magpies, 1 Japanese Bullfinch, 3 Yellow-throated Buntings (*Emberiza elegans*), 2 Japanese Meadow-buntings (*E. cipiopsis*), 1 Japanese Blue Flycatcher (*Cyanoptila*), 7 Japanese Zosterops, 5 Varied Tits, 3 Loo-Choo Robins, 3 Japanese Robins.[73]

Several of these, plentiful in Japan, but rare in Europe, were gifted by Takatsukasa or Kuroda Nagamichi in hospitality extended to the well-traveled French aristocrat. Delacour refers to them, including "M. Matsunaga and F. Mitsui" as "my Japanese friends ... whose kindness to me during my visit ... I cannot acknowledge sufficiently."[74] These pheasants, songbirds, and a Japanese duck found a home in Normandy within his large collections, but some were also given to a man Delacour describes as an "intimate friend" with whom he "spent the third week of each month across the Channel" when in Europe[75]—British bird breeder Alfred Ezra (1872–1955), the Avicultural Society's president whose Foxwarren Park estate boasted one of Europe's finest zoos with rare birds.

In his initial September article, Delacour noted the Japanese tendency to raise small birds rather than larger varieties more common in Europe, and extends praise for the meticulous care in keeping them: "Small birds, however, are the favourites with the Japanese, who can keep and breed them perhaps better than any other people. Both insectivourous and seed-eating birds seem to thrive as well, if not better than in any other country."[76] Much of his prose betrays besottedness with Japan in an Orientalist exoticism prevalent amongst prewar French elites. He also lauds birdcage aesthetics described in superlative comparisons common to European observers fetishizing an Asian culture:

> Japanese cages are simply wonderful; whether they are open bamboo cages or breeding-box cages, they are always pretty and most beautifully constructed. In comparison, our best cages look desperately coarse, unfinished and tasteless ... Open bamboo cages generally rest on a pretty lacquered tray, from which they remain separated by a movable barred bottom. For each cage there is a special case, into which it can be put at night.[77]

Although Delacour portrays the refinement and cleverness of Japanese cages for a British audience primed for exoticism, he also viewed the Japanese as a highly civilized, intelligent, and aesthetic people who possibly surpassed Europeans in craftsmanship and aestheticism. All were common tropes in 1920s-era Anglo-American popular imaginations of Japan which nevertheless betray a standpoint of paternalistic European cultural superiority.

In a subsequent August article on "Japanese Aviculture," Delacour raves about the high quality of Japanese bird-keeping, and views certain Japanese methods of raising birds as superior to European ones. He discovered that Japanese fed their birds a high quality diet far exceeding European prepared bird feed: "In Japan all insectivorous birds, native and foreign, are fed on the same mixture, composed of ground husk of rice and rice itself, salad and fish-meal [sic] I sincerely believe that this food is the best of all such artificial foods."[78] In what he viewed as a practical and highly civilized Japan, bird enthusiasts could avoid dirty and tedious insect procurement for their pets: "One must bear in mind, of course, that no live insects are given in Japan, except in the case of moult or illness: mealworms are not obtainable. Consequently birds thrive on the artificial food and on it only [sic] all there is to do is moisten."[79] This method, again, was Delacour's evidence of a highly civilized people in his exoticized perspective.

In imperial Japan, Delacour also noted which bird varieties were most likely kept, and indicated that native species (including ones from Okinawa) were easily accessible, while those from China and Southeast Asia were common, although rare varieties from the colony of Taiwan still were highly sought after. As small caged birds, "The usual inmates ... are the lovely Loo Choo and Japanese Robins, Blue Flycatchers *(Cyanoptila)*, Zosterops, various Tits and Thrushes, Redstarts, Bush Warblers, Buntings, Orioles, Jays and Magpies, etc. Higher cages are used for Larks. All these native birds are easily obtainable and commonly kept."[80] He also noted how "a good many South American and some East African birds can also be obtained. European, North American and West African birds are extremely scarce."[81] In Delacour's observations, Japanese bird-keeping curiously followed imperial hierarchies that paralleled prevailing European understandings of race and ethnicity. He also toured the country's three main public zoos in Tokyo, Osaka, and Kyoto, indicating the rarest and most remarkable bird species housed at each, which betrayed global collecting sprees supplemented by imperial gifts:

> At Tokyo—Formosan Occipital Pies *(Urocissa careulea)*, Japanese Storks ... Formosan Sibias *(Lioptila auricularis)*, Alcippe *(A. morrissoni)* ... Yucatan and Pileated Jays, Roulrouls, Manchurian, White-necked, Hooded, Australian and Black-necked Cranes, European and Australian Pelicans, beautiful Pelagic Sea Eagles ... At Osaka - Manchurian, White-necked, Demoiselle, Sarus, Common, Hooded and White Asiatic Cranes, Japanese Storks ... Philippine Pelican ... Giant Barbet, small Japanese Woodpecker *(Iyngipicus)*, Cuckoo, Mexican Toucans, Pelagic Sea Eagles ... At. Kyoto - European Pelicans, Manchurian, White-necked, Sarus, Common, Demoiselle and Hooded Cranes, a Condor, various Sea Eagles.[82]

In addition, Delacour was invited to the private estates of Takatsukasa, Kuroda, and Yamashina, all located in central Tokyo's elegant Shibuya and Minato Wards. However, one scientist was anxious to repay the Frenchman's lavish hospitality at his chateau and Paris flat. Delacour first mentions his visit to Takatsukasa's grounds, and refers to him as a prince: "Prince Taka-Tsukasa [sic], who visited Europe last year [1925] and is well known to many of us, keeps a large number of birds in his garden in Tokyo" where "Guiana Parrotlets breed freely."[83] Escapees from Takatsukasa's expansive garden during the war's destructive chaos likely provided precursors to thriving postwar colonies of similar-looking parakeets swarming in central Tokyo.[84] Adept at social customs amongst French elites, Delacour presented Takatsukasa with rare live specimens safely procured during travels in French Indochina before his Japan trip, which included "a fine male Rheinardt's Argus Pheasant, Edwards' Pheasants, Siamese Firebacks, Tantalus and Episcopal Storks, Black-headed Ibises and Edward's Porphyrios."[85] The rare Edwards's Pheasant *(Lophura edwardsi)*, "discovered" only in 1896 by the Paris-based French ornithologist Alphonse Milne-Edwards (1835–1900), and now extinct in the wild, was a particularly welcome gift. Delacours collected several such birds during his 1924, 1926, and 1928 expeditions to Indochina, of which two females and one male were given to Takatsukasa in 1926; these later produced six chicks by 1928, of whom some were given to aviculturists to create a successful Japanese breeding colony.[86]

Such efforts revealed to him possibilities for staving off extinction with superior bird-keeping methods like those practiced by Japanese.

In particular, Delacour lavished his greatest praise upon Kuroda Nagamichi, to whom Delacour refers by his doctorate, also earned at Tokyo Imperial University under Iijima. Notably, Kuroda was the first family heir not educated at Cambridge; his father Nagashige was admitted to King's College in January 1885 and matriculated during Lent Term that year, earning his bachelor's degree in 1887,[87] and in 1891, receiving his M.A.[88] Nagashige's younger brother Nagatoshi (1881–1944) also matriculated at Cambridge nearly two decades later, but never graduated.[89] When the elder Kuroda returned to Japan, he received a plethora of titles and new responsibilities due to his British education: "Master of Ceremonies, a member of the House of Peers, Vice-President of the House of Peers, Privy Councillor and so on."[90] This background of intergenerational connections to Great Britain through his aristocratic father and uncle allowed Nagashige's son Nagamichi an entrée into politics and preferential treatment in Japan's scientific world.

Delacour was most interested in Kuroda's avicultural successes in his large waterfowl collection, including common Japanese duck varieties, plus a vagrant American Wigeon (*Mareca americana*) drake,[91] all captured at his *kamoba* described as a "duck-hunting ground."[92] At the estate, "a fine old house situated in a large garden,"[93] where the *Ornithological Society* held its meetings, "[i]n a courtyard close to the garden and on the way to the museum, there are many aviaries and a bird-room. The jewels are three males and one female of the fine Mikado Pheasant."[94] In 1906, *Syrmaticus mikado* was named for Meiji Emperor Mutsuhito after Taiwan was absorbed into the Japanese empire following the first Sino-Japanese War.[95] Indigenous Taiwanese caught it for long tail feathers to decorate ceremonial dress. Delacour notes how the bird "has quite disappeared from European aviaries, and as the Government rightly prohibits the capture of this rare bird the only chance we have of ever seeing it again in our countries lies in Dr. Kuroda's future success with his birds and new ones he may obtain through an official permission."[96] Wild pheasant genera are only indigenous to Asia and areas in the Caucasus and Balkans, with European varieties usually captive or introduced species.[97]

The Mikado Pheasant was dear to both Takatsukasa and Kuroda, who in the March 1922 edition of *Birds*, proposed a new genus, or *Neocalophasis*.[98] Moreover, in February 1926, Kuroda published a monograph on *The Pheasants of Japan Including Korea and Formosa*, where he described over a dozen types found in Japan and its empire[99]—yet much of his knowledge was gleaned domestically from six specimens collected for him in Taiwan's mountains and three living representatives from his personal aviary.[100] In a beautifully bound folio-sized volume, Kuroda built upon American ornithologist and explorer William Beebe's (1877–1962) superb, and critically acclaimed, *A Monograph of the Pheasants* (1918–22).[101] Kuroda's publication also rectified the fact that an American had once served as an expert on this favored native Japanese game bird, protected on imperial preserves from commoners' predations, and composing an essential imperial identity favoring hunting and dominion over land through stewardship over birds. Witmer Stone (1866–1939), *The Auk*'s editor and the book's reviewer, soundly praised Kuroda's work: "It forms a most important contribution to our knowledge of this interesting group of birds."[102]

Originally produced in the early twentieth century, Beebe's *Monograph* was his most notable work as ornithological curator at the New York Zoological Park (now known as the Bronx Zoo). On January 26, 1941, for the same position, the Bronx Zoo later hired an exiled Delacour as consultant;[103] a decade later, Delacour updated both Beebe's and Kuroda's work when he published a 347-page monograph on pheasants called *The Pheasants of the World* (first edition, 1951; second edition, 1957; third edition, 1964; and fourth edition, 1977).[104] The wealthy philanthropist and businessman Anthony R. Kuser (1862–1929) initially funded Beebe to research the world's pheasants, a then poorly understood bird genera from the *Phasianinae* subfamily, in a project meant to bring greater renown to the zoo.[105] Beebe's observations were gleaned after seventeen harrowing months in the field collecting and observing all known types of pheasants and discovering others, including Kuser's Blood Partridge (*Ithagenes kuseri* Beebe) from China's northern Yunnan Province named for his benefactor, during circuitous travels through Ceylon, India, the Himalayas, Indonesia, Singapore, Borneo, Sumatra, Malaya, Burma, and China from 1910 until 1912.[106]

Following these expeditions, Beebe made a stopover in Japan, where he and his wife née Mary Blair Rice (Blair Niles [1880–1959]) enjoyed an exhaustive social schedule, but waited for a permit to study birds in the Meiji Emperor's Imperial Preserve. This permit arrived sealed and handwritten two weeks later; indeed, Beebe felt that the Japanese ornithologists he met in New York and encountered again in Tokyo "were mired in bureaucracy and making scant progress" with little knowledge gleaned from field observation.[107] In frustration, the couple went to China, which they had fled to outwait a plague outbreak. Beebe returned to study copper and green pheasants in the Imperial Preserve, where the Imperial Household gifted him with two cranes with his promise of a future pair of swans from New York's Bronx Zoo.[108] (Sadly, one crane suffocated in its crate en route to the United States.)[109] Back in Tokyo, Beebe lectured at the Tokyo Zoological Society, with Takatsukasa and other Japanese ornithologists in attendance, but felt odd to "tell them about their own birds."[110]

This lack of broader knowledge on pheasants in Japan was remedied by Kuroda's publication in January 1926, whose copy he gave to Delacour while showing off his own live pheasant collection. Kuroda's expertise on pheasants made a lasting impression on Delacour who later published his own study.[111] Besides the charismatic pheasants, Delacour also mentions the more staid inhabitants of Kuroda's garden aviaries: "There are also ... two species of Godwits, Turnstones ... Gulls (*Larus crassirostris*), a beautiful rare Lory (*Eos rubiginosus*) Japanese Robins and 'Nightingales' (*Horornis c. cantans*) and *H.c. canturians*, Suthoras [Webb's Crow-Tits] ... and different seed-eating birds."[112] Clearly, songbirds were not as fascinating to Delacour, but his tours of these notable Japanese ornithologists' aviaries permitted him a close look into their bird-keeping methods.

Lastly, Delacour visited over a dozen aviaries at Yamashina's Shibuya estate, where he also kept smaller bird varieties like Waxbills. Yamashina's grounds featured "one suite of eight and another of five small aviaries, with roomy houses, each one for a pair of birds. I noticed among others Diamond Sparrows, Sydney and Crimson-rumped Waxbills (*Estrilda rhodopyga*) with nests and young, three birds which do not breed in cages."[113] Rather impressively, Delacour noticed how Yamashina provided such superior conditions for domesticating birds that normally wild birds bred in captivity—an important issue, since the turn-of-the-century saw mass extinctions like the passenger pigeon, but also near-extirpations of rarer species only recently discovered, like the Sheldrake described by Kuroda in 1917 and examined in a 1924 *Birds* article.[114] However, not all Japanese boasted the cosmopolitan connections of these scientists, whose friends brought rare birds from their travels as gifts, nor possessed resources to maintain such impressive collections. Nevertheless, their meticulous care of birds represented a broader trend in Japanese society with deep historical roots.

Even though Delacour spent most of his time amongst professional ornithologists, a more plebian bird-keeping craze also existed in prewar Japan, with antecedents dating back to the Tokugawa period and before. During his visit to Osaka, Delacour noticed that the city "has over one hundred bird shops, fifty of which have been started within the last two years; it shows plainly that aviculture is rapidly increasing in the country."[115] He extends his highest praise to bird-keepers in Japan and their aviphilia; however, within covertly patronizing compliments, he implicitly harbors colonial assumptions of the West's superiority, despite laudatory words: "I hope the above notes will show to many of us, who may think that we are the only good aviculturists in the world, that our Japanese friends have not much to learn from us, and our members should do their best to keep pace with such enthusiastic bird-lovers."[116]

According to *Asahi Shimbun* advertisements, Takatsukasa published a manual on how to raise birds as early as 1923,[117] with subsequent advertisements for revised editions of this same guide printed by the publisher Sôkabô from 1924 until 1926.[118] During his visit, an animated Delacour likely urged Takatsukasa to write more detailed books to document his superior bird-keeping methods, which Takatsukasa later published from 1927 until 1930. Takatsukasa more likely recognized in Delacour's enthusiasm a broader Western interest, and saw this as a timely opportunity to publish methods already endorsed highly by Japanese friends and by an influential European in global ornithological circles. Hence, his 1930 *Collected Works on Bird-Keeping*, complete with color illustrations,[119] was simultaneously published with an English-language version titled *Illustrated, Colored, Bird Collection, with Care and Feeding*.[120] Since Takatsukasa wrote his English-language books by himself, it took him longer to publish, unlike Marquis Hachisuka, who often outsourced editing to "Ms. M. Lawson," a woman typist at the British Museum of Natural History who he hired for the purpose.[121]

Takatsukasa worked on perfecting his observations in exhausting detail, like his later *Birds of Japan* project, whose volumes were published from 1932 until 1943, which never progressed beyond a volume devoted to *Galliformes* (chickens and relatives); the war's impact likely shelved this project due to his growing responsibilities. All of these publications, including a short 1941 tourist booklet on *Japanese Birds*,[122] were meant for an Anglo-American audience familiar with Delacour and curious about his Japanese counterparts, while the country's unique bird life also drew these foreigners to Japan's shores.

Notably, Delacour's 1926 impressions of his Japan visit in *The Avicultural Magazine* provide an intriguing glimpse into how gift-giving customs and social reciprocity functioned in Japanese aristocratic society where these particular forms of sociability involved exchanges of both live and prepared bird specimens. Nevertheless, besides caged birds carefully tended in estate aviaries, birds viewed and encountered in nature and the wild were also important occasions for elaborate forms of sociability with hidden political ramifications amongst bourgeois Japanese.

Homosocial Encounters and Bird Watching: Producing Conservation amidst Gendered Social Networks

Outside of formal Japanese ornithological scientific networks, while often intersecting with them, bird watching served as a leisurely pastime for rising middle to upper classes and allowed them to engage in amateur observation of bird life. The largely homosocial bonding engendered in Japan's ornithological world excluded or marginalized women, which mirrored other contemporaneous groups holding professional weight and political power that kept women in supportive roles. Historian Miriam Kingsberg Kadia emphasizes that "Japan's universities admitted only men. Women could not attain the qualifications, networks, and knowledges expected of full colleagues in the human sciences. Such research accordingly developed as an almost wholly male enterprise."[123] These prewar characteristics of Japan's transwar ornithological circles continued deep into the postwar period and changed little until the global women's rights movements of the 1960s when many male-oriented organizations reflected upon their attitudes toward women. Nevertheless, throughout the transwar period, Japanese and Anglo-American male ornithologists benefited greatly from professional assistance by wives, female companions, or siblings directly aiding their completion of research or publication. Women often served as typists, like Kuroda Nagahisa's sister, who typed his notes to finish an important postwar survey of Japanese birds.[124] In addition, since the 1940s, Hachisuka's female secretary, and alleged mistress, Usui Mitsue served as a stenographer and interpreter due to his increasing deafness.[125] Though of crucial importance in accomplishing their work, these contributions by women went largely unrecognized, except within acknowledgments in frontispieces of books and publications geared to Anglo-American audiences.

This description of a highly masculine gendered organizational structure also resembled practices of the *Wild Bird Society*, founded in 1934 by the poet Nakanishi Gotô, who at his uncle's urging, spent his fifteenth year at the Jindaiji Buddhist temple in Chôfu, a Tokyo suburb, to train as an acolyte, and soon began enjoying bird observation.[126] Women joined bird-watching expeditions as daughters, wives, or mistresses of men engaged in relaxed ornithological observation, thus allowing them to appear in convivial company on outings amidst a supportive entourage. Kuroda Nagahisa noted that Nakanishi organized this group with the assistance of "leading ornithologists such as Uchida, Kuroda [Nagamichi], Takatsukasa, Momiyama, Kiyosu [Yukiyasu (1901–75)], etc."[127] These highly respected men in Japan's scientific world set the organization's tone and sealed its legitimacy despite potentially frivolous connotations. In 1934, the group began publishing *Yachô* [Wild Bird] journal, revealing its serious mission amidst a prevailing public consensus viewing bird-watching as a leisurely elite pastime since classical times.

Notably, the *Wild Bird Society* was also one of Japan's first, and largest, prewar environmental groups, though Hidefumi Imura and Miranda Alice Schreurs posit that "an organized environmental movement did not really take root in Japan in the prewar period."¹²⁸ Japan's ornithologists had long hoped for such an organization. On March 11, 1927, Takatsukasa earlier penned a short *Asahi Shimbun* opinion piece that expressed his views on bird conservation:

> Human activity basically disturbs places for birds to live peacefully; where birds meet with humans, their numbers progressively diminish. How one might do it would be to ban netting and hunting, supplement their being, increase places for them to peacefully rest, and by assisting in doing these things, and foster and raise birds by hand, I wish to cultivate them like children.¹²⁹

Despite initial lack of organized lobbying for bird conservation, the group received support from official political circles and well-known established cultural producers. Notably, Ministry of Agriculture and Commerce bureaucrat and folklorist Yanagida Kunio (1875–1972) belonged, along with celebrated artist Kishida Ryûsei (1891–1929).¹³⁰ Besides enjoying observations of bird life in nature amidst a growing prewar Japanese ecological consciousness, writers like Yanagida, poets, and artists engaged in bird watching to complement their cultural production where avian subjects often appeared as elegant interlopers—linking their activities to historical informal enjoyment of birds in Japan's arts and literature.

Although the *Wild Bird Society* primarily served to allow like-minded men to fulfill their amateur desires to approach birds, it also performed a broader social network linking influential men and a few women. The differences between the two most prominent ornithological associations in Japan, the *Ornithological Society* and the *Wild Bird Society*, appeared in their professionalization, gendered membership, and function. The *Ornithological Society* attracted a largely professional group of individuals identifying as ornithologists, while the *Wild Bird Society* functioned to allow men (and some women) from varying backgrounds to enjoy bird-watching as a hobby. The former group was dominated by male membership, while the latter welcomed small numbers of women as literary figures or in supporting roles. Literary reformer and feminist Hiratsuka Raichô (1886–1971), an early proponent of Zen Buddhist practices like Nakanishi, also briefly joined. Her penname *Raichô* ["thunderbird" or snow grouse] was inspired by birds glimpsed during her Nagano exile after the 1908 Shiobara Incident involving Raichô and her lover, married novelist and translator Morita Sohei (1881–1949), who intended to commit double suicide.¹³¹ Besides a strong literary and artistic contingent, several male scientists actively participated in both groups' activities, like Nakanishi and Yamashina who led postwar bird conservation, and received the Jean Delacour Medal (1977) and the Dutch Royal Family's Golden Ark Award (1978).¹³²

Along with a conservation focus, the *Wild Bird Society* expressed a bohemian atmosphere. Its founder Nakanishi trained in his youth in asceticism and Buddhist practices at Jindaiji and Zen meditation at Sôtô Zen Academy (now Setagaya Gakuen), after which he became interested in anarchism; thus, he held anti-authoritarian attitudes toward politics and bourgeois social relations.¹³³ As a "true" anarchist,

Nakanishi was against caging birds or restraining himself from affairs, since he felt that monogamy was contrary to human nature.[134] Similar to Delacour, Nakanishi initially caught and tamed wild birds to raise them in cages. However, his opinions evolved, along with his increasingly left-oriented politics,[135] and in 1932, he published *Yachō to tomo ni* [(Living) Together with Wild Birds]. Kuroda notes how "the Little Bitterns, Blue Magpies, Crows, and many other birds enjoyed their free life in his room and garden with a cat and playing with him."[136] While enjoyable for half-wild birds, such domestic arrangements plagued those tasked with cleaning up, including Nakanishi's wife. Presumably left at home during outings where the charming anarchist poet met young people under his thrall, she was obliged to clean up after the pet owls he kept in his Western-style villa.[137] After reaching middle-age, Nakanishi notoriously engaged in affairs with young women, and according to his former wife's niece, often met them during his bird watching outings.[138] These excursions began in 1934 with an inaugural multi-day trip to Mount Fuji's slopes.[139] Not surprisingly, the couple later divorced, much like Hachisuka and his wife, who for a while also tolerated ornithological visitors with motives beyond birds.

Notably, a poem of Nakanishi's, "Limestone Mansion," described Hachisuka's Atami beachside estate, and refers to his study with an elephant skin rug as a "private space,"[140] conveniently connected to a bedroom next door, where Hachisuka could stumble into sleep after a long night of writing or invite a lover for a private conversation after showing off specimens garnered in numerous expeditions to exotic locales. Though Nakanishi cultivated a somewhat eccentric image, and provoked gossip not unlike Hachisuka, in his postwar assessment of Japanese ornithology, Kuroda also viewed him as a "man of burning passion, comparable to Audubon in America, [without whom] vast ecological and conservational contributions made by amateurs and scientists would never have been so remarkable."[141] Nakanishi's reputation for unconventionality notwithstanding, he is still celebrated as a father of prewar bird preservation in Japan, along with Yamashina.

Outside of official and informal networks, birds also reinforced aristocratic marriages and enabled couples to harmoniously relate to each other in imperial Japan. Yamashina, like many aristocrats related to Kyoto royalty and former Tokugawa elites, benefited from an arranged marriage. He closely befriended his younger sister Yasuko's (1901–74) husband Asano Nagatake (1895–1969), whose own younger sister Sugako (1905–66) appeared a good match; after designated go-betweens met and exchanged gifts with proper courtesies, she became his wife following an engagement and requisite ceremonies. Yamashina in his autobiography mentioned that he "had no complaints" because his wife was well-behaved [*sunao*] and an excellent artist drawing birds that he observed in the field.[142] Even though Sugako purportedly loved birds like Yamashina, as a proper aristocratic wife, her social position also required active support of his work. Additionally, her art provided important means of interaction with Yamashina, and her indispensability generated positive effects by fostering a harmonious marital union for her. Little evidence reveals Sugako's true feelings about her husband's ornithological obsessions, but she often accompanied Yamashina on trips observing birds in the field, though he mainly relied upon his collector Ori'i Hyōjiri to hunt and acquire desired specimens from the colonies.[143] Despite their rapport, the couple remained childless, with Yamashina adopting his friend Asano's second son Yamashina Yoshimasa (b. 1927) as his heir.[144] After Sugako's December 22, 1966, death, he admitted that he still wondered about her true persona, since she spoke little and with reticence in public; yet, Yamashina acknowledged that she was a perfect companion, especially when accompanying him on trips.[145]

Yet, even when not overtly interested in ornithology, relatives and partners often suffered from or had little choice but to support particular bird-related fixations of brothers, husbands, or lovers. This ranged from tolerating infidelities resulting from ornithological pursuits to cleaning up messes of free-range pet birds. Not only were ornithologists away from loved ones for long periods of time while in the field, they also brought back fruits of their labor, which, if dead and unprocessed, likely distressed their families and household servants. Certainly, the grudging receipt of such gruesome bounties was not limited to female relatives and companions of Japanese ornithologists.

Tony Austin, son of the American ornithologist, recounted an illuminating episode about his father in Japan: "And he was [sic], all of his field trips, he'd collect and bring us stuff back and put it in the freezer, then skin them out at his leisure. He used to skin mammals, too. He was always sending rats and small rodents back."[146] Tony noted how these habits followed his father back to the United States.[147] In their Gainesville home, Elizabeth, Oliver's wife, jumped while opening the icebox and a "dead" pigeon refrigerated for a future meal lifted its head to peer at her.[148] Tolerating omnipresent ornithological endeavors of male partners or relatives evidently came with the territory.

Nevertheless, ornithology enabled these men to engage in a targeted homosocial bonding, conducted formally in organized professional networks and social clubs, and more privately in their own homes. Birds thus ubiquitously appeared not only in aviaries, specimen rooms, home furnishings like screens or paintings, and stuffed mounted displays, but also on plates as meals and decorations amidst an all-encompassing fascination for these collectors.

Conclusion

In his book *Queer Domesticities*, British historian of gender and sexuality Matt Cook describes how interiors and entertaining at home by men in early-to-mid twentieth-century London created a particular form of homosocial domesticity that he describes as "queer," though not necessarily understood as such at the time.[149] Though few of the ornithologists discussed in this book admitted to any kind of identification as queer, Cook's analysis resonates deeply with the personal bird-related obsessions of these scientists, which in terms of contemporaneous Japanese society's expectations for men could be considered "queer" to the degree by which birds defined their lives.

Personal narratives and stories surrounding the scientists investigated in this study evince how completely ornithology overtook their public and private lives. Within the home, personal museum, or estate aviaries was where increasingly intimate interactions between like-minded men took place, and included markers of their fascination, like mounted specimens garnered from expeditions or collectors, and ever-present images of birds within their personal lives, including on tableware for Hachisuka, whose dinner plates at his 1939 wedding reception at Tokyo's Imperial Hotel featured the dodo.[150] Yamashina, whose physique was weakened by childhood illness, believed himself unsuited to military life, and quit the army in his late twenties to focus on ornithology and study at Tokyo Imperial University.[151]

Ornithology in imperial Japan was a pursuit that not only boasted public dimensions, but also enabled more intense private socialization in the domestic spaces of the home, which nevertheless, still clearly maintained politicized aspects and displayed clues to these scientists' all-encompassing interest in birds. Here, public and private arenas intersected, since most of Japan's prominent scientists involved in bird research were also members of the National Diet's House of Peers or circulated in networks of imperial Japan's ruling elites. The domestic spaces of practitioners also served as repositories of their personalities and ornithological passions that strongly resonated with their public lives. Cultural mimesis informed their hybrid living spaces, but more importantly, characterized these scientists' public interactions, born from either international study or travel, and manifested in embodiment of key traits forming a proper Japanese "gentleman of science," like fluency in Anglo-American conversational practices, elite dress codes, and performances of "honorary whiteness." Chapter 3 reveals how such an individual proceeded on a path toward cultivating his authority as a Japanese ornithologist in the Anglo-American world.

3

Cambridge, UK (1925–9)—From "Scandalous Marquis" to Explorer-Scientist: Japanese in Western Imperial Settings

Introduction

Although imperial Japan's leading ornithologists numbered only a few dozen and composed a small network domestically, many of them held disproportionate political power as aristocrats in the House of Peers; they also notably inserted themselves into influential Western scientific networks in the Anglo-American world as transnationally focused researchers while publishing important scholarly contributions. Engaging in scientific research of birds or other fields and corresponding with Western scientists or personally interacting with their circles overseas was common for Japanese aristocrats and imperial elites, including Emperor Hirohito who as a marine biologist exemplified these trends.[1] However, all could be considered cosmopolitan imperial scientists.

As a key illustration of cultural mimesis and the culturally hybrid worlds inhabited by prewar Japanese ornithologists, this chapter details the controversial Marquis Hachisuka Masauji's formation as a "gentleman of science" in 1920s peregrinations throughout the Japanese Empire, Europe, and Africa while based at University of Cambridge allegedly for advanced study. As a wealthy aristocrat from a prominent *daimyô* family who flew his own plane, and whose father modeled their Tokyo villa after estates visited in Britain, Hachisuka symbolizes a cosmopolitan agent of empire with hereditary political power.

Notably, Hachisuka used his Cambridge association to acquire Western scientific legitimacy, along with his first-ever work, *A Comparative Hand List of the Birds of Japan and the British Isles*, published in 1925 by Cambridge University Press, where he elevated the two island empires' avifauna to similar levels of importance.[2] However, while school records indicate his 1924 matriculation at Magdalene College, he never formally took courses or graduated.[3] Such unsubstantiated formal study at top Western institutions was nevertheless quite common for wealthy Japanese male students, as were challenges to practices and academic traditions entrenched in these institutions.

For example, Hachisuka's mid-1920s spat with luminary professor of Chinese Herbert A. Giles (1845–1935) symbolized several key aspects of his struggle to maintain a particularly Japanese imperial masculinity. Giles believed that Classical Chinese formed an East Asian classical canon's foundations on par with Western Greco-Roman traditions, and thus, patronizingly assumed that Japanese could easily substitute Classical Chinese for one of Cambridge's required classical language exams—yet, Hachisuka failed the exam and then retaliated by refusing to fund a publication of Giles'.

This incident and others amongst overseas Japanese show that wealth and aristocratic class status failed to insulate them from elitist and racist attitudes in the Anglo-American world. "Honorary whiteness,"[4] or conferral of certain privileges usually only garnered by those deemed "white," only went so far in England and the United States. However, social trappings acquired by Japanese there, including alleged degrees, informally earned scientific knowledge, and savviness of foreign social practices, aided them immensely upon return to domestic Japan and the empire to build social capital, scientific credibility, and global connections.

Hachisuka's Time at Cambridge, 1921-7

Even for those with aristocratic titles, England's highly stratified society was difficult to penetrate for foreigners without allies and benefactors—a feat that was even more complex for non-Europeans not deemed white. In the early twentieth century, individuals in Great Britain with an "Oriental" background, including those of Middle Eastern or East Asian descent, even if not disrespected or treated with disdain, were often exoticized or viewed with curiosity, and most certainly, treated as "other."[5] Education historian Hilary Perraton notes that, by the 1920s, "In Cambridge, Japanese, Siamese and Chinese students were better received than Indians, where 'The latter were reported to be regarded as 'black men' and the others merely as 'yellow men.'"[6] From 1902 until 1921, Japanese also benefited from their empire's alliance with England, even though it was formally abrogated in 1923 due to American pressure. Yet, Perraton indicates that these students faced engrained racist perceptions in Great Britain: "For much of the twentieth century the degree of prejudice, or the warmth of welcome, towards overseas students was a function of the darkness of their skin."[7]

However, honorary whiteness could be purchased with elegant dress or acquired through the careful performance of gestures and behaviors reminiscent of powerful Anglo-Saxon elites. The "enclothed cognition"[8] provided by a properly tailored suit aided many prewar Japanese in engaging in a self-curated performance of British masculinity acquired by observation. This learning process was assisted with concomitant trappings of Western sartorial modernity bought at London's Savile Row tailors, like Henry Poole and Company, a favorite of Crown Prince Hirohito during his 1921 visit, and of his brother Prince Takamatsu (1905–87) who visited the clothier on his 1930 honeymoon for a state visit to King George V (1865–1936).[9] Such aspects are embodied in journalist, businessman, and postwar politician Shirasu Jirô (1902–85), who notoriously drove a Bentley during his early 1920s studies at University of Cambridge's Clare College. Fortuitously for Shirasu, the Seventh Earl of Strafford, Robert Cecil Byng (1904–84), also a Clare student, took the wealthy Japanese under his wing, and taught him essentials of British gentlemanly behavior.[10] Even though Shirasu was two years older and not an aristocrat, it helped that he came from an affluent industrialist's family and provided elegant transportation by automobile. From the mid-to-late 1920s, Hachisuka also used his money to commemorate his Cambridge sojourn by funding book publications and a building project. Then, as now, British institutions of higher learning were cash-strapped, lacking plush endowments like their American counterparts, and while stingy with lecturers' salaries, also sought alumni donations for new buildings or even repairs.

Nonetheless, Hachisuka benefited from a privileged aristocratic lineage and immense wealth prompting easier admittance to University of Cambridge, though he appeared a lackluster student at the state-run *Peer's School*, a Tokyo-based institution established in 1877 to educate Japanese nobles, which taught them English for a cosmopolitan, well-rounded Anglo-American-inspired education. Graduating in 1895, Hachisuka's father Masa'aki studied at Cambridge's Trinity College, arranged through machinations of his own father Hachisuka Mochiaki (1846–1918), a former student at Oxford University's Baliol College during the early Meiji period who boasted connections to illustrious mathematician and educational administrator Kikuchi Dairoku (1855–1917), a distinguished Cambridge alumnus.[11] Additionally, Hachisuka enjoyed a storied pedigree as eighteenth-generation family scion related to the Matsudaira clan, nephew of the last shogun Prince Tokugawa Yoshinobu (1837–1913), great-grandson of Tokugawa Ienari (1773–1841), and son of the National Diet's vice president, Hachisuka Masa'aki, also former head of the Tokushima clan. He stemmed from illustrious forebearers, but now bore added burdens to reach expectations demanded by family and aristocratic peers: studying abroad at a leading British university to perform as well as English counterparts, and impress them as a representative of Japan's ruling class.

Much of Hachisuka's recorded history reflects tensions between his aspirations and complicated realities. When Scottish ornithologist Norman Boyd Kinnear (1916–2009), the London-based Natural History Museum's retired director, wrote Hachisuka's obituary in *Ibis*, the BOU's leading peer-reviewed journal equivalent to the American *The Auk*, he indicated that "Masauji Hachisuka, who died on May 14, 1953, after a brief illness, came to this country at the age of nineteen to complete his education and for five years was at Selwyn College, Cambridge, studying zoology."[12] Yet, Kinnear's hagiography of Hachisuka's life at Cambridge, while revealing inconsistencies in dates, degrees, and titles, also conceals his difficulties assimilating into British society. Nevertheless, Hachisuka formed helpful connections with sometimes-reclusive zoology scholars through a burgeoning interest in birds, which eventually subsumed his original motives for studying abroad.

Replete with grandiose plans to specialize in political science befitting his lineage and distinguished position, aged only seventeen, Hachisuka left Tokyo by train on July 18, 1920, and continued to Kobe, where he took a ship and arrived in London nearly six weeks later on September 4, 1920.[13] In 1920, London harbored a vibrant community of 1,638 Japanese residents, whose numbers dwindled from financial difficulties following the 1918 Rice Riots and Siberian Intervention-related (1918–21) inflation.[14] In 1920, the Hachisuka family experienced strikes by Hokkaidô tenant farmers, but losses were absorbed by revenues from other land holdings, and they still sent a scion abroad in style.[15] In London with plans to matriculate at its university, Hachisuka felt tremendous social freedom distanced from entangling obligations and duties at home. In the British imperial capital, he realized his true vocation: ornithology.

In 1919, a year before his journey abroad, and only sixteen years old, Hachisuka joined the Tokyo-based *Ornithological Society* upon his cousin's recommendation, well-connected ornithologist and House of Peers member Duke Takatsukasa Nobusuke, described as "Prince" in most English-language writings. In Japan and Europe, the field attracted eccentric and colorful personalities, where amateur interest in zoology intersected with organized study amongst politically powerful men. In a letter to his mother, Canadian ornithologist James Henry Fleming (1872–1940) described the flamboyant atmosphere at the 1905 International Ornithological Congress (IOC) held in London: "The congress is frightfully noisy[.] Like the tower of Babel any old language goes here apparently the louder the better ... We have got the most extraordinary collection of freaks possible ... Giraffes and barons and freaks some in cases and some outside."[16] Similarly, on a 1924 Paris trip, Takatsukasa introduced Hachisuka to French ornithologist Delacour, whose Normandy chateau boasted a zoo and aviaries of unusual birds. During Hachisuka's time in London and Cambridge, his ornithological passions connected him with enthusiasts from the British and French aristocracies and German and Yankee elites. Hachisuka's following in his father Masa'aki's footsteps represented a trend of overseas study or scientific research at England's top institutions for Japanese aristocrats since the Meiji period.

This included Crown Prince Hirohito, who engaged in scientific studies under his Chamberlain since childhood, and was taught by *Peers School* microbiologist Hattori Hirotarô (1875–1965).[17] Keen interest, and the desire to evade imperial household handlers in field expeditions (like on Sagami Bay) and laboratory work, led Hirohito to conduct marine biology research with Hattori as his "scientific proxy" in corresponding with Western scientists.[18] Hattori's mentorship sparked Hirohito's lifelong interest in hydrozoans, solitary, and colonial microscopic sea- and freshwater creatures related to jellyfish and coral. This included correspondence and specimen exchanges with Belgian researchers in the 1930s.[19] From 1908 to 1914, the Crown Prince attended the *Peers School*, followed by instruction at the *Akasaka Palace Royal Academy* until 1921[20]; a biological laboratory run by Hattori was built in 1925 on Akasaka Palace grounds to further Hirohito's scientific studies.[21]

In March 1921, the Crown Prince embarked on a six-month European tour, stopping in London, Oxford, and Cambridge in mid-May before serving as imperial regent for his father, Emperor Yoshihito (1879–1926), upon return. Like Hachisuka, in England the Crown Prince could indulge his scientific interests; in London, he attended the Linnean Society, a venerable natural history organization researching taxonomies and other issues, where he signed its Roll and Charter Book.[22] Nevertheless, before Japan's future emperor quickly learned proper British etiquette, his table manners were considered atrocious, and lacked his English counterparts' social finesse despite noble birth.[23] Hirohito still enjoyed his visit, dining daily on a full English breakfast including eggs and bacon even following his return to Japan,[24] and in a later interview, remarked that he felt "like a bird sprung from a cage" due to relatively informal social interactions amongst British aristocrats with members of their own class.[25] As a recent arrival to England related to the former Shogun, Hachisuka was rumored to have met the Crown Prince with his motorcycle before a Regent's Park garden party, much to the chagrin of Japanese students accustomed to holding the imperial family in awe.[26] Despite unequal social positions, the two youths appeared on friendly terms and enjoyed common interests in scientific research; when Hirohito became Emperor in late 1926, Hachisuka served twice in his multiple enthronement ceremonies in 1928.[27]

Despite Hachisuka's distinguished lineage, the path toward enrollment at Cambridge was difficult for a non-native English speaker unversed in British classical education. In fall 1921, before Michaelmas term (running from the Feast of Saint Michael to Christmas),[28] Hachisuka arrived in Cambridge, where he sought tutoring to pass the university's entrance exams, and three years later, matriculated at Magdalene College. Curiously, Magdalene never housed Hachisuka, whether lack of room or administrative issues prevented this remains unknown; upon arrival, he lodged at Selwyn College, considered a Public Hostel from its 1882 founding until 1926. Its reputation was as a school where Church of England-affiliated overseas missionaries, clergy sons, and cash-strapped students could be educated.[29] There, he was looked after by Reverend Dr. John Owen Farquhar Murray (1858–1944) and his wife, who warmly hosted Japanese students for decades during his position as Master (1909–28).[30]

Curiously, records, or lack thereof, on his University student card indicate Hachisuka as having matriculated on October 22, 1924, but never officially taking classes, though he possibly audited them. According to Noboru Koyama, curator of University of Cambridge's Japanese collection, this was common for titled Japanese, including dilettantish Marquis Tokugawa Yorisada (1892–1955), a distant family member, who aspired to study music during Hachisuka's stay.[31] To prepare for exam series at University of Cambridge, Japanese students sometimes studied at Leys School, a religiously tolerant Methodist-founded public academy built upon the twenty-acre Leys Estate and opening on February 16, 1875.[32] Students were mostly aged eleven to eighteen, and their studies served as secondary preparation for University of Cambridge or another school.[33] The Leys School was near to the university's colleges, especially Selwyn College, which can be seen across the lawns, so students could view their aspirational focus on a quotidian basis.

Prior to 1945, University of Cambridge required residence of at least nine terms, or three years, for completion of a bachelor's degree equivalent, with ordinary or honors degrees awarded. Students remained at colleges, where they lived, studied, ate, and interacted with professors; the university merely conducted examinations, registered

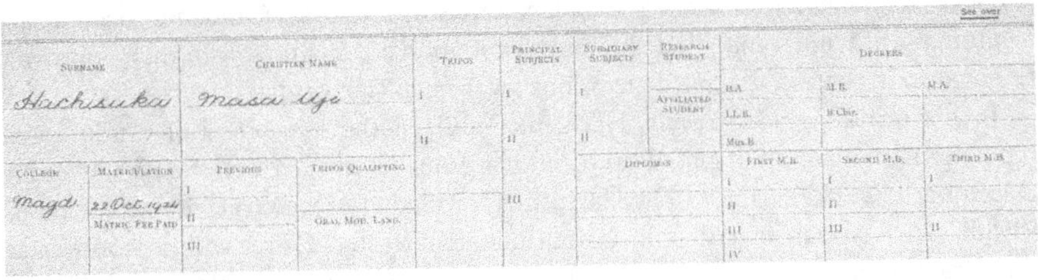
See over
SURNAME: Hachisuka
CHRISTIAN NAME: Masa Uji
TRIPOS
PRINCIPAL SUBJECTS
SUBSIDIARY SUBJECTS
RESEARCH STUDENT
AFFILIATED STUDENT
DEGREES: B.A. M.B. M.A. LL.B. B.Chir. Mus.B.
COLLEGE: Magd.
MATRICULATION: 22 Oct. 1924
MATRIC. FEE PAID
PREVIOUS
TRIPOS QUALIFYING
GRAD. MOD. LANG.
DIPLOMAS
FIRST M.B. SECOND M.B. THIRD M.B.

Figure 3.1 Hachisuka Masauji's matriculation record (October 22, 1924). © Published with permission from the University of Cambridge Library.

students, awarded degrees, and administered fee payment. College and university exams occurred every six months, plus two more important examinations for degrees. For an honors degree, students needed to pass two examinations: the "Previous Examination" (known in slang as "Little-Go"—a quasi-entrance examination whose first part focused on Greek and Latin, and second part featured Classics and a specific study emphasis, like natural sciences or mathematics) and the much harder "tripos" ("three-legged stool") specialist areas examination. Students failing the latter could still sit for an ordinary degree, whose next step was the General Examination (a more challenging version of "Little Go") and Special Examination divided into special categories relating to a study field. For undergraduates, one third received an honors degree, another third the ordinary degree, and the last third, no degree despite residence.[34] The latter apparently was Hachisuka's experience.

Although Kinnear's obituary noted that Hachisuka took a degree at Cambridge in 1927,[35] he more likely left Cambridge (and Selwyn) then; Koyama indicates that Hachisuka "appears never to have taken the preliminary examination. At Cambridge he ignored all the exams and without doing unnecessary things concentrated on his ornithological research."[36] Hachisuka spent his time at Cambridge busily exchanging notes with British colleagues showing off their collections and interacting with distinguished local scholars like Arthur Humble (A. H.) Evans (1855–1943), a British economic historian and former *Ibis* editor from 1901 to 1912, and Francis Henry Hill (F. H. H) Guillemard (1852–1934), a bird of paradise expert who edited the Cambridge Geographical Series, and whose journey with the *Marchesa* throughout Asia and the South Pacific from 1881 to 1882 led to the amassing of 3,000 bird specimens and publication of an 1886 best-seller.[37] In his preface to *The Birds of the Philippine Islands* (1934–5), Hachisuka thanks Guillemard, who recently died, and Ernst Hartert (1859–1933), curator of birds at the Natural History Museum at Tring, of whom he notes, "My close association with Guillemard during my academic terms at Cambridge in the study of books and specimens, and my explorations in North Africa with Hartert, will be a very inspiring memory to me for many years to come."[38] Upon Evans's and Guillemard's recommendations, Hachisuka likely joined the BOU around 1923. In London, he attended dinners of the BOC,[39] the BOU's exclusive offshoot where his aristocratic status and Cambridge associations certainly helped him to make useful contacts amongst like-minded upper-class bird enthusiasts. Hachisuka also formed a close working relationship with Lord Rothschild at the museum on his Tring estate, and in the mid-twenties, claimed that he was a Fellow of the (Royal) Zoological Society (FZS) in his publications.[40] The young Japanese student so determined to become an ornithological scholar made a deep impression upon his mentors; with their approval, he engaged in three collecting expeditions away from Cambridge.

In his ornithological travels, Hachisuka utilized the University of Cambridge's reputation in furthering global connections amongst British ornithologists. From December 1923 to February 1924, Hachisuka traveled to Alexandria with Mohammed Riyadh, an Egyptian friend from Cambridge, and visited Cairo, with a stopover at the Giza Zoological Gardens to meet its retired director Major Stanley Smyth Flower (1871–1946) and his former assistant, now director, Michael John Nicholl (1880–1925).[41] Flower, incidentally, returned to England in April 1924 to live at Spencer's

Green near Tring to continue research at Lord Rothschild's personal museum, where he also mentored young ornithologists including Hachisuka[42]; in 1934, in ill health, he moved permanently to Tring village for closer proximity to Rothschild's collections.[43] Hachisuka later traversed the Sahara by car during an expedition surveying Egyptian birds, whose 1926 publication garnered him notice in Anglo-American ornithological circles.[44] In early 1927, Witmer Stone, *The Auk*'s influential American editor, praised the handsome volume identifying 455 species and subspecies, but lamented the text's majority use of Japanese.[45]

In 1925, Hachisuka and two students, L. J. Turtle and J. Cadbury, left Cambridge for an Iceland expedition, which resulted in a 1927 publication in English with numerous photographs, including the men perched atop Iceland's celebrated white horses.[46] While casually dismissing German and Danish research, Hachisuka's preface to *A Handbook of the Birds of Iceland* notes how he updated the only two English-language book-length works on Iceland's birds from 1863 and 1901,[47] and writes that "To the ornithologist, and in particular to the British bird-lover, Iceland is of considerable interest, in that many of our rarest British breeding-birds are to be found there in abundance during the summer."[48] Hachisuka's interest in rare or extinct flightless birds like the Great Auk and Dodo began under Lord Rothschild's mentorship, with his book including a photograph of a Great Auk specimen from Cambridge's Zoological Museum.[49] Iceland, the bird's last breeding ground, saw it extirpated in 1844 by fishermen who raced to catch its last pair, still incubating an egg, which they crushed in a frenzy to bag them.[50] A 1928 review in the British journal *Nature* praises Hachisuka's contribution:

> The Honourable Masa U. Hachisuka is a very young ornithologist, but his work—his first, we understand, of any magnitude—shows that he is keen, industrious and methodical, and, though it is an ambitious attempt for a first work, there is little doubt it will prove most useful. We congratulate the author, not only on his pluck in undertaking it, but also on the result itself.[51]

Such boldness and ambition expressed by the young Japanese student already engaged in high-level scholarship certainly impressed notable figures surrounding him. In March 1927, Hachisuka embarked with Tring's curator, German ornithologist Ernst Hartert, on a nearly three month-long expedition to North Africa by car on Rothschild's behalf, and returned via ship from the Straits of Gibraltar to Marseilles on May 26.[52] Ever peripatetic, in June, Hachisuka left for a journey to the United States, Canada, and Hawaii, with French ornithologist Jean Delacour, whose Normandy estate Chateau Clères hosted a similar zoo and museum like that of Rothschild. Evolutionary biologist and ornithologist Ernst Mayr (1904–2005), Delacour's later colleague at the AMNH, where the Frenchman served as technical advisor, described this zoological garden in his obituary as an Eden where "Gibbons, gazelles, kangaroos, flamingos, cranes, numerous kinds of waterfowl, and other wildlife roamed freely through his park; carnivores of course were excluded. The smaller and more delicate birds were kept in aviaries."[53]

These outward forays into underexplored fields aided in establishing Hachisuka's reputation as an "authentic" ornithologist and scientist on par with, and potentially surpassing as a prodigy, his British counterparts. However, the most productive part of his sojourn was establishment of a close working relationship with a much older Lord Rothschild, also a former alumnus of Magdalene College where he studied zoology from 1887 to 1889, and whose name opened doors.

Lord Rothschild and the Young Japanese Bird Enthusiast

Lacking his other relatives' charisma, following studies at Cambridge, Rothschild obtained a berth at the family bank N. M. Rothschild and Sons, which he barely tolerated until his departure at age forty. His father Baron Nathanial Mayer Rothschild (1840–1914), Britain's first Jewish peer, promptly disinherited him, but arranged a substantial settlement to keep him comfortably entrusted for life to allow him to concentrate fully on natural history endeavors. Like Hachisuka, Rothschild was an outsider to British high society due to his Jewish background, whose father literally purchased the family's entrance into England's aristocracy with immense wealth and an inherited title from his paternal great-grandfather Nathan Mayer Freiherr von Rothschild (1777–1836), granted by the Austrian Kaiser in 1822. Though socially awkward in public, the younger Rothschild enjoyed hosting friends with whom he exchanged ideas on ornithology and zoological pursuits.[54]

Rothschild's vast estate at Tring nestled in Hertfordshire thirty miles from London housed his own zoological museum built by his father Nathanial as a twenty-first birthday gift; this he opened to the public in 1892 as the Walter Rothschild Zoological Museum, where a team of domesticated zebras pulling a carriage traversed its grounds and guests likely dodged dozens of kangaroos or cassowaries[55] and a morose Galapagos tortoise named Rotumah.[56] In 1895, during a visit, the British artist and illustrator Louis Wain (1860–1939), known for anthropomorphic drawings of cats, described his impressions within Tring's gates in *The Windsor Magazine*, a popular Victorian monthly:

> Beyond, the gardens and the orchards–magnificent in the resplendent wealth of posy, and replete with gorgeous carnation bloom endless in tint of every variety of colour-matching–are sweet with ineffable delicacies wafted insidiously from the wealth of ripening fruit, while the air resounds with the contented pipe and chatter of many lusty throats to the sibilant accompaniment of swarming bumblebees.[57]

In torpid prose, Wain described the Museum as "a fine pile, standing out like the wing of some moated grange covered with clematis."[58] Rothschild's interest in extinct species and passion for flightless birds like cassowaries led Hachisuka to research a similarly flightless Dodo, a key study published right before Hachisuka's untimely 1953 death.[59]

When Hachisuka arrived in England in 1920, the enterprising teenager soon appeared at Tring, and scoped out its Natural History Museum, befriending an avuncular Rothschild, and thereafter, regularly visited him on Sunday afternoons to talk shop.[60] There, he met Hartert, who curated Rothschild's ornithology collection and had contributed to a *Hand-List of British Birds*, published in 1912.[61] These meetings were clearly fruitful, since in 1925, Hachisuka published his own contribution, *A*

Comparative Hand List of the Birds of Japan and the British Isles, where he compared avifauna from the two island archipelagos, and in whose foreword, he first thanked Hartert, and then, extended gratitude toward Momiyama Tokutarô and his Japanese mentor Takatsukasa for helping revise his manuscript.[62] Curiously, at the forward's very bottom appears "Magdalene College, Cambridge. October 1924."[63] This location and date were meant to seal Hachisuka's authority firmly within this British college's orbit where he enjoyed an academic lineage through his father, while he also dedicated this book "to my sister Viscountess Yasuharu Matsudaira," who sometimes appears as "Viscount" or male in English-language ornithological literature apparently published under her husband's name. Indeed, in the *Hand List's* bibliography of "Some Notable Works on the Ornithology of Japan," instead of as Toshiko (Hachisuka) (1896–1970), she is listed as "Y. Matsudaira."[64] Her name appears last after Takatsukasa, Uchida Seinosuke, and Kuroda Nagamichi, whose own father Nagashige was a student at King's College who received his B.A. in 1887 and M.A. in 1891, after which the Emperor nominated him the House of Peers' Deputy-Speaker.[65] Thus, this bibliography of Japanese contributions to ornithology reveals strong representation by the aristocracy and imperial elites. Clearly, the context for Hachisuka's endeavors was his need to impress numerous illustrious relatives and mentors.

Hachisuka's mentor Rothschild never married, and despite a shy and retiring personality led a turbulent life as would Hachisuka, later known as "the scandalous marquis" for alleged and real personal affairs and financial issues. Nevertheless, he took an avuncular interest in Hachisuka, and encouraged him to invite French ornithologist and aristocrat Jean Delacour to Japan. In 1927, following a several months-long tour from June to mid-September of the United States, Canada, and Hawaii with Delacour, an enamored Hachisuka held tea ceremonies and introduced him to friends, including Kuroda Nagamichi, whose laboratory and famed specimen collection they visited. In late September, the two, along with Delacour's mother, embarked on a trip to Korea and Northeast China, both sites harboring Kuroda's beloved rare crested shelduck (*Tadorna cristata*), first described by him as a new genus in 1917. On October 27, 1927, possibly to assuage gossips' suspicion of his obsessive devotion to ornithology and its practitioners, the Tokyo *Asahi Shimbun* announced news of Hachisuka's engagement with Princess Kitashirakawa Mineko (1911–70), which his father Masa'aki brokered through imperial go-betweens.[66] However, Hachisuka's sister Toshiko remembers that their father abrogated this when "unsavoury aspects about his life" became known, whereupon he threatened to disinherit his son.[67] The Princess later married Tachibana Tanekatsu (b. 1912–unknown) in 1933,[68] while Hachisuka eventually married a commoner in 1939, a Japanese-American woman he met in Pasadena, California while recovering from illness.[69] His decision was likely because her wealth and ordinary class status could potentially insulate him from his father's lasting posthumous grip on his personal life, both socially and financially.

In a similarly complicated personal situation, until 1908, Rothchild kept two mistresses simultaneously: actress Maria Barbara Fredenson-Walters (unknown) with whom he had a daughter, Olga Alice Muriel Walters Yarde-Buller (1906–92), and another, Lizzie Ritchie (unknown), who most likely suffered from manic depression. Once the two women discovered each other, and Ritchie divulged the affairs and illegitimate child to Rothschild's mother, his brother Charles (1877–1923) helped settle his accounts by providing each woman with a house and 10,000 pound yearly stipend while paying off the museum's mortgage.[70] Concurrently, Rothschild conducted a doomed affair with a ruthless married peeress, who constantly blackmailed him for money for nearly four decades and whose name was never publicly disclosed. By 1930, Rothschild was rumored as near-broke. Details failed to surface until his niece Miriam Rothschild (1908–2005) published her uncle's biography *Dear Lord Rothschild* in 1983, which still desisted in naming his blackmailer, described as "ruthless" and "cunning."[71]

Most pertinent to ornithological history, Rothschild's nemesis was finally silenced by a huge payment, forcing him to sell a considerable portion of his collection, or 280,000 bird skins, to the AMNH on February 13, 1932, for the relatively paltry sum of $225,000.[72] The wealthy widow Gertrude Vanderbilt Whitney (1875–1972) offered funds in an expertly brokered deal by the New Haven-based surgeon and amateur ornithologist Leonard Cutler Sanford (1868–1950) who served as an ANMH trustee and as the strongest patron of its ornithology department ostensibly competing against Harvard University's MCZ.[73] Ernst Mayr, just hired in 1931 to serve as the AMNH's ornithological curator, happily accepted the task to categorize and catalogue the collection due to its quality and rarity (though unpacking of 185 cases sent from Tring would only begin in 1935); he never suspected Rothschild originally considered him for a job as Hartert's successor as Tring's curator of ornithological collections after Hartert's 1930 retirement.[74] Mayr's hiring, along with the coup of acquiring Rothschild's collection, precipitated the decline of British ornithological dominance and shifted it toward the United States,[75] a process that Hachisuka likely understood when he made a foray into the furthest American colony in Asia—the Philippines.

Notably, in Hachisuka's first 1931 book on birds of the Philippines, the first person he thanks in the Acknowledgments is his mentor Lord Rothschild.[76] Here, Hachisuka also complains that, in several months of research at Tring, following his own 1929 expedition to the Philippines, he could make his way "through the Philippine series only as far as the Timelidae [babbling thrushes] before the collection was disposed of to the American Museum of Natural History, New York."[77] Negating the import of earlier collections formed by Americans in the late nineteenth and early twentieth centuries cached at the University of Minnesota Academy of Natural Sciences and University of Michigan, Hachisuka asserts that the best extant bird specimens gleaned from the Philippines were instead found in two places: the British Museum and Rothschild's Museum in Tring Park. These collections initially arose from contributions from Danish collector Johannes Waterstradt's (1869–1944) September to November 1903 expedition to Mount Apo's slopes, where he only reached Balek village at 6,000 feet,[78] and British zoological collector Walter Goodfellow's (1866–1953) April 1903 foray and 1905 expedition. Rothschild clearly whetted Hachisuka's appetite to collect birds in a species-rich area attracting numerous explorer-scientists; this expedition would compose materials for a monumental four-volume set, discussed in Chapter 4. However, not all connections that Hachisuka made in Cambridge were positive or productive.

The British Sinologist and Japanese Zoologist Clash and Part Ways, 1927

The University of Cambridge Library caches a curious document still revealing handwritten notes by its former owner in a fountain pen's sepia-inked cursive. "H. A. Giles" appears scrawled on the upper right-hand corner of a publishers' announcement on high-quality parchment paper for an allegedly forthcoming *Record of Strange Nations: From the Chinese of 1392 A.D.*[79] The abbreviated name penned on the announcement refers to Herbert A. Giles (1818–95), Professor of Chinese at Cambridge, who succeeded Sir Thomas Wade (1818–95), the University's first professor of Chinese from 1888 until 1895, and improved upon his Romanization system, now known as the Wade-Giles system used in Mainland China until pinyin replaced it in 1958; it was still used in Taiwan until 2008. Intended as a co-production between Giles and Hachisuka, the book would feature their particular expertise in an elegant illustrated translation of a medieval Chinese bestiary worthy of a fine library. "Consisting of about 392 pages, containing about 176 full page reproductions of woodcuts including Chinese text, each faced with type page of descriptive matter, handsomely bound," it would have boasted a modest size of 12.5 by 8.5 inches and would have sold for "Three Guineas Nett," or three full guineas.[80] Until 1971, British retailers advertised exclusive luxury goods in guineas rather than pounds to make their commodities seem more elite, since few could easily access the sums implied by guineas: "three guineas nett" equals three pounds and three shillings, or the considerable price of 199.19 British pounds in 2020 currency.[81] Their target audience would have been wealthy educated elites seeking an unusual antiquarian tome for their personal libraries.

Further notes penned by Giles on this early 1927 document reveal the degree to which Hachisuka clashed with a man initially serving as his mentor. A few pages into the publishers' announcement, and at the very top of a section entitled "Record of Strange Creatures" extracted from Hachisuka's introduction to the text, lies scrawled Giles' sharp criticism, along with his scribbled underlining of "Hon. Masa. U. Hachisuka's introduction."[82] In a remark meant for his private consumption only before it was archived after his death, Giles left the following blithe statement belittling Hachisuka's achievements, but likely reflects feelings of threat from this youthful Japanese upstart: "This would-be scholar tried to get into the University through the Previous Exam in Chinese, but failed miserably."[83] Obviously, this assessment was far off the mark, at least for Hachisuka's scientific work. By 1927, though only in his early twenties, Hachisuka had already produced two publications, with one receiving acclaim in the key US ornithological journal *The Auk*. Both men possessed strong personalities, and quite often imposed their opinions. In Hachisuka's defense, though known as charming and convivial, Giles conflicted with many of his colleagues and students, and he never became a full-fledged fellow of any Cambridge colleges, despite a distinguished 35-year teaching career from 1897 to 1932.[84]

Initially hired by the University where he would "receive no stipend," Giles lived off his diplomat's pension from earlier postings in China, later supplemented by a modest 200 pound yearly salary reluctantly initiated in 1899 which never rose in his entire career[85]—he was initially paid the contemporary 2020 equivalent of 25,869.17 pounds

(28,308.63 US dollars) and ended his career with a yearly equivalent of 14,052.39 pounds (17,661.75 US dollars).[86] Koyama posits that "It may also have been the case that the demand from the Japanese students for a classical Chinese examination was one of the stimuli for Giles coming to Cambridge."[87] In 1887, for the Preliminary Examination, Japanese students gained an exemption from examining in Greek, and were allowed testing in English instead, though Latin still remained as requirement.[88] In 1906, after a three-year proposal holding Giles's support, and previous interventions by Sir Ernest Satow (1843–1929), a British diplomat in China and Japanologist, along with Lord Cromer (Evelyn Baring) (1841–1917) based in Egypt, the university now allowed "Oriental students to take English and one classical Oriental language (to be chosen from Arabic, Chinese and Sanskrit) instead of Greek and Latin. This policy was officially adopted by the Council of the Senate on June 16, 1906."[89]

Giles's distinguished, though underpaid, position also accounted for his needing financing to publish this rather peripheral early Ming dynasty (1368–1644) work, called *Yiyu tuzhi* [literally, "Map and Chronicle of Exotic Places," which he translated as *Record of Strange Nations*], that he began translating in 1916.[90] He first sought its publication in 1924 by University of Cambridge Press, which required from him 750–1000 pounds (an astronomical subvention for a poor academic) for reproduction of the original with its translation, notes, and plates, despite the advocacy of academic luminaries from the university, including anthropologist Alfred Cort Haddon (1855–1940), classicist and archaeologist Sir William Ridgeway (1853–1926), and philologist Peter Giles (1860–1935), who was also Master of Emmanuel College from 1911 until 1935.[91] Hachisuka's addition to the project, along with his zoological expertise to identify animals described in the medieval Chinese bestiary, was financially as well as zoologically helpful for its completion. But what led to their mid-1920s falling out that ultimately scrapped its publication?

Giles had advocated strongly for pushing through Chinese, as opposed to Latin, to replace the classical language requirement for East Asian students in the Previous Examination to be passed as qualification to enter the University. Yet, this purportedly positive development resulted in a fierce battle with none other than Hachisuka himself. The young Japanese aristocrat, though a lackluster student at the *Peer's School* in Tokyo which favored study of English over *kambun* [literary Chinese], initially concurred that Classical Chinese works matched the Greco-Roman canon, and therefore, proficiency in Classical Chinese, or *yuyan*, should be accepted as equal to knowledge of Greek or Latin in a Western Classical tradition. Unfortunately, Hachisuka never met the grade for this exam, with Giles undoubtedly writing a purposefully difficult exam inspired by over two decades of careful poring over Chinese Confucian classics and other texts.

In addition, Giles possessed a long history of picking fights with former mentees, including Edmund Backhouse (1873–1944), a Sinologist later found to have forged much of his work. However, in his memoir, the Cambridge professor also left statements expressing virulent contempt toward Japanese in his personal interactions and loathing of their nation's international relations with China, like the 1915 Twenty-One Demands finally rescinded in 1921. Yet, what Giles seemingly most abhorred about Japanese was that he found them arrogant. In November 1923, he reviewed a collection of Li Bo's (701–762) poems translated by diplomat and translator Obata Shigeyoshi (1888–1971) for the journal *The Chinese Student* and vituperously writes about the experience:

> *The Japanese are notorious for their inaccurate commentaries,* and the work of Mr Obata confirms this opinion; not to mention that, although his English is fairly good for a foreigner, it is hardly adequate to the requirements of a difficult Chinese poem. *Add to these short-comings a boundless faith in himself amounting almost to impudence, and a broadly expressed contempt for everybody else's work,* which I hope will receive a check when he sees that in one poem of about 40 lines I have pointed out some 25 mistakes, and you have the chief characteristics of Mr Obata's volume.[92]

Naturally, such a damning review from a Cambridge-based Sinologist could harm Obata's career; so, the journal failed to print it or correspond further with Giles, until Obata asked for his article's return in February 1924, which he received only after a long delay.[93] However, the greatest irony is found in Giles's own words in his 1925 memoir where he mentions liberties taken in translations and describes Chinese in Orientalist terms as an enthralling, yet chaotic, language:

> *Close grammatical analogy never had any charms for me.* I managed, however, by wide as opposed to meticulous reading, to imbibe, and in later years to enjoy, something of the spirit which informs the mythologies, histories, and poetical literatures, of ancient Greece and Rome, *before being finally engulfed in the maelstrom of Chinese.*[94]

Initially, the two men clearly collaborated, with an arrangement whereby Hachisuka would foot the considerable bill for the book's printing, and have his name associated with the storied British scholar of Chinese and former diplomat. But Giles's disparaging and unjustified comment of Hachisuka as a "would-be scholar" indicates that relations had already soured by the late 1920s.

Though nearly two years lapsed since the publisher's announcement for the proposed book, Giles must have felt taken aback when receiving a letter from publisher Norman Parley, dated January 14, 1929. His ire still appears in penciled marks left on the prospectus with a bracket around the F.Z.S. following Hachisuka's name, as if he doubted his credentials,[95] and below Haddon's name, he wrote "See publishers' [sic] letter" with the French word "*merde*" [shit] written below it.[96] Parley's letter blames the printing's scuttling on Hachisuka's remaining incommunicado and interprets lack of forthcoming cash as his abandoning the project: "Mr. Hachisuka not having replied to any of our correspondence, evidently does not intend to go forward with the proposal to print this book, [sic] we are therefore destroying the printing which we have held in stock in connection with it, and I am now returning to you the MS, which you committed to my care."[97] However, Parley continues his letter in a palliative tone: "I am very sorry that the project has not gone forward and am sure you will recognize that it has been through no fault of ours. I enclose a copy of the Prospectus and take it that it would be of no interest to you to have some of these?"[98] Almost as an insult, enclosed within the publisher's small parcel containing the letter was a single copy of the publishing announcement for the now-aborted text upon which an irascible Giles had devoted over a decade of scholarship.

Nevertheless, at a time when Europe and Asia were not easily connected via telecommunications or quick travel times, mail delays or slow transportation could still be blamed for an individual's lack of communication. Indeed, in spring 1927, Hachisuka embarked on an expedition to North Africa, followed by a summer journey to North America. At the letter's date, Hachisuka was in Manila, and would soon commence scaling Mount Apo's peak in the Philippines, where he remained until April 15, 1929.[99] After returning, Hachisuka went back to Tring to use its libraries and specimen collections—no doubt to compare what had been previously collected and written—and could have repaired the rift with Giles (or at least replied to the publisher). According to contemporaneous Japanese social customs, likely similar to those of the British upper-classes, a reticence existed toward expressing direct personal conflict. Thus, a person might choose to break off contact without exchanging words of rancor. Clearly, Giles had committed an unforgivable slight toward Hachisuka, which was not uncommon for Japanese at Cambridge. Koyama also indicated that Giles had kept one of Hachisuka's books on the Dodo and never returned it[100]; this valuable text was possibly gifted by his mentor Lord Rothschild from his own personal library.

Nevertheless, to entrench himself into the university's storied history and surmount his falling-out with Giles, Hachisuka made sure to perpetuate his connection with Cambridge in stone, simultaneous to his 1929 return to Tring when he might have repaired his relationship with the famed Chinese professor, but chose not to. Hachisuka and two other former Japanese students provided funding for an archway built during the global Great Depression's (1929–39) early years. This donation, along with their presence at one of England's top academic institutions in the 1920s, provided much sustenance for their feelings of national legitimacy and pride. From 1929 to 1930, construction of the Selywn College Library was funded by the War Memorial Fund and donations of these three Japanese aristocrats, including Hachisuka, Tokugawa Yorisada, and Sawada Kiyoshi (1903–94).[101]

On an arch connecting Selwyn College Library's upper floor to the main building's "C" staircase appeared Hachisuka's distinctive swastika, or *manji*, of his *kamon* [family crest], forever memorializing his sojourn and sealing his mark upon the college as a Japanese and non-white Asian.[102] The swastika's implied counter-clockwise revolution and parallel setting of its arms placed within two concentric circles intimates an ancient East Asian Buddhist religious symbol. In this non-confrontational and philanthropic manner, Hachisuka and the two other benefactors combated the era's exclusionary racial politics by making their names visible; however, the mysterious symbol unwittingly generated decades of postwar speculation due to the swastika's later connotation with the Nazi regime (1933–45).

Nevertheless, Hachisuka's immortalization of his family crest in stone symbolizes how he left a clear imprint upon the University, even though Selwyn was not yet a true college and only his matriculation at Magdalene College was indicated on his student record card. Giles with his purposefully difficult exam in Classical Chinese and recalcitrant attitude toward Japanese obviously put a huge roadblock in the young student's path, but he circumvented this difficulty through informal study with mentors more amenable to his interests. Hachisuka's passionate love for ornithology, desire to collect the rarest birds in expeditions during his time away from the university, and obsessive attention to detail in describing them allowed him to engage with similarly zealous mentors like Takatsukasa, Guillemard, Rothschild, and Hartert, who fostered his abilities and admired his youthful energy. From the mid-twenties to mid-thirties, he produced multiple monographs and articles in Japanese and English detailing his explorations and connections with the globe's most important practitioners in the field and in the process carved out a name for himself in the Anglo-American world's ornithological endeavors.

Conclusion

Hachisuka's experiences in Great Britain provide an intriguing glimpse into the formation and early career of a Japanese scientist who enjoyed the privileges of an aristocratic position and wealth, which all aided in establishing useful connections with the British, and later, American ornithological worlds. However, he also met local prejudices as a subaltern, who tried to shield himself from carefully concealed racist sentiments within the trappings of honorary whiteness while overcompensating in his field's particular qualifications, like university degrees or memberships either personally attained, garnered honorarily, or even feigned. Such actions represented important survival steps in Hachisuka's formation as an internationally recognized ornithologist and national representative in the 1920s when popular Anglo-American conceptions of Japan oscillated between two common tropes, the menacing "yellow peril" and the exotic "cherry blossom," represented by threatening samurai and hospitable geisha, respectively.[103]

On a broader level in international affairs, Imperial Japan trounced Tzarist Russia in 1905 and aided in the Siberian Intervention to stop the Bolshevik Revolution from 1918 to 1921, but was also booted from the Anglo-Japanese Alliance by the Americans in 1921 and no longer enjoyed welcome of its immigrants to the United States after the Immigration Act of 1924—sometimes referred to as the "Asian Exclusion Act." These broader events in foreign policy certainly percolated down into the personal lives of Japanese abroad and influenced their reception as professionals on a global level amidst a contemporaneous hierarchy of nations. Other scientists experienced similar trials as Hachisuka, whose story revealed people of color's struggles in earning scholarly reputations in a largely Western, Anglo-Saxon scientific world. Within its racial hierarchies, Japanese asserted their place as honorary whites in an arena alternately enthralled and threatened by them as they occupied this space. Nowhere was this more apparent and contested than in the Philippines, an American colony since the late nineteenth century, and a place where American racist notions came to regulate quotidian life.

4

The Philippines (1929–31)—A Japanese Ornithologist Encounters the American Empire

Introduction

Prevailing transwar expressions of elite Japanese masculinity surrounding Western science arose out of strong associations with imperialism. Garbed in adopted Western dress and practices while hunting for specimens, Japanese ornithologists like Marquis Hachisuka Masauji engaged in mimetic performances of imperial masculinity as explorer scientists in forms of collecting imperialism.[1] This chapter focuses on Hachisuka's exploration of Mount Apo during a 1929 expedition in the Philippines, with his findings documented in four (intended as five) volumes in the early-to-mid 1930s. The Japanese and American empires touched waters in the Philippines, where Japanese scientists like Hachisuka and American military personnel like raptor expert Captain Lloyd R. Wolfe first crossed paths. While outfitted in military dress of the American colonizers, Hachisuka encountered and employed Indigenous peoples, carved his name upon a rock atop Mount Apo, and discovered new species, while circulating in American colonial spaces and performing a (waning) honorary white imperial masculinity just before the Great Depression's (1929–39) global assault on Anglo-American financial power.

In the Philippines, Hachisuka also met a cast of characters uncannily reappearing during Japan's postwar Occupation. In addition, when the US-based Harvard University MCZ exchanged specimens with Yamashina Yoshimarô and others in Japan, shipments traversed ports like Manila as key American halfway points to other Asian destinations. The Philippines, with multiple lush island environments, functioned as a liminal space between two rising empires and later rivals, with many quasi-uncharted areas hosting rich species diversity. These interactions are examined in a broader context of international relations and social space within Mindanao, the Philippines' second largest island and an imperial contact zone for Americans, Japanese, and Indigenous peoples.

The Explorer-Scientist: "Hunting for Empire" and the Thrill of Collecting Imperialism

In the late nineteenth century, a Social Darwinist orientation expressed in ruthless imperial geopolitics translated into Japanese science.[2] Notably, specimen acquisition mirrored territorial acquisition with the spread of political influence, a process accelerating amongst Western European imperialist powers with the 1884–5 Treaty of Berlin's subsequent "Scramble for Africa" and early twentieth-century American expeditions into Latin America. Yet, as US historian Matthew Karp argues, American imperialism began much earlier in the nineteenth century.[3] The 1846 US invasion of Mexico was preceded by previous imperial attempts to protect slavery in the Republic of Texas, Cuba, and Brazil. While Japan avoided this particular Scramble, its foreign policymakers' collective late nineteenth-century obsession with the Korean peninsula, termed *Seikanron* ["Rectify Korea Discourse"], and with early twentieth-century China after victory in the Sino-Japanese War (1894–5), provided an apt parallel, whereby military theorists like Yamagata Aritomo (1838–1922), a Meiji-Era *genrô* from Chôshû, interpreted Prussian theorist Franz von Liszt's (1851–1919) legal concepts of "zones of sovereignty" and "zones of advantage" as increasing spheres of influence in Korea and Manchuria.[4] These areas provided Japan's zoologists opportunities for scientific exploration and collection by attaching themselves to surveying crews mapping such territories of interest—often to counter or revise earlier work by Western scientists.

Imperial competition amidst such power projections also played out in scientific exploration, where biologists and zoologists sought new species in allegedly "uninhabited" places in a process of naming and claiming where accrual of avifauna under one's nation or personal designation translated into mastery of colonies' wilder portions. Many of these scientists were systematists, adherents of a branch of zoological and biological study emphasizing classification and taxonomies, later attaining greater sophistication with genetic research. Historian of science Nadin Hée reveals strong connections between Japanese scientific knowledge production and colonial control in Japan's colony Taiwan, where violence supported scientific endeavors.[5] This also pertained to the Philippines, where an American zoologist's writings inspired a 1890s colonial takeover, photographs from scientific explorations led to imposition of imperial power, and expeditions received military protection.[6] Japanese, whose colony Taiwan neighbored the Philippines, circulated there as diplomats, businessmen, and scientists, benefiting from rising numbers of expatriates running plantations and trade in Mindanao's Davao area in the late 1920s when the islands were firmly under American colonial control. As political scientist Reo Matsuzaki argues in *Statebuilding by Imposition*, though both colonies were governed by imperial powers, Japanese successfully implemented control in Taiwan by co-opting resident Chinese elites and imposing rules and norms, while in the American-run Philippines, Filipino elites jockeyed for power and engaged in corruption by establishing localized power centers contributing to unrest.[7] Thus, the Philippines contained pockets of politically contested areas far into the 1920s, like in Mindanao, where imperial Japan's business concerns progressively gained strength. Japanese scientists joined these informal extensions of imperial power.

Amidst this political backdrop, Japanese ornithologists, including Hachisuka and Duke Takatsukasa Nobusuke, head of Japan's hunting association, engaged in a social process that Gregory Gillespie calls "hunting for empire" where "hunting narratives tell of specific cultural attitudes, historical circumstances, and ways of seeing."[8] Others, like former Prince Yamashina, hired collectors like Ori'i Hyojiri to venture into Japan's far-flung imperial frontiers like the Daitô Islands in the early twenties, and employed Nakamura Yoshio (1890–1974) in Manchuria in the mid-thirties. Their amassing of exotic trophies during safaris or gathering specimens in expeditions operated amidst social conventions inspired by Great Britain and emulated by aristocratic Japanese. John MacKenzie reveals how these practices enacted imperialism domestically and overseas through exploratory conquest: "in the high noon of empire hunting became a ritualized and occasionally spectacular display of white dominance."[9] For imperial agents displaying imperial masculinity, hunting involved politically charged traversal of space, where dominance over animals and natural environments conferred white (or honorary white) hunters power over areas where they encroached and, often, over Indigenous peoples.

During the late nineteenth and early twentieth centuries, colonized peoples lost access to hunting rights on game preserves and parks amidst conservation's exclusionism, beginning amongst imperial powers like Great Britain. Japan also engaged in similar actions in early twentieth-century Taiwan and Korea, while in 1910, the Americans created a Forest Reserve for the Philippines on Mount Makiling,[10] where Hachisuka hunted birds with Captain Wolfe. In contrast, the Sahara desert, familiar to Hachisuka and his British or French counterparts, saw its megafauna decimated from lack of conservational attention resulting from perceptions of desert biomes as "empty" and uninhabitable, despite impoverished Indigenous inhabitants' ingenious survival efforts.[11] Japan extended similar disregard toward colonial Korea, where Japanese-administered forests were extensively planted to replace those cut for fuel in the late Yi dynasty (1392–1910); after Korea's 1910 Annexation, these were transformed into extractive natural resource production where cubic meterage of harvested timber rose from 700,000 in 1910 to 2.8 million in 1939, often exported to Japan.[12] Historian David Fedman underlines how harnessing Korea's natural world extended colonial control: "Japanese forestry bureaucrats cleaved farmers from materials essential to their livelihood and, come Korea's long and frigid winters, survival. By outsourcing the heavy lifting of reforestation, they bled-off forestland to Japanese corporations, capitalists, and settlers."[13] Koreans were thus forced to scrape up ground cover and sticks for fuel, while "illegally" trespassing into once-available forests.

In prewar domestic Japan, imperial forests or *Goryôrin* composed 5 to 6 percent of woodlands as reserves.[14] This included exclusive duck preserves, where only aristocrats or imperial elites could afford upkeep for high-class hunting receptions. After 1945, national land was democratized for popular usage of forests and green spaces in tourism instead of resource extraction or elite hunting. Yet, until the mid-twentieth century, privileges to hunt in certain spaces correlated to socio-political status and positioning in nested imperial hierarchies, where hunting allowed practitioners to enact imperial masculinity.

The explorer imperialist was originally a primarily British trope—of the great white hunter or pith-helmeted scientist exploring colonial territories on collecting expeditions in Africa or South Asia. Scientific expeditions involved ritualized hunting forays similar to safaris, including sending home rare animal trophies and, often, dead creatures' complete skins. Cosmopolitan American and Japanese imperial scientists were informed by British explorers and appropriated aspects from original British models.[15] Hunting and shooting avifauna in collecting enterprises also followed rules of British upper-class hunting practices and sociability.

Such "cultural mimesis" resembles Barbara Fuch's analysis of practices in New Spain, where colonized peoples' cultural performances increased their social capital in a stratified colonial society, where "the deliberate replication of Spanish ideology, provide[d] a powerful rhetorical weapon for writers marginalized by the same ideology."[16] Centuries before, Britain was archenemy to rival Spain, whose former colonies Americans and Japanese now explored. Moreover, from 1902 until 1921, Japan enjoyed nearly two decades of prestige through the Anglo-Japanese Alliance, which disappointingly lacked renewal to avoid alienating the United States. In the 1920s and early 1930s, imperial Japanese explorer scientists appropriated British or American gear, and similarly pith-helmeted, traveled through Philippine jungles and mountains commanding native porters to find unknown avifauna. In Manchuria and China, they also adopted this western style or the Kantô Army's militarized khaki clothing, with Chinese coolies transporting materials, as investigated in Chapter 5. Yet, not until wartime rupture in late 1941, American scientists continuously traveled and exchanged intellectual information, while sometimes competing with British counterparts as formerly colonized newcomers in the global scientific community, while Japanese ornithologists like Hachisuka, Takatsukasa, and Kuroda Nagamichi also engaged with Anglo-American colleagues. Prewar American and Japanese scientists converged in the Philippines with common interests in birds and enacted forms of avian imperialism connected to ownership of scientific knowledge.

Exploration in the Philippines, Late Nineteenth to Early Twentieth Centuries

By the late nineteenth century, Japan, as a newly emerging empire in East Asia, abutted waters with American imperial possessions in the Philippines around its colony Taiwan, and Guam near the Marianas (once part of the Philippines sold by Spain to Germany, who lost them to Japan following a post-First World War Mandate). The United States, a similar parvenu on the imperial world stage, had by 1898 projected its Manifest Destiny beyond its Pacific west coast to include territories of Hawaii and Guam, and in the same year, after victory in the Spanish-American War under President William McKinley (1843–1901), "purchased" the Philippines, ceded by Spain to the United States for a $20 million settlement. These islands soon became areas ripe for exploration by American scientists often directly engaged in political affairs and colonial administration. Similarly, the Philippines attracted Japanese explorers in the 1920s, like Hachisuka, who acted out imperial (and personal) desires through scientific pursuits.

From the beginnings of American rule, colonial authorities used the islands to fulfill ornithological obsessions. From 1901 to 1913, American zoologist Dean Conant Worcester (1866–1924) served as the Philippines' Secretary of the Interior, where his first year in office saw his establishment of the Bureau of Science's Division of Ornithology.[17] Worcester knew what scientific marvels he could anticipate; in 1887, as a University of Michigan undergraduate, he assisted a collecting expedition in the Philippines under ornithologist Joseph Beal Steere (1842–1940), Chair of the Zoology Department, who sought specimens for the university's museum. Worcester recounted how the islands fascinated the scientists and students, despite difficulties including Spanish authorities' suspicions:

> The close of our trip found us with health seriously impaired by hardship and exposure. Bourns (Dr. Frank Swift Bourns [1866–1936]) and I were firmly convinced that we should never again wish to risk such an undertaking. But unpleasant experiences became enjoyable in retrospect, and as we worked over our material and realized what had been accomplished and what remained to be done, the old fever came back upon us.[18]

In 1890 to 1893, Worcester returned with the Menage Expedition (sponsored by real estate tycoon Louis F. Menage [1850–1924]), collecting over 3,000 specimens of avifauna, plus material objects wrested, purchased, or cajoled from Indigenous peoples.[19] Using Manila as a base, and garnering Spanish goodwill through the US State Department, with safe passage granted by Governor-General Valeriano Weyler (1838–1930)—known in 1898 as "the Butcher of Cuba" for pioneering concentration camps [*reconcentrado* policy]—he and Bourns conducted six expeditions to amass these acquisitions. Upon his US return, Worcester's status rose when University of Michigan hired him as lecturer in 1893; he quickly advanced in 1895 to a position as Assistant Professor of Zoology and museum curator.[20]

In 1898, Worcester scored a coup when *The Philippine Islands and Their People*,[21] detailing the archipelago's history, ethnography, geography, and zoological wonders, became a popular best seller enthusiastically read by President McKinley, who invited him to the White House and recommended his service in the First and Second Philippine Commissions.[22] Having solidified his reputation as an expert, on January 20, 1899, Worcester joined the Schurman Commission to recommend American policies for colonial rule over the Philippines.[23] This post perfectly positioned him to continue ornithological work.

In 1906, Worcester and ornithologist Richard Crittenden McGregor, a Stanford University-educated COC member who would run Manila's Bureau of Science, published *Hand-list of the Birds of the Philippine Islands*,[24] for which McGregor had served as Worcester's collector from 1901 to 1905 while a colonial bureaucrat. In 1909, McGregor published *A Manual of Philippine Birds*, which described 739 species with common English, Latin, and local names[25] in a work Hachisuka extensively consulted for his 1931–5 handlist.[26] American zoologist Joseph Grinnell (1877–1939), editor of the COC's key publication *The Condor*, and director of University of California at Berkeley's Museum of Vertebrate Zoology, notes in McGregor's obituary that "[i]ts lasting authority rested in part on the exhaustive collections McGregor and his native assistants had industriously and discriminatingly gathered during the eight preceding years since his first arrival in Manila."[27] These two studies updated Worcester and

Bourns's 1898 *Distribution List* of Philippine birds, which originally "enumerated 243 genera and 596 species for this same area," but now raised "this total to 284 genera, 691 identified species and two species (*Oceanodroma* sp. and *loriculus* sp.) not identified,"[28] and included Palawan Islands species. These omitted and "undescribed" birds now joined the "known" Anglo-American natural world, despite remote islands with seemingly impenetrable vegetation.

Notably, the 1906 *Hand-list*'s introduction indicates that a considerable number of Americans in the Philippines were AOU members,[29] a trend revealing intimate connections between colonial bureaucrats, military authorities, and ornithologists amongst Americans posted in Asia, and their acquisition of privileged positions allowing fulfillment of passions for birds. In a 1911 letter to Grinnell, McGregor in characteristically flamboyant prose described his vehement refusal of a plum Brooklyn Museum position offer, and waxes rhapsodic about the Philippines in suggestive verbiage: "Think of the unexplored parts of the Philippine Islands, and the ripe, juicy regions nearby! No U.S. for me."[30] Clearly, birds, politics, and exploration were nested intimately together in this lush American colonial space, as for Japanese in geographically diverse locations like Manchuria, the Pacific, and Occupied Japan.

Hachisuka's Masterpiece: *The Birds of the Philippine Islands*

McGregor's conclusive 1906 and 1909 studies were surpassed in only two decades when Hachisuka one-upped his results with a four-volume set of findings gleaned from his 1929 expedition, published in the early-to-mid 1930s by Cambridge University Press, well-respected in the Anglo-American global scientific and academic community. The four books reveal a detailed, systematic account of bird species that he encountered and categorized, updated, revised, and described. In the second volume until the fourth, Hachisuka classifies Philippine bird life into the latest taxonomies, including several color plates. He planned a fifth volume, but Lord Rothschild's sale of extensive collections prevented this.[31]

On each volume's first page, a surprising plethora of titles appears behind Hachisuka's name; in print, he is referred to as "The Hon. Masauji Hachisuka, F.Z.S. (Fellow of the Zoological Society of London), F. R. G. S. (Fellow of the Royal Geographical Society), Member of the Ornithological Society of Japan; Member of the British Ornithologists' Union; [and] Corresponding Fellow of the American Ornithologists' Union"—new designations and memberships acquired since his late-1920s publications, including a new corresponding membership to the US-based AOU in 1930. Senior Japanese ornithologists, like Kuroda Nagamichi, in a two-volume text on Java's birds, listed fourteen impressive organizations and associations joined before 1936, including the *Ornithological Society*, BOU, AOU, and BOC.[32] For many people of color, imposter syndrome, defined as a sense of "intellectual phoniness,"[33] could lead to considerable collections of titles and qualifications to overcompensate in justifying their authority due to discrimination, both perceived and imagined. The two ornithologists hailed from the world's only non-white empire besides Ethiopia. Hachisuka additionally adopted a masculine culture of national competition and come-uppance, and correspondingly, adopted trappings of Anglo-American imperial masculinity in a world requiring such assurances by joining the "proper" scientific and professional organizations.

The collected volumes of *The Birds of the Philippine Islands* span considerable transformations in Anglo-American relations with imperial Japan after the latter's military aggression in northeast China and growing business dominance in Mindanao. The first volume was published on March 16, 1931, and the second on September 14, 1932, sandwiching them between the Japanese Kantô Army's 1931 conquest of Manchuria—an organization where Hachisuka's collector Nakamura had served— and the Japanese government's controversial diplomatic recognition of Manchukuo in March 1932. On October 2, 1932, the Lytton Commission, headed by Cambridge graduate British Earl Victor Bulwer-Lytton (1876–1947), concluded in its report that imperial Japan served as the Incident's aggressor and determined Manchukuo was an illegitimate state, with the region to be returned to China.[34] Thus, the Anglo-American world's political favor slowly turned away from imperial Japan, which disregarded the Commission's findings.

Notably, Hachisuka's early 1929 Philippines trip and AOU corresponding membership since 1930 soon oriented his ornithological connections firmly toward the United States. These American connections eventually helped him immediately after the war. However, after briefly returning to Japan via Kobe with his expedition team, he returned to catalogue his collections at the British Museum of Natural History in England, where he benefited from warm friendships with Lord Rothchild at Tring and Delacour, who often visited.

Not surprisingly, Hachisuka's diligence in establishing his reputation and overseas ornithological networks, coupled with meticulous scholarship, resulted in a favorable September 22, 1934 review in the British publication *Nature*, commending Volume II, published in 1932:

> Considering the work as a whole, we can only congratulate the author on its excellence and completeness. Every work of this nature merely forms a basis for future work. Macgregor's [sic] "Handlist of the Birds of the Philippines" appeared in 1926 [sic][35] and since then merely additional notes by the same author and a few odd notes by other ornithologists have appeared from time to time in different publications. The present work brings all our information on this group of islands completely.[36]

The British reviewer clearly remained unfamiliar with American ornithological circles, since he misspelled McGregor's name and erred in the date for his key *Handlist*–by two decades! Ornithological studies in systematics were often revised as new discoveries arose, and the reviewer notably praised Hachisuka's contribution as the field's latest standard. It included a comprehensive list of the volume's weaknesses, plus still-debatable classification matters:

> On the whole, the author freely accepts genera though he reduces a good many names to synonyms and certainly cannot be accused of being a wholesale splitter.

He has, however, frequently been ruthless in refusing to accept sub-species as definable, especially when he does not consider their stability proved for any given area. On the other hand, in a few cases he accepts a greater number of races than we ourselves would recognise, as for example those of the common moorhen.[37]

Hachisuka's dismissal of sub-species populating certain areas is labeled as "ruthless," which possibly references larger systematics issues—McGregor's earlier work also failed to acknowledge sub-species. Nevertheless, mountain slope microclimates, and other geographical diversity, prompted difficulties in clear categorization, and Hachisuka possibly erred on the side of caution. Regardless of shortcomings, *The Birds of the Philippine Islands* was a tour de force for the young Japanese ornithologist who made his mark in the Anglo-American scientific world.

A Hint of Scandal Provides Pretext for an Expedition

Remarkably, Hachisuka dedicated *The Birds of the Philippine Islands (Volume 1)* to his father Masa'aki, who nearly disinherited him after dissolving an engagement to Princess Kitashirakawa upon his return to Japan, allegedly because of unsavory conduct at Cambridge. To curry favor with Masa'aki, he named the *Turnix sylvaticus masaaki*, a tiny buttonquail subspecies found only in the Philippines, for his father, describing it in 1931.[38] In 1926, Masa'aki greatly welcomed aristocratic French ornithologist Jean Delacour as Hachisuka's close friend during his first Japan visit, while Masa'aki constructed a new, Cambridge-inspired villa "to suit his son's taste" on his large property,[39] built to anticipate his impending marriage. Still in Europe, Hachisuka returned to Japan in 1927 with Delacour and his mother, following two months-long expeditions.

For Delacour, his mother's accompaniment on travels was extraordinary, since he habitually spent fall and winter on tropical expeditions—usually in Indochina. His AMNH colleague Ernst Mayr's (1904–2005) obituary for his friend delicately explains his choice to remain a confirmed bachelor: "Delacour never married, but, as the last surviving son, he was devoted to his mother and took care of her until his death at the age of 94."[40] Delacour likely became increasingly intimate with Hachisuka and brought his mother to deflect from their relationship's possibly controversial aspects. Regardless of their friendship's nature, in prewar times, Delacour's mentorship certainly helped Hachisuka to integrate further into global Anglo-American ornithological circles as a non-white scientist.

The trio's arrival in Japan was met with great fanfare. At the Yokohama docks, Delacour remembers an imposingly large group of relatives, retainers, Peers, and Buddhist priests in "local costumes" (traditional formal wear) extravagantly greeting the ship hosting the men and older matron. Many arrived from Tokushima in Shikoku, a day-long train journey of 507 kilometers, joining those who met Hachisuka at Yokohama harbor with a train engaged for the occasion, including his uncle, Prince Tokugawa, President of the House of Peers, and descended from the last Shogun. This gathering of dozens of high-born family members, aristocratic friends, clergy, and numerous household servants undoubtedly preceded celebrations for Hachisuka's engagement to an imperial Princess, including an elegant reception and banquet.

Nevertheless, Delacour misread the pomp and circumstance, and believed that Masa'aki organized this welcome because "[h]is son had shared our life in France, so he wanted us to be a part of theirs in Japan."[41] He exoticizes the reception and banquet at the family's estate, including when "Hachisuka was taken to a small chapel for prayers."[42]—likely, the *yuino* ["extending of (customary) gifts"] engagement ceremony, where both families privately exchanged ritual gifts and placed them on the altar, where Hachisuka burned incense for ancestors at his family's temple on the grounds. In his memoirs, the Frenchman opined, "[i[t was most interesting, and I am thankful to have witnessed something so touching and so refined, which is not to be seen again."[43]

Most importantly, in Japan, Delacour met Hachisuka's ornithological friends and gained entrée into exclusive Japanese aristocratic circles. He undoubtedly stayed at Hachisuka's estate with his mother as was customary amongst close friends of wealthy Japanese and European elites at the time; for Masa'aki to host overseas French aristocrats showcased his cosmopolitan urbanity. Delacour thoroughly enjoyed meeting highborn companions similarly besotted by birds, with leisure devoted to their passion: "I had excellent friends in Japan. It was lucky that many of the men interested in natural history, particularly in birds, belonged to the old aristocracy. They still had the means of pursuing their interests efficiently and they also had the proper background for their studies. Zoological research, in those days, was almost entirely in private hands."[44]

After his Japan visit's conclusion, Delacour and his mother accompanied Hachisuka to "North America, China, Korea, and Manchuria," after which Hachisuka remained in Japan from 1927 into 1928 to fulfill family duties, including participate in his friend Emperor Hirohito's enthronement. Later in 1928, the Delacours traveled back to Japan, from whence Hachisuka accompanied them to Europe.

However, during Delacour's second return to Japan, Hachisuka's relationship with his father soured, and he broke his engagement with Princess Kitashirakawa. Beyond Hachisuka's alleged Cambridge indiscretions mentioned by his oldest sister Toshiko,[45] a break occurred with Masa'aki during Delacour's second Tokyo visit. In a 2015 interview, Hachisuka's protegé, Nakamura Tsukasa, Yoshio's son, described his personality as *jiyu honpô* ["free-spirited" or "reckless"],[46] which, for a Japanese aristocrat, could cause serious ramifications if his actions surpassed social norms. Nakamura indicated that Hachisuka's reputation in Japan arose as a "womanizer" or "scandalous Marquis" [*sukiyandaru no kôshaku*] because of jealousy against a rich, strong-willed young man who piloted his own plane and whose name frequently appeared in newspapers.[47] Though often seen in company with young women, widows, and aristocrats, called "new lovers," the media never pinpointed names.[48] Nevertheless, the act or situation apparently threatened the family's honor and social standing.

Hachisuka notably waited almost a full decade to return permanently to Japan following years overseas in England and the United States, including expeditions to the Philippines, Belgian Congo, Latin America, and other areas. He only came home when Masa'aki's sudden December 31, 1932 death[49] forced him to fly his plane to Tokyo via Siberia to administer his inheritance and other duties as heir.[50] His next extended return after years in the United States was prompted by the September 16, 1937 death of his second oldest sister Fueko (1898–1937), a writer and literary scholar divorced from Baron Matsuda Masayuki (1892–1976) living in Hachisuka's Japanese-style Atami villa.[51] By 1938, with the Second Sino-Japanese War's (1937–45) intensification, and Japan's political climate disengaging from the Anglo-American world, Hachisuka failed to freely travel throughout the globe and indulge his cosmopolitan research pursuits. Added were health concerns, worsened by tropical diseases like malaria plus tuberculosis, and worsening hearing loss, possibly congenital, which his older sister also suffered before death.[52]

In Hachisuka's obituary, Kinnear mentions that "On his way home [in 1935, presumably from Mexico City] he was taken ill at Los Angeles and was forced to remain in California till he was cured in 1938 [actually the end of 1937]."[53] Though a nervous breakdown in LA is possible, Hachisuka likely contracted tuberculosis in Mexico, where many immigrants to the United States with the disease were deported, which resulted in further outbreaks,[54] and before streptomycin's advent in 1946, took an extended "cure" in either the LA-based Barlow Sanitarium for respiratory diseases, or Las Encinas in Pasadena, specializing in nervous disorders, housed in a comfortable "craftsman-style bungalow."[55] Historian of Public Health Emily K. Abel asserts that "Although notorious for its polluted air today, Los Angeles once billed itself as a health resort, especially for people with 'lung troubles.'"[56] From late 1936 into 1937, Delacour visited Hachisuka for several months in Pasadena prior to his return home,[57] but in his friend's obituary tersely stated that, afterwards, he "never saw him again. In the course of these years in California, he had made many close friends."[58] The French ornithologist also mentioned Hachisuka's 1939 marriage to the Japanese-American Chiyeko Nagamine, from a wealthy agricultural developer's family, who he met in Los Angeles,[59] and his 1938 return to Japan, where Hachisuka remained until his sudden 1953 death from an alleged angina attack. At age thirty-six, producing an heir was expected of Hachisuka, though marriage outside of the aristocracy still required the Imperial Household Agency's approval.

Though homosexuality in Japan held less stigma than in the Anglo-American world, with the Meiji-era advent of capitalist modernity, Japan's aristocracy adopted Western attitudes and social norms, including heteronormative notions casting same-sex intimacy as unacceptable.[60] No historical records clearly indicate such intimacy between Delacour and Hachisuka, but Delacour's memoirs mention their close relationship, beginning when he met a seventeen-year-old Hachisuka, newly arrived for study in England, when Takatsukasa, visiting European ornithological collections, took his relative to dinner at the French ornithologist's Paris home:[61] "That was the beginning of a long and close friendship with Hachisuka, which ended in 1950 [1953] with his death at the age of fifty."[62] Moreover, the French aristocrat's memoirs often included homoerotic descriptions, especially when applied to interests in rare plants and birds. Delacour euphorically described a defining moment in early youth, admiring a suggestively shaped Mexican orchid (*Stanhopia wardi*) with the family gardener, Gaston, whose careful tending of tropical plants turned the boy's desires toward the tropics: "I shall never forget the shock of pleasure I felt, as a heavenly vision was revealed to me, for the first time, the true splendour of nature ... The scent was sweet, but powerful."[63] He also noted that "a unique spot remains in my heart for an

Englishman who had an exceptional influence on my younger years," or his "best friend" Hubert Delaval Astley (1860–1925), President of the Avicultural Society, who had "exquisite taste" and whose "affection" was "wholey reciprocated" despite a thirty-year age difference.[64] Delacour possibly combined scientific mentorship with physical interest in the promising, and quite attractive, young ornithologist.

In prewar Japan, homoerotic friendships and relations amongst elite schoolboys at Tokyo's First Higher School and the *Peers School* were extremely common,[65] with little social stigma for same-sex school friends or older men engaging in physical intimacy with younger men still studying at university; yet, beginning in their mid-twenties, most Japanese men were still expected to marry and produce an heir.[66] Amongst former *samurai* or *daimyô* elites like Hachisuka, traditions of sexual mentorship [*shudô*], couched in bonds of "brotherhood,"[67] continued until the 1898 Meiji Civil Code instituted compulsory heterosexuality for Japanese subjects, and reformers cast aspersions upon male-to-male love as "unmodern" or "uncivilized,"[68] rather than "immoral" as in Great Britain, which maintained sodomy laws where homosexuality was criminalized until partial legalization in 1967.[69]

The aristocracy was concerned with lineage, and so, arranged marriages were common. For example, Prince Takamatsu (1905–87), one of Emperor Hirohito's younger brothers, freely expressed feelings toward younger naval cadets in his personal diaries, but inwardly chafed against imperial duties, nevertheless marrying Princess Tokugawa Kikuko (1911–2004) in 1930.[70] Though friendly as spouses and active in multiple arts and health charities, they never produced children and led separate lives.

Such relationships resembled that of the childless fictional character *Sensei* ["Teacher"] with his wife and *Watashi* ["I"], his besotted mentee, in the celebrated writer Natsume Sôseki's iconic 1914 serialized novel *Kokoro* [Sincerity]. Literary scholar J. Keith Vincent pinpoints Sôseki as having "straddled the modern homosocial divide" between the Tokugawa and Meiji periods, and views the story's themes as quasi-autobiographical.[71] In a story set in 1912, the Meiji era's final year, two male protagonists grapple with "a world where romantic love between men and women has eclipsed the world of men"; as revealed in a long letter to Watashi, Sensei commits suicide due to his inability to reconcile his values with the modern era.[72] Like Watashi, Hachisuka came of age during the Taishô period, when gender norms and expressions of love amongst men were steered toward compulsory heteronormativity in a Japan obsessed with modernity.

Regardless of the scandals attached to his name, by the late 1920s, Hachisuka temporarily distanced himself from both Great Britain and Japan, and considered an exploratory trip to the Philippines, a US colony renowned for bird life. He soon began to write American leaders in the field for references. On August 23, 1927, Hachisuka's efforts bore fruit when Joseph Grinnell enthusiastically wrote a letter of introduction to Richard C. McGregor,[73] and noted how "ornithologically inclined visitors to the Philippines were always warmly and helpfully welcomed by McGregor."[74] Hachisuka planned to be the first man to fully traverse Mount Apo and hoped to discover new species on its slopes—actions potentially restoring an image of masculinity hopefully admired by his outraged father.

Historian of sexuality Robert Aldrich notes how "Explorers channelled energies into expeditions and homoerotic friendships rather than 'normal' married life,"[75] which complicated their social expectations, especially as aristocrats, but also provided outlets where they could travel and enact professional and personal passions in liminal colonial spaces where their privilege afforded leniency toward certain behaviors deemed deviant in their own societies. Noting characteristics possibly pertinent to Hachisuka, Aldrich argues that "sexual ambivalence and the desire of certain men for emotional and physical union with male partners produced a direct and identifiable influence on their public lives, whether in political actions, philosophical beliefs or artistic and literary creations."[76]

Individuals in the Philippines were quite receptive to Hachisuka, enabling him to meet with McGregor, and Lieutenant Wolfe, indulging in common obsessions for birds together prior to setting off on his Mount Apo journey. Traversing a dangerous mountain peak in Mindanao and discovering new species while besting earlier American ornithological surveys were important ways for Hachisuka to prove his imperial masculinity to both his father and the profession. A passion for ornithology allowed all kinds of men to socialize freely together, which sometimes, became a metaphor for more intimate friendships.

Hachisuka's Stopovers in Colonial Ports and Fateful Philippines Arrival

Prior to Hachisuka's voyage to the Philippines, he had already traveled with two of Europe's top ornithological authorities as an apprentice in his early twenties. In 1927, when Hachisuka returned to Japan from Cambridge, he had experienced numerous expeditions that year when he journeyed with Tring's curator of birds, celebrated German ornithologist Ernst Hartert, to North Africa, and subsequently with French ornithologist Delacour to North America, Hawaii, and Northeast China. Prior to leaving Japan in late 1928, he founded the *Nihon seibutsu-chiri gakkai* [Biogeographical Society of Japan][77] with Watase Shôzaburo (1862–1929), a highly respected zoologist educated in Japan and the United States at Sapporo Agricultural College and Johns Hopkins University (1886–1890), where he studied animal morphology, and taught at University of Chicago (1892–5) and Tokyo Imperial University (1901–24).[78] Historian of science Janet Browne indicates that, for Britain, "biogeography ... was one of the most obviously imperial sciences in an age of increasing imperialism ... the conceptual framework, methodologies, and practical techniques developed to deal with foreign animals and plants during this period [late eighteenth and nineteenth centuries] took their tone directly from those used in imperial expansion."[79] Japan, though an imperial latecomer, shared similar attributes for practitioners in the geography field. Thus, as an imperial agent furthering Japanese scientific interests, Hachisuka was well-positioned to lead his own expedition, despite relative youth at twenty-five.

In December 1928, after embarking by ship in Yokohama, Hachisuka left Kobe and traveled with his household retainer Itô S., along with Ministry of Agriculture official Nakamura Yoshio, an experienced collector, by steamer to the Philippines via Shanghai and Hong Kong, when Japan began to accelerate informal influence over the islands. Hachisuka remained in the Philippines until April 5, 1929, and returned to port at

Kobe on May 16, less than six months before the New York Stock Exchange crashed and seriously impacted American fortunes.

The results of Hachisuka's expedition were quite fruitful. After returning to England in spring 1929, Hachisuka spent several months in London and Tring working with Lord Rothschild's personal ornithological collections while making his own observations. With assistance of "Ms. M. Lawson of the Bird Room" as his typist,[80] whose voice often intrudes into Hachisuka's narrative in particularly British assumptions,[81] he began organizing notes from his journey in the library of London's British Museum of Natural History to form multiple volumes of *The Birds of the Philippines*. To substantiate previous international research on the region's bird life and provide fundamentals for his own contributions, he compiled *separata* of papers from 1890 to 1934 chiefly on birds of the Philippines and later cached them in the Museum's library; these *separata* were composed of twenty-one articles reprinted from various journals[82] and included a 1932 letter sent to him from Canadian ornithologist James Henry Fleming [1872–1940], who collected some 32,267 specimens.[83] Hachisuka's intellectual production reveals how a Japanese aristocrat and Peer sought out a British symbol of imperial power where he worked on his masterpiece as an imperial agent cataloguing the latest observations on an American colony's avifauna

With H. F. & G. Witherby, a press run by Hachisuka's friend from University of Cambridge, and the same publisher for Audubon's and William Beebe's (1877–1962) works, he ultimately published four volumes entitled *The Birds of the Philippines*, in Parts I to IV, issued in 1931, 1932, 1934, and 1935. Hachisuka's first volume included a history of ornithology in the Philippines, and descriptions of perambulations throughout Mindanao, plus his historic expedition traversing Mount Apo. The four handsome green leather-bound volumes were illustrated by 166 gorgeous color plates, "by Gronvold, Kobayashi, Keulemans, Frohawk, and Horsfield; other plates were taken from the works of Elliot, Gould, Sharpe and Marshall."[84] Unfortunately, the last two parts anticipated to complete Hachisuka's notes on the Passeriformes (perching birds or songbirds) never came to fruition—possibly due to Lord Rothschild's 1931 selling of a sizable portion of his collection.

The first volume of Hachisuka's set includes an ethnographically rich travelogue in the introductory chapter, "A Short Account of the Author's Journey to the Philippines," which denoted people he met and places frequented from an elite explorer's viewpoint.[85] It also outlined the Philippines' geographical setting, including maps and descriptions of vegetation, animal life, and diversity of birds,[86] which preceded detailed accounts of prior explorations by earlier European scientists.[87] Also, before Hachisuka's setting off into Mindanao's rugged interior, he described a voyeuristic traversal of numerous exoticized spaces when the S.S. Wilson docked at port en-route to its final destination, taking fascinating jaunts into dark alleys of Chinese areas in colonial cities like Shanghai, Hong Kong, and Manila to observe bird-sellers.

While in Shanghai, Hachisuka immediately headed for quaintly narrow streets where birds were sold, where "the cages were hung up on the roofs and entrances, leaving hardly any space to enter the shops."[88] In Hong Kong, he met Mr. Kinoshita, manager of the N. Y. K. shipping line, characterized as a "keen local sportsman" who took him to play golf—a then still Anglo-American pastime making inroads amongst

Japanese elites. The zoologist in Hachisuka noted with relish the golf course's location near a swamp abutting Repulse Bay, which promised wildlife like "pythons, bamboo-snakes and tigers in that neighbourhood."[89] Then, the two Japanese visited a local museum and bird shops, but also entered with curiosity a nearby butcher shop and outdoor public market featuring Canton's favored delicacies, indicating for them seemingly indiscriminate southern Chinese appetites for wildlife, including pangolins and domesticated dogs. Hachisuka left a pithy account of wanderings into side streets where he gleefully ogled exotic zoological wares:

> armadillo[90] [pangolin] vertically cut in half in the same way in which we cut mutton at home.[91] It is hung by a nail from the roof and sticks out into the street, affording one of the many choice materials used for the famous Canton chop-suey! I am told the Cantonese eat rats, but that is not by any means at all! By the way, the commonest dog in Hong-Kong is a fat, black-tongued, Teddy-bear-looking animal, known to be very faithful to its master. It is a native Chow dog, the word "Chow" meaning "to eat." The name bestowed on this much-favoured animal therefore means "a dog to be eaten." We then strolled into a big fish-market, where vegetables and birds were sold in one corner. Domesticated Chinese geese (*Cygnopsis*), with a big knob, and Canton duck were carelessly packed in cages.[92]

Stereotypical descriptions of crowded cages in narrow streets with "careless" vendors represented Hachisuka's colonial gaze toward his Chinese neighbors, and mirrored common Japanese (and Anglo-American) impressions of allegedly bizarre Cantonese tastes in this Chinese treaty port. Following this short Hong Kong layover, the American ship steamed toward its last stop in the Philippines.

Finally, in early 1929, after changing into "white summer kit" as his ship approached, Hachisuka landed in Manila amidst New Year's Day festivities, where he "caught a glimpse of the city in the morning mist and then very soon afterwards white buildings became visible, reflected under the strong sun."[93] The bay lapping at the mirage-like white colonial capital revealed water hyacinths choking waterways with this prettily ornamental, but invasive, plant described as "a native of tropical America" brought over in 1912.[94] Before Hachisuka and his two assistants, a household retainer and specimen collector, disembarked from their ship, local press members, and the Japanese consul, Mr. H. Okamoto, met them on board for interviews and assisted them with bureaucratic formalities involving importation of firearms and ammunition for their party, formally called the Mount Apo Expedition.

Soon thereafter, Lieutenant (later Captain) L. R. Wolfe,[95] a raptor enthusiast and fellow-member of the COC,[96] and his wife, received Hachisuka and brought him to "the Aquarium, the Botanic Gardens, and finally to the Bureau of Science to meet Mr. McGregor, the author of 'Birds of the Philippines,' who has a beautiful collection of bird skins, including all his types and many of Mearns' co-types."[97] Grinnell, as McGregor's correspondent, noted that "sequestered most of the time as he was, from ornithological contacts, greatly valued such visits,"[98] while in the late 1920s, "Wolfe was another American whose sojourn in the Philippines meant much to McGregor."[99] Zoologist Wilfred Osgood (1875–1947), McGregor's intimate friend at Stanford

University and a COC member, remembered that "He was given to strong attachments in which he displayed more downright affection than is usual with boys. He never gave up a friend, but always had someone who took a particular place."[100] McGregor's warmth toward like-minded men passionate about ornithology was enthusiastically extended to Hachisuka, who benefited from his expertise after bringing the fruits of his expedition to McGregor's Bureau.

Due to his aristocratic background, impressive ornithological knowledge, and excellent English perfected in Britain, both the expatriate American and Japanese communities appeared to warmly welcome Hachisuka. He thus befriended all the "right" people, and arrived during the social season's full swing, with Western New Year's festivities soon merging into the Chinese Lunar New Year. However, these social connections were also a pretext to garner maximum assistance for his scientific expedition. Serious ornithologists throughout the world dreamed of exploring the Philippines' multifarious tropical environments hosting remarkable bird species with many still "unknown" to science. Moreover, in Manila, McGregor was the leading authority on the islands' bird life.

Amongst the Philippines' American colonial authorities were an astounding number of AOU and COC members. The fateful meeting with Lieutenant Wolfe for ornithological reasons enabled Hachisuka to build helpful connections with Americans. Prior to a morning birding foray with Wolfe in Lake Taal's deep crater, Hachisuka was set up with his military tailor, who provided him with the proper outfit for his expedition: "it being the most practical wear in the tropics."[101] As a Japanese literally clothed in trappings of American power, his uniform performed the enclothed cognition of colonial safe passage. Moreover, to protect the health of Hachisuka and his assistants, McGregor arranged for inoculation against cholera, typhoid, and dysentery by a Dr. Schöbel, who provided them each with one syringe of cobra antivenin.[102] Armed with the requisite explorer's kit and tropical disease countermeasures, Hachisuka and his crew provisioned themselves for a trip to Mindanao, and readied themselves for the coming forays into illusive forests on the legendary Mount Apo's mountainsides, as the first Japanese to officially traverse its peak.

A Short Colonial History of Mindanao

Popular British poet and novelist Rudyard Kipling (1865–1936) evocatively described the American "civilizing mission" in the Philippines in the first stanza of his notorious 1899 poem, "The White Man's Burden: The United States and the Philippine Islands": "1 Take up the White Man's burden / 2 Send forth the best ye breed / 3 Go bind your sons to exile / 4 To serve your captives' need; / 5 To wait in heavy harness / 6 On fluttered folk and wild / 7 Your new-caught, sullen peoples, / 8 Half devil and half child."[103] Here, the poem lauds the nobility of sending out young white men to selflessly devote themselves to American imperial aims in the Philippines, while its hortatory tone echoes in a 1900 speech by US Senator Albert Jeremiah Beveridge (1862–1927), urging Americans to annex the Philippines. Literary scholar Ann Kaplan asserts that "Beveridge implied tautologically that the empire provided the arena for American men to become what they already were, to enact their essential manhood before the eyes of a global audience."[104] This aim certainly characterizes what Hachisuka attempted as a young Japanese scientist. However, Kipling's poem also forebodingly hints at potential lack of cooperation in this self-serving American endeavor, and insinuates likely insurrection by colonized subjects, characterized as juvenile, savage, and potentially dangerous. The United States acquired the Philippines as a colony in 1898, but its military soon became entrenched in a brutal war until 1902 to subdue independence movements.

These tropes became pronounced on Mindanao, the colony's second largest island, hosting diverse Indigenous peoples, including the Moros, followers of Islam maintaining trade and cultural connections with a broader Southeast Asian Islamic maritime world. Sociologist Stuart Schrader indicates a lengthy history of global policing by Americans, who began this process quelling early twentieth-century Indigenous insurgencies in the Philippines, an initial imperial acquisition.[105]

Later, patrol of lands based on colonial expeditions was applied elsewhere throughout the world, including in Japan during the Allied Occupation, with use of subalterns to effectuate imperial goals.[106] For American governance to succeed, beyond establishing patrols implanting colonial control, the area's overlords needed to attract Anglo-American settlers to make the US presence ubiquitous and visible while creating new land-use patterns with employment of native laborers. Nowhere was this more pronounced than in Mindanao.

For colonists who desired to control the Philippines, Mindanao first required subjugation as the colony's second-largest island with easy access to southeast Asia and rich agricultural potential.[107] Historian Olivier Charbonneau indicates how the early twentieth-century American press portrayed areas around Davao in a largely Muslim province as a fantasy "New West" to recruit white settlers—complete with Indigenous parallels to American Indians as potentially pliable workers once defeated.[108] In America, expansion in the West was declared closed in 1890 after the US Army won a brutal victory against the Lakota people whose resounding defeat followed the Wounded Knee massacre, perpetrated against largely unarmed elderly men, women, and children.[109] According to historian Patricia Irene N. Dacudao, similar campaigns occurred in the Philippines, sometimes with the same US actors, but afterwards, Americans permitted fair degrees of autonomy.[110] After Indigenous groups were quelled with force by 1903, US military proconsuls governed then-Moro Province until 1914, when a "Filippinization" campaign led to increased Catholic Filipino settlement and official oversight[111] by subalterns of mixed Southeast Asian, Spanish, and Chinese descent.

Like many imperial narratives, including those of Europeans in late nineteenth-century Africa, and Japanese in Manchuria after 1931, the Philippines' land was portrayed as "empty" and ripe for colonization with ample exploitable natural resources.[112] Hachisuka's late 1920s description of the Cotabato plain reinforced this view: "In this corner of Mindanao are over 500,000 acres of splendid virgin land, and it is practically uninhabited."[113] However, prior to colonial development's implementation, the area's natural wealth needed assessment to scientifically catalogue lands and demystify peoples, flora, and fauna. Charbonneau asserts that "Military intelligence-gathering expeditions trekked into the hinterlands, collecting demographic, agricultural, and epidemiological information; scientific data flowed in from Spanish sources; and government bodies like the Bureau of Lands made the region knowable through assiduous mapping."[114] American scientists like McGregor in the Bureau of Science codified assemblages of colonial knowledge, where ornithologists became integral to this apparatus, including his predecessor, interior minister, and bird specialist Dean Worcester. Moreover, "Authorities encouraged enterprising individuals to explore mining, forestry, and plantation agriculture, with the hope that these activities would make the Muslim South 'one of the wealthiest areas in the world.'"[115] Hachisuka accomplished the same as a Japanese imperial agent, arriving when Japanese economic dominance now characterized Mindanao, especially near Davao.[116]

By 1911, during experimentation's heyday to make the colonial tropics fruitful and productive through industrial farming, Americans or white Europeans, including Worcester, owned eighty-nine plantations using Indigenous labor to raise cattle or grow abacá (known as "Manila hemp," harvested from fibers of a banana plant used for rope cordage), agricultural products, and coconuts.[117] Hachisuka reported kind hospitality from both Worcester's son Frederick L., who shared his father's remarkable photo collection, and the Scottish McLarens (written as M'Laren), who owned a rubber plantation and hosted the explorer and his assistants in Kabacan, "the centre of the plain of Cotabato, and in this fertile soil hemp and coconuts grow like weeds."[118] Anglo-Americans described their business endeavors as "uplifting" and "civilizing" the Moros through productive work and providing children with plantation school education.[119]

Despite white plantation owners' kind hospitality, their efforts at "improving" local people's lives were limited. Hachisuka's impressions of plantation schools were unsettling, where "nearly half the schoolchildren were suffering from many tropical skin diseases, mostly yaws and 'double-skin,'"[120] hinting at malnutrition and sub-standard living conditions. The explorer and his crew responded by offering rudimentary medical treatment: "They found our small medicine-box very useful, and most of them received treatment for the first time."[121] Yet the scientist complained of these new duties: "Our skinning hour was considerably shortened owing to the patients who waited outside for medical treatment."[122] Here, Indigenous children from white-run plantations were accorded mercy from neighboring imperial agents, the Japanese. In only a decade, what Hachisuka and other imperial agents witnessed in travels in Anglo-European colonies in Southeast Asia transformed into anti-colonial rhetoric against white imperialists, denoting Japanese as better "stewards" of exploited Asian neighbors.

Nevertheless, what Charbonneau calls "the potentially deracinating effects of the environment," often named "Philippinitis," or a mutable illness encompassing malaria, home-sickness, depression, and debilitation from tropical heat, plagued the project described in Kipling's "The White Man's Burden," and limited further expansion of hoped-for white settlers.[123] Concurrently, Moro and other Indigenous peoples resisted with force the American-run government's various bureaucratic schemes.[124] By 1914, Filippinization of government agencies promoted greater Catholic Filipino migration into Mindanao, while Japanese emigration, especially to Davao, grew quickly from the mid-teens into the 1940s.[125]

In areas near Davao, Japanese planters began growing abacá hemp for the cordage trade, and eventually replaced Americans. There, they worked with Filipino officials and "cultivated important patronage relationships with northern politicians and tapped into Japanese trade networks to move their cash crops, in effect succeeding where white settlers had failed."[126] Dacudao underscores abacá's pivotal role in Davao's socioeconomic and cultural development: "By 1940, Davao supplied 53.3 percent of total abacá production in the Philippines, a far cry from its 3.4 percent share just 25 years earlier."[127] Because this region in Mindanao provided a vital resource grown only in this American colony, when tensions with Japan grew in the 1930s, US policy-makers feared a Japanese monopoly over a vital natural product. Thus, Dacudao asserts that "abacá production brought once-peripheral Davao, or the frontier, into the epicenter of government—the White House—in a matter of just a quarter of a century."[128] In 1941, Japanese settlers fully controlled Davao's business world with numbers reaching over 17,000.[129] This Japanese economic and cultural presence felt overwhelming to American observers, who called the area *Davao-kuo*, or "Little Japan"[130]—a pun on Manchukuo, or Japanese-occupied Manchuria.

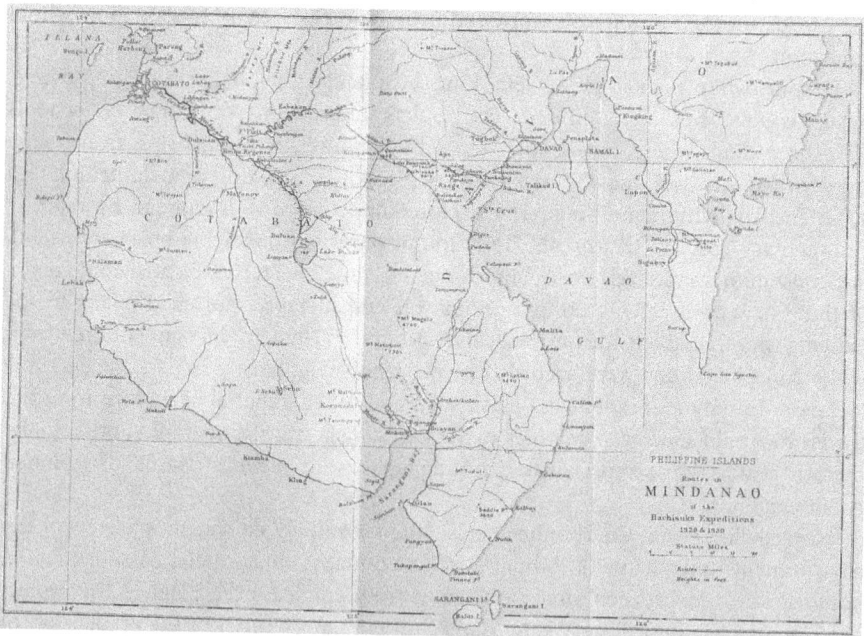

Figure 4.1 Map of the "Routes in Mindanao of the Hachisuka Expeditions 1929 & 1930," in Hachisuka Masauji, *Birds of the Philippine Islands: With Notes on the Mammal Fauna: Volume I, Parts I & II, Galliformes to Pelecaniformes* (London: H. F. & G. Witherby, 1931–2), insert. © Reproduced with permission from the University of Cambridge Library.[131]

Fully aware of such economic and political tensions, Hachisuka used his ornithological expertise and knowledge of Anglo-American social norms to profit from the American colonial presence through McGregor's Bureau of Science and Lieutenant Wolfe's military connections, while he also tapped into Japanese business and planter networks near Davao via Misters Koike and Yamaguchi. From an imperial perspective, Hachisuka needed to explore the area since it was never fully subdued by either Spain or the United States, which acquired increasing global prominence due to its Japanese population involved in near-monopolization of trade in an essential commodity. Therefore, Hachisuka's exploration of a strategically important area in the Philippines meant far more than birds, much about empire, and rising Japanese interests in Southeast Asia, which increased visibility of imperial Japan's informal power.

The Perilous Ascent up Mount Apo

As the Philippine Island's highest volcanic mountain peak, Mount Apo attracted many climbers amongst colonial settlers and visitors, though certainly, they were by far not the first to cross or climb it. Indigenous peoples believed that the mountain was sacred, so they treated entrance into areas around it with fear and respect: Apo meant "grandfather" or "elder."[132] Spanish colonizers' initial attempts to reach the peak were elusive, with one unsuccessful 1859 expedition led by Davao's first Governor, José Oyanguren (1800–59), followed by an 1870 expedition by Réal, a Davao military officer who also failed to reach it; in October 1880, another governor, Don Joaquin Rajal, finally ascended to the top. In 1882, German naturalists Alexander Schandenberg (1852–96) and Otto Koch followed these forays, along with early twentieth-century ornithological expeditions of Danish collector John (Johannes) Waterstradt (1869–1944) in 1903, American field naturalist Edgar Alexander Mearns (1856–1916) from 1903 to 1904 and 1905 to 1907, British zoological collector Walter Goodfellow (1866–1953) in 1905,[133] and lastly, US Army officers in 1928 who named a lake near the peak.[134] From the mid-1920s onward, ten Japanese reportedly scaled Mount Apo, but left no records; Hachisuka's journey was unprecedented since he first fully traversed the mountain.[135] Reaching Apo's peak clearly represented a rite of imperial power for agents of old and new empires alike.

The Mount Apo Expedition headed by Hachisuka, lasting from January 1 to April 5, 1929, allowed him to collect 243 specimens, with support by local Japanese networks of businessmen and intermediaries and dozens of Indigenous peoples serving as guides, porters, and outfitters for field camps. Organized in-country by "Mr. Koike," the expedition's general manager with extensive knowledge of the area following a previous climb to the mountain's peak, Hachisuka also acknowledged three Japanese crucial to the expedition's success: "Messrs T. Hoshi, Uchimura and Ishii, were the gentlemen who took part in the trip, and we owe much to their valuable assistance."[136] The team reaching this peak included Hachisuka; Koike, representing

Davao's Japanese community; Nakamura, the expedition's collector; Ito, Hachisuka's retainer; Ichikawa and Takamori, photographers; and two Bagobos, including Ouvah, a young hunter.[137]

"Mr. K. Miyasaka, of the Ohta Development Co." also provided introductions to local authorities,[138] whose connections revealed intersections between officialdom, economic development, and imperial scientific endeavors. Notably, "This gentleman has shown particular interest in scientific research, and … he presented me with a rare Monkey-eating Eagle on a previous occasion."[139] The rare "Monkey-eating eagle" (*Pithecophaga jefferyi*), or Philippine Eagle, a large raptor with a characteristic brown frill of feathers behind its neck, was most commonly found on Mindanao.[140] Such sociability involving gift-giving of unique specimens set Miyasaka within a history of Japanese retainers offering gifts of raptors to their lords in highly political exchanges involving future granting of favors in implied reciprocity. In turn, Hachisuka would provide valuable publicity for Japanese economic and commercial endeavors in Davao, perhaps mentioning them in the National Diet's House of Peers after reaching the requisite age of thirty for office. Naturally, should unrest occur in Mindanao, this could allow for speedy dispatch of Japanese troops to protect civilians.

Provisioned with ample personnel and advice, Hachisuka headed up Mount Apo with his retinue of two Japanese assistants, plus numerous traditionally garbed Indigenous porters, who required their own forms of protocol. A highlight of the

Figure 4.2 "At Lunch," a photo of the Mount Apo Expedition Team in Hachisuka Masauji, *Birds of the Philippine Islands: With Notes on the Mammal Fauna: Volume I, Parts I & II, Galliformes to Pelecaniformes* (London: H. F. & G. Witherby, 1931–2), Plate 9. © Reproduced with permission from the University of Cambridge Library.[141]

expedition was setting a stone upon Apo's peak, inscribed with Hachisuka's family name in *katakana* for other Japanese to see—another means of honoring Masa'aki after their falling-out. The stone would "claim" territory in a language that American colonists likely could not understand. Climbing and traveling through peculiar alpine slopes of mossy primeval forests and treacherous volcanic landscapes at Mount Apo's summit seemed a rite of passage for many explorers, including Hachisuka, with particular difficulties enabling harrowing recountings of the journey's perils. Many referred to the feared mountain as the "Devil's landscape" because of sulfurous eruptions issuing from vents upon slopes. The peak looked much like a snow-capped mountain from yellow sulfurous powder.

At the peak, owing to poor weather conditions, the Japanese explorers failed to find evidence of the previous year's (1928) military expedition by a joint American and Spanish-Filipino team with their Indigenous guide, undertaken by Lieutenant Colonel E. C. Bruns, Governor D. Gutierrez, Captain L. L. Gardner, Lieutenant A. Montera, Major A. S. Fletcher, and Manobo chief Datu Apang. According to a notebook given to Hachisuka by (now) Colonel Fletcher to consult before his journey, their names appeared on a piece of paper headed, "FIRST TO CLIMB MOUNT APO FROM THE COTABATO SIDE," followed by February 16, 1928, which they put into a sealed quart

Figure 4.3 "The Author at Lake Faggamb," Hachisuka Masauji, *Birds of the Philippine Islands: With Notes on the Mammal Fauna: Volume I, Parts I & II, Galliformes to Pelecaniformes* (London: H. F. & G. Witherby, 1931-2), Plate 10. © Reproduced with permission from the University of Cambridge Library.[142]

bottle, and placed at the believed highest peak surrounding the volcano's crater.[143] Hachisuka and his team searched for the elusive bottle, hidden by driving rain and thick fog. Instead, he commemorated their exploration by quoting Fletcher's account in *Birds of the Philippine Islands* to explain the area's unique name. According to the US military officer,

> As our little valley and lake seemed to have no name, we thought it would be the proper thing to give them one. We decided to make a pronounceable word from the first letters of names of the seven members of the party. This resulted in FAGGAMB—the initials of the names of Fletcher, Apang, Gutierrez, Gardner, Aduk, Montera, and Bruns. Let us hope that Faggamb Lake and Valley will be noted on future charts and maps about two miles from Apo's peaks, with a compass bearing about north fifteen degrees west.[144]

Notably, Hachisuka used Fletcher's notes as references in a method similar to consulting a handlist to update previous systematic findings on birds.

Figure 4.4 "Inscription on a Rock at the Summit of the Nameless Peak in the Apo Range," Hachisuka Masauji, *Birds of the Philippine Islands: With Notes on the Mammal Fauna: Volume I, Parts I & II, Galliformes to Pelecaniformes* (London: H. F. & G. Witherby, 1931–2), Plate 13. © Reproduced with permission from the University of Cambridge Library.[145]

Other members of the Japanese community also reached this area, and placed a small rock on the peak to commemorate this feat. When the 1928 military expedition investigated a westerly peak, they found evidence of a four-foot square post carved with Japanese characters, indicating a previous expedition where the "Japanese of Davao" placed a marker to indicate Mount Apo's highest peak; however, Fletcher's group of Americans and Filipinos discovered with relief that this was "thirty feet less than the highest peak."[146] Finding the highest peak was difficult, because Mount Apo "is really about ten peaks superimposed on one base about two miles in diameter and about 1500 feet below the highest peak," with multiple live and extinct craters created when an eruption blew the mountain's top off.[147] Nevertheless, Hachisuka and his team attempted to find it, where their complete traversal of the mountain would also assure the Japanese a competitive advantage in exploring the area as scientists and imperial agents.

While perched atop Mount Apo sipping *sake* [Japanese rice wine], at exactly 11:00 a.m., Hachisuka and his men festively burst into "the national song as it was a national day, and drank [to] the health of the Mikado."[148] This holiday, held on February 11 since the Meiji Emperor's 1873 blessing, was known in prewar times as *Kigensetsu* ["Festival of the Accession of the First Emperor and the Foundation of the Empire"], and more recently, *Kenkoku no hi*, or National Foundation Day; it commemorated Japan's mythical founding by Emperor Jimmu in 660 A.D. as indicated in the *Nihon shôki* [Chronicles of Japan, 720].[149] To commemorate this momentous occasion and link it with the Japanese empire's achievements, Hachisuka discovered a four-foot high flat rock to serve as a marker, and ordered it chiseled with his name in *katakana*, along with the date his group reached the peak, in a laborious process taking several hours with an Indigenous bolo knife.[150]

After returning from collecting specimens on February 7 to 13 and resting at Lake Faggamb, a lake near Mount Apo's peak, their euphoria attenuated when Hachisuka and his team headed for Quinatilan. To him, a disturbing switch in command occurred amongst Bagobo porters, headed by a young guide named Ouvah. Hachisuka complained that "It was to my regret that Ouvah, my useful native boy hunter, refused to come with me any further. He liked his work, and enjoyed our company, but as we were going into the country of another tribe (the Manobos), he was averse from going with us there."[151] Traversing the peak brought the Japanese explorers into Manobo territory, a group that Ouvah publicly hoped to avoid—a revelation manipulating facile American and Japanese understandings of Indigenous people. Ouvah's youth and lack of full accession to manhood amidst male tribal leadership likely prompted him to pass the traversal's honor to a more-deserving respected elder: the venerable Manobo chief Datu Apang, also referred to as "Datu," a respectful title used by Major (later Colonel) Fletcher in his expedition notebook. While Hachisuka grumbled about losing a valuable specimen collector, "So now we had not a single native from the Davao side, and the entire outfit was entrusted to the hands of Apang and his men,"[152] he was left in highly capable hands to negotiate the physical and spiritual perils of the mountain with this man, whose leaving of offerings and recounting of the 1928 military expedition's exemption from a usually required human sacrifice for a safe climb up the mountain reveal that he was also a shaman or *balyan*.[153] Most importantly, Datu Apang had served as the previous expedition's "native guide," and intimately knew the territory; he also owned Quinatlan's largest home where he could provide the Japanese expedition with superior hospitality.[154] Hachisuka celebrated his twenty-sixth birthday on February 15 at the Datu's residence, where the family's women regaled him with a feast complete with rice cooked in bamboo tubes over a fire, along with foods representing Spain's colonial empire, like tomatoes and sweet potatoes brought from the Americas.[155]

An interesting peculiarity of Manobo villages near Mount Apo's lower slopes was the intriguing possibilities of past connections to Indigenous peoples from areas conquered by Spain now known as Mexico and Guatemala.[156] Hachisuka noted that the mysterious village of Quinatilan, with a possibly Indigenous Nahuatl name, was "situated 1,800 feet above sea-level. It is the last village before going up the Apo and is not shown on any map."[157] The Spaniards, during their sixteenth-century conquest of the Philippines, employed Indigenous men and women, like Central Mexico's Nahuatl-speaking Tlaxcaltecs, as armed auxiliaries and camp staff, of whom some apparently never returned home.[158] Quite possibly, they absconded into mountainous areas near Mount Apo that possibly reminded them of their homelands' volcanic slopes, bringing with them tomato and maize seeds to plant in their new environments. At nearly 5,000 feet, Hachisuka found tomatoes as big as gooseberries growing on bushes, which he understood as "originally introduced by the Spaniards from Mexico," while he noted that some villages cultivated and ate maize.[159] Near Quinatlan, he remarked that villagers ate *camote* [sweet potato, *camotli* in Nahuatl].[160] Remarkably, the clothing of the Manobo girl from Quinatlan depicted in Hachisuka's first volume resembled the traditional garb of Guatemala's Xinka with her white top and black-and-white checked wrap skirt, although no records exist of Xinka auxiliaries accompanying Spaniards to the Philippines.[161] Clearly, Mount Apo's slopes were an important imperial contact zone for multiple cultures and empires for over four hundred years.

Hachisuka's descriptions and photographs of Indigenous people reveal an anthropological fascination with their diversity, a trait shared by earlier American colonial administrators like Worcester, who left behind a copious archive of 15,000 images.[162] He was even shown such photographs by Worcester's son Frederick L., a plantation owner who was also a naturalist.[163] While Hachisuka certainly granted native peoples humanity in his descriptions, he also found them highly exotic, and thus replicated the American authorities' colonial gaze, where he envisioned anthropological diversity similar to his approach to the island's diverse bird species.

In Santo Tomas near Baguio, Hachisuka entered with impunity a sacred "mummy" cave "discovered by Japanese not many years ago."[164] Within, Hachisuka complained of a "peculiar smell, sweet and honey-like,"[165] and described nests of small swifts, live bird inhabitants of the cave he disturbed. Yet he also desecrated bones of local peoples' ancestors by examining Igorot-like tattoo patterns on the leg of a "perfectly preserved" woman's corpse, with others cached within "crude" wooden coffins carved with images of crocodiles or lizards, or just "laid on the floor, with their legs tucked up close to their bodies."[166] Hachisuka noted that fabric found in the caves resembled that of the "Igorot, a head-hunting tribe of a mountain region of Luzon, who closely resemble the Formosan aboriginals,"[167] also citing possible connections between Indigenous

Taiwanese and Filipinos, in an observation indirectly justifying for him increased Japanese control over this area of Mindanao.

The island's Indigenous peoples were descended thousands of years ago from similar Pacific ancestors as aboriginal Taiwanese, rendering them more familiar to Japanese who interacted with them as colonial peoples. In fact, successful negotiations with Indigenous porters required knowledge of their myriad beliefs plus cultural and religious practices, but also a respectful attitude, which colonial authorities rarely extended, and even targeted what they viewed as "superstition" for elimination.[168] For American colonists and their visitors, Muslims were viewed as "intelligent, but deceitful," while Bagobos had a reputation for being fearful and loathe to leave home for long. Of course, these subaltern individuals sometimes manipulated such particularities when clients were led to assume that ethnic, religious, or cultural particularities motivated their actions should they refuse a task or request. Some looked more Asian to Hachisuka, while others, disparaging called "Negritos" by the Spanish in a term that stuck, possessed physical traits more like Sub-Saharan Africans. The Japanese, whose nation came under semi-colonial control by Americans in 1854-8, and who only shed the so-called unequal treaties with the United States in 1911, or less than two decades ago for Hachisuka and his retainers, also undoubtedly felt kinship with Filipinos they met as Asians and fellow members of fragile representative nations—one with suppressed independence subsumed into a "democratic" empire's colony, and the other that experimented with greater democracy under imperial rule.[169]

In fact, after his Mount Apo traversal, Hachisuka personally visited several members of the 1928 US military expedition, some now promoted to higher positions in the colonial hierarchy. While entering the Cotabato plain, he realized his suffering from malarial fevers, and felt obliged to return to Zamboanga. Here, he recuperated in the home of the head local Filipino authority, Governor Gutierrez, under his wife's ministrations and a local doctor's regular attendance. The Governor, in fact, helped lead the previous year's American expedition scaling Mount Apo from the Cotabato side, whose contribution was as "the first to conquer the Apo from this province and to return the same way."[170] During Hachisuka's recovery, he enjoyed viewing the Governor's large collection of weapons and implements amassed from Lanao and Cotabato Moros.[171] He noted that all Moros lost their distinctive swords, signs of masculinity like for premodern Japanese samurai, because "The Americans had made a regulation forbidding the Moros to wear these swords"; yet they were allowed to retain "a special thin-bladed working bolo, which they carried above over their shoulders without a shield. These bolos proved to be more dangerous, and finally this regulation became useless."[172] Bolos, large knives resembling machetes, were viewed by colonizers as working implements used to harvest plantation products in the fields, and therefore, were more an emblem of colonial subservience. As Dan Inosanto (b. 1936), Bruce Lee's (1940-73) former training partner, and a modern practitioner and master of *escrima* [a Filipinization of the Spanish word for fencing], or Filipino martial arts including knife fighting, has discussed, the bolo served with deadly effectiveness in rainforest fighting during the Philippine Insurrection (1899-1902).[173]

In Zamboanga, after Hachisuka's recovery, he also visited (now) Colonel Fletcher at the Petit Barracks, the area's military garrison.[174] Fletcher was the 1928 expedition's leader, honored with the first F in the Faggamb name. Two women, "Mrs. M'Laren" (McLaren) and her sister "Miss Young," "being good shots and horsewomen,"[175] had "accompanied Fletcher's expedition to Kidawapan in the previous year."[176] Hachisuka's last social call, on March 2 to 6, was to the Basilan-based Atong Atong Plantation, owned by Mr. I. Yamamura, described by Hachisuka as hosting "a family of naturalists and the first Japanese to have opened up the field of natural history in the Philippines."[177] Then, he traveled by boat to neighboring islands to collect birds and bats. Captain Wolfe also took Hachisuka on a tour of "Fort M'Kiling (actually Makiling Forest Reserve surrounding Mount Makiling) in Rizal near Manila," which boasted large open spaces harboring quail, francolin, and grass owls.[178] Through such social calls, planned and contingent, Hachisuka made visible his scientific and exploratory efforts to both Filipino and white American authorities, including local American and Japanese plantation owners, like Worcester's son.

Although Hachisuka took credit for the Mount Apo traversal's performative aspects, his employee Nakamura, a former Kantô Army member, collected specimens in forays amassing 784 specimens in Mindanao's five main localities, all while suffering from malaria.[179] Nakamura was also the first colonizer to ascend Mount Tumadgopt, a 6,000 foot peak in Mindanao's more isolated eastern division featuring a humid, tropical climate spared from typhoons, and rumored to harbor myriad unique species.[180] The birds Nakamura collected numbered 105 species, whereas certain birds endemic to Mount Apo were absent.[181] As for the mountain's until-then "undescribed" nature, Hachisuka explained this based on environmental and cultural factors gleaned from Nakamura's descriptions: "From the middle of the mountain to the summit is a matted tangle of undergrowth, almost impossible of penetration. The natives are very superstitious, and will not enter this uninhabited territory."[182] Clearly, the Japanese explorer misunderstood Indigenous beliefs and spiritual practices, involving purification rituals and gifts or sacrifices left for mountain's spirits to ensure a safe journey through sacred space.

To catalogue and categorize his burgeoning collection of specimens, which included Nakamura's acquisitions, Hachisuka returned to the Bureau of Science, where McGregor happily helped sort his latest specimens.[183] Back in England, the Japanese ornithologist was immensely grateful: "His [Nakamura's] valuable notes, setting down descriptions of birds and their habits, were recently sent to me, and the systematic parts of all these will be printed in full; while his collection is now being worked through in the British Museum."[184] In this bastion of British imperial power, a Japanese imperial agent's cataloguing of an American colony's bird life largely collected by a former member of the Kantô Army proceeded apace before the Great Depression's advent significantly dented Anglo-American imperial control.

On the ship *Empress of France*, Hachisuka left the Philippines and returned to Kobe on May 16, where "it was still not too late to admire the cherry-blossom," an iconic harbinger of spring and national symbol for Japanese.[185] He brought with him many live specimens, including a "new species of *Loriculus* and a magnificent cuckoo," which, unfortunately, failed to survive the journey.[186] Hachisuka's nostalgic reminiscences of Mindanao and Mount Apo mask the traversal's difficulties, and perpetuated fictions of his penetration into a pristine Edenic environment timelessly removed from American rule:

> The tranquillity and solitude of Faggamb, the bugle-call of a beautiful fruit-pigeon, the melody of *Orthotomas* or a herd of black deer will not be disturbed again in the Apo for many years to come: the Bagobos we had taken with us were too superstitious and would never return again to any height. Many treasures to naturalists will once more remain undisturbed in the heart of the mountain of heaven, Apo.[187]

Development eventually came to most Indigenous villages on Mount Apo's slopes, but such a view of a prelapsarian environment was informed by Hachisuka's imperial gaze perceiving the Japanese as the area's proper paternalistic curators.

Conclusion

By the late 1920s, American economic control slowly transferred to local Filipinos, and by 1941, was completely replaced by Japanese planters in Mindanao near Davao at the Pacific War's advent. Hachisuka's historic expedition, along with earlier informal Japanese forays up Mount Apo, represented a growing Japanese presence in the Philippines; though initially friendly with the Americans, tensions grew as the 1930s progressed. Hachisuka's most experienced collector, Nakamura, had worked as a translator and patrolman for the Kantô Army in Manchuria, in an interesting foreshadowing of future Japanese colonial control, which reached the Philippines with a direct attack on Manila in early December 1941 with imperial Japan's declaration of war against the Americans and their colony.

Long-standing connections between ornithology and colonial control continue to haunt birding's more modern incarnations, such as in current threads in the Cornell Ornithology Lab blog where enthusiasts still wax rhapsodic about Philippine birds, and note how the best place for watching them is at Subic Bay, a former American military base decommissioned in 1992 after Mount Pinatubo's violent volcanic eruption that same year.[188] Deeply intertwined connections between colonial control, scientific expeditions, and academia characterized Japanese endeavors in colonies or satellite states—like Manchukuo, whose lands, flora, and fauna were mapped out by Japanese scientific explorers in a key 1930s expedition. As in the Philippines, Japanese ornithologists would play an important role in that colonial region.

5

Manchukuo and the Japanese Empire (1932–40)—Deploying Avian Imperialism in the Media, Military, and Scientific Expeditions

Introduction

Located in Japanese-occupied northeast China, or Manchuria, imperial Japan's client state of Manchukuo (1932–45)[1] provided a strategic laboratory for Japanese exploration and scientific research, including ornithology. Domestic and overseas media publicized such endeavors, where broader publics consumed reports of Japanese scientific expeditions and avifauna in a new state under imperial Japan's tutelage and stewardship when Manchukuo's founding was still controversial in the Anglo-American world. Ornithologists from the imperial capital interacted with those in colonial peripheries to share specimens, data, and political information in areas where scientific knowledge mobilized mechanisms of control. The Kantô Army even mobilized and militarized carrier pigeon squadrons to collect data and surveille territory. In Manchukuo, mobilization of ornithologists and the birds they studied prefigured broader domestic mobilizations in wartime Japan from the early 1940s onward.

From June to October 1933, celebrated Waseda University Professor of Geology Tokunaga Shigeyasu (1874–1940) led the first Japanese state-funded overseas scientific expedition[2] to Manchukuo, whose complete report for the nearly three-month-long investigation included a bird volume.[3] The historic Tokunaga Expedition involved data collection and fieldwork with twelve other scientists, with some, including Tokunaga, garbed as colonial explorers in British gear sporting incongruously tropical pith helmets despite Manchukuo's dry and temperate summer weather. The scientists covered mostly Jehol (Rehe) and Manchuria's southwestern areas previously colonized by Japanese, resulting in a 25-volume compilation of reports in 3,937 pages of scientific information underlining imperial Japan's claims to a contested region.[4]

Zoology formed the collection's most prominent section, with "Birds of Jehol" featuring contributions by Hachisuka, Takatsukasa, Kuroda Nagamichi, Yamashina, and Uchida. Beautifully illustrated with color images, their research notes appeared in English, with abridged Japanese portions, illuminating the intent to address a global, or at least Anglo-American audience.

Besides as research venues, Manchukuo's "wilds" also became coveted settings for patrons of collectors like Nakamura Yoshio and Ori'i Hyojiri, who sought specimens for Japan's military or wealthy private benefactors. Nakamura worked for the Kantô Army before the region's Japanese takeover and in the late 1920s served as Hachisuka's specimen collector in the Philippines, while Yamashina entrusted Ori'i to collect bird and mammal specimens in far-flung colonies like Korea and the Ryûkyû (Okinawan) and Daitô Islands near Taiwan.[5] Such expeditions or collection forays resembled imperial land surveys, with establishment of taxonomies and lineages within a systematics process mirroring colonial control where technocratic influence supplanted formal rule in Manchukuo.[6] Such modern expeditions capturing or shooting birds resembled Tokugawa-era samurai retainers' travels over vast territories to collect live falcons to gain intelligence over lands they traversed.

Discovering and documenting new species, especially in contested locations like Manchukuo and around colonial borders, lent legitimacy to imperial Japan's claims, where science aided in crafting the Japanese into "worthy stewards" of areas under their control and in liminal imperial spaces like late-1920s Mindanao. Land surveying and specimen collecting, for state-funded expeditions and private individuals, directly supported Manchukuo's political fate and legitimacy. Imperial Japan's most prominent ornithologists played key roles, lending expertise to codify avifauna from the new state's borderlands. Some, like Takatsukasa, Yamashina, and Hachisuka held influential House of Peers offices in Japan's National Diet, highlighting their pivotal importance in expanding avian imperialism with international impact. In Manchukuo, ornithology and empire clearly intersected, with Japanese ornithological projects presaging Pacific War-era mobilizations.

The Media and Manchukuo as a "Laboratory of Modernity" for Avian Imperialism

By the 1930s, Japan's imperial ambitions reached Manchuria via machinations of the Kantô Army, tasked after the Russo-Japanese War (1904–5) to guard strategically and economically important South Manchuria Railway (SMR) routes from Dairen to Changchun. On September 18, 1931, the so-called Manchurian Incident erupted when Kantô Army General Ishiwara Kanji (1889–1949) and Colonel Itagaki Seishirô (1885–1948) instigated a plot to invade all of Manchuria where loyalists exploded tracks near Shenyang and blamed Chinese conspirators; thus began a satellite state to imperial Japan later known as Manchukuo in 1932.[7]

While the Kantô Army served as Manchukuo's military garrison and putative security detail for Japanese colonists and their endeavors, the South Manchuria Railway Company (SMRC) dominated transportation, hospitality, and commercial industries of southeastern Manchuria in areas largely settled by Japanese. Historian Prasenjit Duara characterizes Manchukuo as "a laboratory for what was not possible to achieve in Japan itself."[8] Under Japanese tutelage, Manchukuo also became a utopia for converted leftists, effectuating artistic and literary ambitions to become informal propagandists.[9]

For scientists, the country became an important locus for Japan's imperial undertakings, like ornithologists, whose contributions described, codified, and even militarized Manchukuo's bird life by establishing a "carrier pigeon army" within the Kantô Army.[10] Japanese scientists also led forays into Manchukuo under their protection, which even boasted specimen collectors. These included Nakamura, who admirably served his patron Hachisuka in the late-1920s Philippines and understood Manchuria's territory from Kantô Army patrols. Through expeditions and new publications, ornithology directly supported Japan's empire.

A broader public also followed science in the satellite state when Japanese media publicized Manchukuo in a plethora of topics. Historian Louise Young describes the Manchurian Incident-sparked media phenomenon as "war fever" infecting domestic Japan and its empire.[11] The *Asahi Shimbun*, a liberal paper often criticizing imperial Japan's past militarism, soon enthusiastically supported Japanese adventurism in China's northeast, and optimistically described forays including scientific expeditions like Tokunaga's in 1933. Contemporary *Asahi Shimbun* journalist and former Editor-in-Chief, Funabashi Yoichi (b. 1944), argues that this seminal moment began with an October 1, 1931 Osaka *Asahi* editorial lauding the Manchurian Incident, indicating that "[t]here is no reason we should oppose celebrating this event."[12] This generated articles and opinion pieces portraying Manchuria's desirability to the general public who consumed imperial Japan's allegedly most reputable news source and contemporary literature venue. Certainly, *Asahi Shimbun* was instrumental in highlighting Japanese scientific feats and other news to readers throughout the empire.

Issued in Osaka on January 25, 1879, with the first copy published in Tokyo on July 10, 1888, the *Asahi Shimbun* developed into a leading paper, hiring as contributors news luminaries like establishment writer Natsume Sôseki (1867–1916) on April 1, 1907, who penned serialized novels including "The Poppy," 1908; "Sanshirô," 1908; "And Then," 1909; and his iconic "Sincerity," 1914.[13] In autumn 1909, Sôseki made a landmark Manchuria tour, later described in *Here and There in Manchuria and Korea*, arranged by a student friend from Tokyo Imperial University, SMRC President Nakamura Zekô (1867–1927).[14] Beset by gastrointestinal issues, Sôseki grumbled during the whole trip, and undoubtedly understood his travels as the SMRC and *Asahi Shimbun*'s advertising tool. He thus described the area in superficially poetic prose to avoid fraught politics, pointing out "various natural sights such as the flights of northern birds," when the Korean independence activist An Jung-geun (1879–1910) assassinated the Korean protectorate's Japanese governor-general Itô Hirobumi (1841–1909) at Harbin Railway Station.[15] Despite Sôseki's few enjoyable bird-watching moments in Manchuria, political unrest and ill health prevented him from Korean travels.

Sôseki's privileged background resembled *Asahi Shimbun*'s many newspaper readers, and though dyspeptic, his experiences resonated with a general public interested in Japanese imperial initiatives. By the 1920s, the paper's subscription also became a marker of high social status; in contemporary Japan; *Asahi Shimbun* is still preferred by business, culture, and opinion leaders, with 23.8 percent of current readers enjoying household assets of over 100 million Japanese yen[16] (911,882 US dollars).[17] Imperial Japan's cultural, economic, and political elites, including scientists, both consumed the *Asahi Shimbun*'s news and manufactured it.

While *Asahi Shimbun* rapidly evolved into a reputable media source for Japan's bourgeoisie, and an organization supporting imperial enterprises like Tokunaga's, the

Dairen-based *Manchuria Daily News* from its origins served as the Japanese colonial government's mouthpiece following Japanese emigration to the area after Japan's 1905 Russo-Japanese War victory. Such a role continued after Manchukuo's 1932 formation. This state-funded newspaper originated in October 1908 as an English column in the Japanese-language *Manshû nichi nichi Shimbun* [Manchuria Daily News]. Tamura Kaîchirô (unknown) initiated the newspaper's English incarnation in Dairen. Housed within SMRC's Dairen headquarters and assisted by the SMR Research Department, the paper served as the Japanese government's propaganda arm which financed its publication. Beginning September 1, 1936, it appeared daily, excepting Mondays, while on November 1, 1939, its offices moved to Shinkyô (contemporary Changchun) where publication continued until 1940.[18] The *Manchuria Daily News* forms an important window into how the state portrayed Japanese activities in Manchuria, including scientific ventures and ornithological contributions to Japan's imperial state formation.

In the early 1930s, the *Asahi Shimbun* and other leading papers supported Japanese government initiatives in Manchuria and informally expanded imperial Japan's propaganda.[19] Young posits that their management soon became assured of the Kantô Army's wisdom in occupying the region: "In short, editors committed themselves to unifying public opinion behind the occupation because they were convinced that the army was right."[20] In domestic Japan, the Manchurian Incident sparked massive government crack-downs on leftwing dissent; this prompted anxieties amongst intellectuals and media producers, leading to *tenkô* [political conversion] supporting the imperial regime.[21] For scientists and ornithologists, as imperial elites embedded in imperial institutions, *Asahi Shimbun* and *Manchuria Daily News* would become important venues for publicizing their work.

Despite publishing constraints for journalists and the media, plus rising validation of militarism bolstered by commercialism, newspapers remain a key source of information on contemporaneous Japanese scientific forays. Here, Tokunaga's Expedition and other ornithological endeavors were presented to the general public. The *Asahi Shimbun* also helped to fund imperial Japan's first overseas scientific expedition to Manchukuo, and therefore, carefully detailed its activities. Concurrently, the *Manchuria Daily News* revealed general interest in birds rising in popularity amongst early twentieth-century Japanese news readers, where ornithologists imparted scientific and political knowledge on Japanese-occupied northeast China.

Avian Imperialism: The *Hand-list of Japanese Birds* and Birds in Manchuria and the Empire, 1932–40

On September 18, 1932, the one-year anniversary of the "Manchurian Incident," *Manchuria Daily News* initiated "On a Few Birds Peculiar to Manchuria,"[22] a multiple-part series by amateur ornithologist Mizuno Kaoru (1895–1980). Born in 1895 near Hiroshima's military port, Mizuno majored in natural sciences at Hiroshima Normal School, and in 1925, accepted a position as junior high school science teacher in

Manchuria, where he remained for two decades until repatriation after the war.[23] He also collected specimens for elite Japanese ornithologists like Kuroda Nagamichi.

In Mizuno's first article, published on September 18, 1932, and based on July 1930 studies, he indicated that 450 birds are endemic to the "vast and deforested Manchurian plains" whose numbers cannot compare to ones in a Japan "thickly dotted with woods and groves."[24] Establishing the paucity of Manchukuo's bird life amidst allegedly uninhabited and poorly stewarded environs, he revealed to readers his intentions to describe birds "peculiar to Manchuria" and "seldom met with in Japan proper," and briefly compared "Manchurian" birds to their Japanese counterparts. Mizuno described their habitats as "empty" or denuded in rhetoric echoing Japanese state-builders.[25] Such comparative descriptions highlighting occupied northeast China's ostensible inferiority inscribed birds within Japan's imperialism borne from incipient militarism justifying quasi-colonial Japanese rule over Manchukuo.[26] Japanese media publications in English also notably addressed Anglo-American audiences challenging imperial Japan's tutelary role in a perceived illegitimate takeover of Manchuria.

Mizuno's first two articles mainly featured popular game birds, including quail and grouse varieties as attractive bounties for sport hunting from a hunter's and specimen collector's perspective. Offering detailed descriptions of birds' habitats, plumage, and behavior, his essays appealed to Anglo-American sport hunters and bird-watchers and others interested in Japanese stewardship over Manchukuo's natural environment. Such articles in a longer-running series likely stimulated tourism to the newly minted state by these same foreign readers.

The initial September 18 article features the Korean Barred Buttonquail, or "Burmese hemipode" (*Turnix tanki blanfordii* of the *Turnicidae* family, also "yellow-legged buttonquail") where Mizuno details the bird's migration patterns and nesting habits in areas where imperial Japan established colonies (or invaded later): Siberia, Mongolia, Manchuria, the Liaodong Peninsula, Port Arthur, Jinzhou, the Gulf of "Pechili" [sic] [Gulf of Chihli, or Bo Hai gulf], the Shandong Peninsula, South China, and Burma.[27] Early twentieth-century Japanese ornithology's intersections with sport hunting appear when he signals to readers, "People interested in quail hunting can tell the approach of the Port Arthur season by the appearance of quail at Chinchou [Jinzhou]."[28]

Mizuno's next September 21 article highlights the Black Grouse, or Ussurian Black Cock (*Lyrurus tetrix*), described as a favored quarry shot in winter by "tens of thousands" that "are shipped, frozen, to Europe, packed in straw bags."[29] These ubiquitous birds composed the bulk of meat for Japanese expeditionary forces during the 1918–21 Siberian Interventions.[30] Such general columns written by local educational authorities and amateur ornithologists like Mizuno revealed growing popular interest in avifauna from Japan's empire, shared with Manchukuo's foreign audience and Anglo-Americans reading *Manchuria Daily News*. They also evince a strong imperial flavor where Japanese writers like Mizuno positioned themselves as honorary whites.

Mizuno's writings were supported by elite Japanese from the imperial capital and informed readers of Manchukuo's ornithological conditions beyond birds. Between the late 1920s and early 1930s, Mizuno befriended the older, Tokyo-based Kuroda Nagamichi, whose acquaintance grew from ornithological correspondence and concerns over sociopolitical conditions along the Korean border affecting the SMR. In 1930, stationed at the border town of Andong, Mizuno first sent Kuroda seventy bird specimens for identification.[31]

Their collaboration was further strengthened by communication over a 1932 *Hand-List of Japanese Birds* to which Kuroda contributed, coinciding with general Japanese and global interest in Manchukuo, declared an "empire" in 1934. From 1930 until 1934, Mizuno continuously sent Kuroda specimens he collected, which prompted six *Birds* articles by Kuroda.[32] In turn, Mizuno benefited from Kuroda's patronage in privately publishing his hand-list of Manchurian birds in 1934. This extensive correspondence between an eminent Japanese ornithologist from Tokyo with a younger Japanese mentee in Manchukuo, whose informal research informed broader Japanese ornithological studies, exemplifies how avian imperialism functioned in relationships between the imperial center and its periphery. Such quasi-symbiotic, though by no means equal, relationships gained each scientist benefits needed to advance his career.

Like for Mizuno and Kuroda, the *Ornithological Society's Hand-List of Japanese Birds*, updated each decade, provided periodic occasions for correspondence, collaboration, and critique amongst larger Japanese ornithological circles throughout the empire and beyond. Since 1912, the *Ornithological Society* published a new *Hand-List* every decade, its publication dates bookending important developments in imperial Japan, with 1912 corresponding to the Meiji era's completion and *Ornithological Society's* establishment, while 1932 ushered in Manchukuo and the *Ornithological Society's* twentieth anniversary. For these scientists, "Japanese birds" noted in the 1932 *Hand-List* also meant those endemic in "Korea, Sakhalin and Formosa," included in 856 total species and subspecies described; the appendix "Birds of Micronesia under Japanese Mandatory Rule" listed 169 species and subspecies from the Marianas, Palau, Carolines, and Marshall Islands.[33] Avian imperialism surfaces in authors' claiming bird species throughout Japan's burgeoning twentieth-century empire, and publishing a hand-list mostly in English as the Anglo-American world increasingly criticized imperial Japan's encroaching militarism in China.

In January 1933, *The Auk* editor Witmer Stone published a mixed review of the *Hand-list* in the AOU's leading journal a month before Japan's League of Nations' departure on February 24, 1933.[34] This action followed the Lytton Commission's damning October 1932 report blaming Kantô Army aggression on Japan, denying Manchukuo's legitimacy, and demanding Manchuria's return to China. Stone still positively reviewed the *Hand-List*, produced by Japan's top ornithologists, including Hachisuka, Kuroda (Nagamichi), Takatsukasa, Uchida, and Yamashina, all later contributors to *Birds of Jehol*: "The work is well printed with an excellent index and is a credit to all concerned."[35] Yet he expressed veiled criticism after lauding a format like American models: "The arrangement is much like that of the A. O. U. [sic] 'Check-List', species and subspecies being treated alike, but there are no references to place of publication of genera and no mention of type species."[36] Stone implies lack of due diligence and possible obscuring of histories for descriptions and range. Such flaws indicated by Stone were remedied by Yamashina in a 1939 article on Manchurian bird species in *Birds*, carefully listing collecting locations for specimens. For the *Ornithological Society*, new hand-lists of Japanese birds published each decade stimulated debate and research highlighting the empire's latest ornithological work.

In 1932, Takatsukasa, as a *Hand-List* author, expanded upon its findings in *The Birds of Nippon*, an unfinished multi-volume project continuing until 1943, when wartime duties dominated his attention. Only seven volumes were completed; these editions were printed and publicized by the Tokyo press Yokendô and promoted overseas by British press H. F. & G. Witherby, which published Hachisuka's previous books; both presses were likely owned by the scientists' friends. On February 28, 1943, Takatsukasa self-published the eighth volume amidst domestic publishing pressures and imperial Japan's designation of Great Britain as an enemy.[37] In 1967, over two decades later, a 623-page revised and expanded edition of Takatsukasa's exhaustive study was published posthumously by the English-language bookseller Maruzen, and included sections on "Galli, Turnices, Limicolae, Alectorides, The Birds of the Island of Bonin and Micronesia, and Order Psittaci."[38] Initial reception of *The Birds of Nippon* was positive for Takatsukasa's first volume. In 1932, in *The Auk*, Stone indicated "Prince Taka-tsukasa's [sic] book bids fair to be our authoritative work on the Japaneses [sic] avifauna and we wish him success with his great undertaking."[39] He also praised its high-quality materials with thick paper and magnificent color images.

In 1934, to promote imperial Japan's endeavors in codifying colonial avifauna in Taiwan, Takatsukasa published two articles in *Birds*, one featuring birds of Red Headed Islet or Orchid Island,[40] followed by another on birds of Bonfire Island or Green Island, all Taiwanese islets rich in rare birds.[41] He co-authored these articles with Kano Tadao (1906–45), a Tokyo Imperial University geography graduate student serving as the Governor of Taiwan's staff member.[42] This position allowed Kano direct access to bird observation in the field and enabled him to collect specimens for Takatsukasa and send them to domestic Japan. Kano also wrote the 1935 report on insects for the 1933 Tokunaga Expedition. While embedded in Japan's colonial apparatus of control over Taiwan, and as a leading Tokyo ornithologist's protégé, Kano investigated avian behavior in the wild and enabled the imperial center to conveniently connect with the colonial periphery through science.

Another *Hand-List* contributor, Kuroda Nagamichi, published a key mid-1930s study of Indonesian birds. In 1925, Kuroda had attended the Fourth Pacific Congress in Batavia, where he acquired core collections of Javanese birds, to add specimens "by exchange and purchase,"[43] revealing reciprocity with political import amongst Japanese ornithologists and Western networks. His 1926 work on pheasants covered varieties from Japan's colonies of Korea and Taiwan; in 1933, he published *Birds of Java* while collaborating on *Birds of Jehol* for Tokunaga. In 1933 and 1936, on high-quality paper and burgundy cloth-embossed covers, Kuroda self-published 200 copies of the 794-page two-volumed *Birds of the Island of Java*, his best-known work, with one volume on *Passeres* [songbirds and small, perching birds] and the second on *Non-passeres* [other bird varieties].[44] Each included a folding map and illustrations by Japan's top illustrator of birds, natural history painter Kobayashi Shigekazu (1887–1975), later famed for his illustrated *Birds of Japan in Natural Colors* (1956).[45]

Like Hachisuka, on both volumes' frontispieces, Kuroda lists all his academic credentials and global ornithological memberships. Arranged chronologically, they date from 1907 until 1936, ending with "etc., etc." as if he will continue joining relevant organizations. Kuroda's memberships comprised twelve organizations in nine different countries, listed in English, German, or French.[46] The second volume adds two more organizations: life member, Academy of Natural Sciences, Philadelphia (1935); and member, Ornamental Pheasant Society (1936).[47] Acquisition of multiple memberships in global ornithological associations, societies, and clubs was common amongst contemporaneous Anglo-American ornithological authorities like S. Dillon Ripley (1913–2001). In 1950, *New Yorker* staff writer Geoffrey T. Hellman (1907–77) notes of Ripley that "To help him feel at home on his travels, he has joined ... some fifteen other social clubs and ornithological or zoological societies."[48] Japanese ornithologists as cosmopolitan imperial scientists also joined every club and organization they could. Below this impressive list, both volumes' frontispieces include his family crest, enclosed within the embrace of two white peacocks facing each other. Kuroda's signature study of Javanese birds is thus circumscribed within a long aristocratic family history, symbolizing Japanese ruling elites' informal control over an area where scientific activities predated formal imperial takeover.

Stone's 1934 review commended Kuroda's study: "The book is excellent in its plan and execution and will be our standard work of reference for the birds of this island for many years to come. Ornithologists owe Dr. Kuroda a debt of thanks for his energy and labor in providing this much needed publication."[49] He also mentions that "[o]ne of the plates, representing the Pittas and the Broadbill, is by the author's fourteen-year-old son, Nagahisa Kuroda,"[50] whose artistic contribution presented his early adolescent debut into Japan's ornithological world, "while the others are by Shigekazu Kobayashi, generally acknowledged to be the leading bird artist in Japan."[51] Stone's only complaint against Kobayashi's renditions was their crammed placement: "His figures are, as usual in such works, necessarily crowded—often ten to fifteen to a plate—but are excellent representations of the birds."[52] These two volumes, well-received in Anglo-American ornithological circles, marked Kuroda globally as a Japanese authority on birds in neighboring Asian territories later absorbed into imperial Japan's Greater East Asian Co-Prosperity Sphere through military conquest.

The year 1934 marked significant collaboration between Japanese ornithological experts in the colonial periphery and imperial metropole. In the same year, under auspices of his mentor Kuroda Nagamichi, Mizuno published a scholarly oriented *A Distributional List of the Manchurian Birds*.[53] In 1940, while a schoolteacher,[54] he published *The Great Illustrated Encyclopaedia of Manchurian Birds in Original Colours*.[55] Matsuda Michio, a wild-bird song specialist[56] and blogger in Japan's contemporary ornithological world, indicates that it resembled his previous volume's format and others Kuroda published.[57] Below Mizuno's name, the brown cover shows a "2600" publication date, calculated from Japan's founding by mythical Emperor Jimmu in 660 B.C., and features an illustration of two raptors in a desert landscape overshadowed by a large ancient Chinese tower against a sky-blue background.[58] Notably, the landscape lacks people, where raptors dwarf an arid terrain framed by a grassy plant and archaic building. The book's foreword indicates support from Mantetsu (SMRC) Scholarship Funds for this illustrated encyclopedia.[59] The Kuroda and Yamashina families were primary SMRC stockholders, closely supporting this funding source, while Kuroda mentored Mizuno and inspired its creation.

Prior to collaboration with Tokunaga and Tokyo-based ornithologists on scientific reports from his expedition, younger scientists like Mizuno and Kano were thus enmeshed within colonial networks and codified bird species throughout Japan's empire and environs. These amateurs or junior researchers in colonial peripheries strove to build reputations in imperial Japanese circles; thus, they sought patronage of well-positioned scientists and organizations in Tokyo. In turn, these scientists required in-country expertise to produce authoritative volumes on colonial bird life. Such quasi-symbiotic reciprocity allowed Tokyo-based authorities to acquire needed knowledge, while supporting their mentees to further their careers.

Notably, the 1932 *Hand-List*'s same Tokyo-based contributors contributed to Tokunaga's 1933 survey of Manchurian birds in Jehol. Like Kuroda's patronage of Mizuno, they relied on local collectors to procure bird specimens from Manchukuo. Despite imperial science's logistical challenges, the early 1930s thus marked fruitful years for Japan's leading ornithologists. They revised earlier handlists of domestic Japan's birds, codified those in colonial areas from afar, and interacted with younger counterparts from colonial peripheries. Products of their labors were broadly disseminated into Anglo-American ornithological networks, despite growing distrust toward imperial Japan after Manchuria's military takeover. This atmosphere of accelerating domestic knowledge exchange amidst imperial Japan's distancing from diplomatic ties with the United States and Great Britain characterized scientific endeavors from the early 1930s onward.

Here, ornithology intersected with politics; for Japanese scientists, knowledge creation was imperial, amidst describing, codifying, and naming species in areas under Japanese domination or influence like Manchukuo, Taiwan, and Java. In Manchukuo, birds themselves joined Japan's imperial apparatus, amidst the increasingly militarized use of science for empire.

The Tokunaga Expedition, June to October 1933

After Manchukuo's contested establishment, domestic Japan's public became acutely interested in the hard-won new state's imperial ventures, referenced in articles, advertisements, serialized novels, reprinted round-table discussions, and detailed accounts of scientific expeditions. Press surrounding the newly minted polity vaunted the alleged progressiveness of those heading a brighter future under scientific principles involving proper stewardship of natural resources.

Notably, on March 8, 1933, *Asahi Shimbun* published an article describing a scientific round-table discussion on Manchuria and Mongolia's natural resources, held in its Tokyo head office the previous day.[60] The empire's leading media organization organized an informational seminar with authorities in geology, zoology, forestry, agriculture, paleontology, archaeology, animal husbandry, and the military—including Ministry of Agriculture official Iwata Hisayoshi (unknown); Tokyo Imperial University botanist

Nakai Takenoshin (1882–1952); artillery captain and Army Scientific Research Institute engineer Notô Hisashi (unknown); and Waseda University geologist and paleontologist Tokunaga. After a half-decade media hiatus, Tokunaga reappeared in *Asahi Shimbun* since a 1928 discovery of a rare four million-year-old shrimp fossil.[61] Amidst these experts, Tokunaga contributed knowledge to diverse aims like introducing horse- and-cattle raising to Manchukuo, improving forestry, and studying the "frightening" [*osoroshii*] "Hida" people—an Indigenous group likely related to the Kitans, a proto- Mongolian group that founded the Liao Dynasty (916–1125).[62] On March 26, a special

Figure 5.1 Photograph of the expedition leader, in Tokunaga Shigeyasu, *Dai'ichi ji ManMô gakujutsu chôsa kenkyû-dan hôkoku* [Report of the First Scientific Expedition to Manchoukuo (and Mongolia): Under the Leadership of Shigeyasu Tokunaga, June–October 1933, Section 1] (Tokyo: Office of the Scientific Expedition to Manchoukuo, Faculty of Science and Engineering, Waseda University, October 1934), 50. © Image from book in author's private collection.[63]

one-page feature on the round-table honored its eleven participants like celebrities with a row of photographs captioned with their names.[64] Such formal displays of Japanese imperial science's role in harnessing Manchukuo's natural resources set the stage for Tokunaga's well-publicized expedition of an under-studied Manchurian border area.

The Manchukuo expedition served as capstone for the geologist's career, arriving nearly four decades after his graduation from Tokyo Imperial University. After Tokunaga's 1940 death from pneumonia-induced heart failure, his obituary eulogized him "as a martyr for science and the nation."[65] Like Kuroda Nagamichi, Takatsukasa, and Uchida, Tokunaga hailed from a distinguished academic lineage with prominent connections to American and German zoologists through Morse and Iijima, who pioneered paleo-malacology in Japan, with later interests in ornithology. After graduating Tokyo's *First Higher School*, Tokunaga originally studied zoology, and received his 1897 degree under Iijima, with professional training resembling his later ornithological collaborators in *The Birds of Jehol*.

The Manchukuo expedition was state-funded under the Tokyo Foreign Office's "cultural department" with SMRC collaboration, while the Foundation for the Promotion of Scientific and Industrial Research, a precursor to the Japanese Society for the Promotion of Science, also contributed, with the *Asahi Shimbun* publishing company providing an airplane, and the Kantô Army furnishing wireless communications.[66] According to science historian Morris Low, "The expedition's journey served a variety of purposes. It can be seen as a way of nurturing budding colonists and heroicizing Japanese imperialism."[67] In the expedition's October 1934 general report, Tokunaga indicates "with mutual assistance Japan and Manchoukuo are exerting themselves to enlighten the people of the new state, and to explore the natural resources of the land," while he criticizes alleged Chinese backwardness: "It can be safely said, however, that the general lines of Manchoukuo civilization are still in a crude condition; the people can be active in no proper way to develop productive industry, mining industry, and other important undertakings of the country."[68] Clearly, Manchuria "required" Japanese guidance to render it productive and mobilize an unmotivated populace.

These aims to "civilize" Manchukuo's purportedly benighted inhabitants through science were widely publicized in international media to justify the scientists' mission. Aside from *Asahi Shimbun*, Tokunaga's expedition received extensive coverage from Manchukuo-based official news outlets like the English-language *Manchuria Daily News*, catering to an Anglo-American audience and other Westerners reading English. In 1933, when imperial Japan left the League of Nations, the empire became *non-grata* in the Anglo-American world; thus, many Japanese tried to remedy this situation by publicizing their nation's allegedly "peaceful" scientific endeavors. Tokunaga's general report frames imperial Japan's mission to reap natural resources and industrialize the region in a language of scientific inquiry:

> Here, from the angle of purely scientific observation, we may think it timely and important to explore the land in the hope of rendering contributions in the cause of Asian civilization. The devotional assistances the present expedition offers to this kind of exploration will, we hope, not merely serve the purpose of developing

Asian civilization, but also will render the industrial development of Manchoukuo practicable. In this lies the chief object and mission of the expedition."[69]

During Tokunaga's Dairen visit to organize his crew's arrival, and after a brief Japan return for preparations, he notified news organs, including the Shanghai-based *The China Journal*, published in English from 1923 until 1941 by the influential British-owned *North-China Daily News and Herald* for the China Society of Science and Arts' largely Anglo-American membership.[70]

On June 8, 1933, in *Manchuria Daily News*, Tokunaga indicated "The area covering Western Jehol and Eastern Mongolia, however, is still unknown to the scientific world."[71] *The China Journal* reveals his June 21 "*Rengo* [Japan Associated Press][72] telegram" from Japan's port city of Moji for advance publicity, indicating "Outer Mongolia had been covered by Dr. Roy Chapman Andrews [US explorer (1884–1960) from 1922 to 1928 and future AMNH president][73] but Inner Mongolia and Jehol is still unknown to the scientific world."[74] One-upping this renowned American paleontologist echoed Tokunaga's Dairen press conference.

Nevertheless, in July, *The China Journal* refuted the area's scientifically undescribed status, and countered that "not only has Dr. Andrews himself and his associates visited and secured specimens from Inner Mongolia and what is now called Jehol, but other British and American scientists and explorers, including the Reverend George Dr. Wilder (ornithologist), Mr. F. R. Wulsin and Mr. A. de C. Sowerby, have explored these areas more or less thoroughly."[75] The article minimizes Tokunaga's accomplishments by highlighting Anglo-American exploration; yet, it concludes charitably: "This does not, however, mean that a scientific expedition into the region will not be able to gather a lot more new data and new specimens; unquestionably there remains much to be done."[76] Japanese scientists encountered such detractions, prompting them to prove findings to skeptical foreign audiences with particular urgency.

An almost daily log of their journey appears in Japanese-controlled media, like *Asahi Shimbun* and *Manchuria Daily News*, saturating the news with exploits, however mundane. On June 5, Tokunaga arrived by ship in Dairen, Manchukuo's top commercial port, ahead of twenty-four other Japanese university professors and assistants to arrange their mid-July arrival and three-month-long expedition.[77] First, they would travel in a two-week study trip where "tourism was central to the maintenance of empire itself," and exemplified what historian Kate McDonald terms "the spatial politics of empire," or "the use of concepts of place to naturalize uneven structures of rule."[78] Into the contemporary period, study trips, combinations of tourism and serious investigation, are still popular for Japanese scientific researchers and other professionals.

Upon arrival, Tokunaga likely checked into the Yamato Hotel, SMRC's flagship for Japanese and foreign visitors, boasting an impressive high-ceilinged entrance hall flanked by pink marble and gold accents illuminated by electric lighting, and rooms arranged like suites, with parlors abutting bedrooms featuring large floor-to-ceiling windows.[79] The hotel's six stories grace the city's focal point, a large circle with streets rayed outwards like Baron Haussman's boulevards characterizing Belle-Époque Paris, which local Japanese called *Dai hiroba* [Great Plaza]. Within was a statue of Count Gotô Shimpei (1857–1927)

SMRC's first director from 1906 until 1908, who previously headed Taiwan's civilian affairs. The Yamato Hotel's construction began during Gotô's tenure, and was completed in 1914. The expedition departed on SMRC railroads from such "civilized" surroundings toward Jehol's remote areas, followed by trucks on track-less routes trailing dry riverbeds, and employed an aeroplane for aerial photos of the region.

Tokunaga's expedition was thus framed by modernity and militarism, including in transportation on a ship serving as military troop transport, while a heavy Kantô Army detail protected the scientists during surveys of southwestern Manchuria. The June 8 *Manchuria Daily* article stated that a Kantô Army military troop accompanied them, along with botanist "Dr. Nakai of Tokyo Imperial University, and Dr. Shimizu [Saburô] (unknown) of the Epidemic Prevention Institute in Shanghai."[80] A geologist's expertise helped survey the terrain, a botanist discerned agricultural potential and which plants thrived in its ecosystem, and a medical researcher assessed epidemiological risks for human populations that potentially hampered productivity. This large colonial operation involved land surveying and cataloguing of "native" flora, fauna, and anthropological specimens while scientists and their associates avoided locals and the diseases they might carry.

Clearly an imperial endeavor, the expedition symbolized Japan's military reach into remote border regions valuable for harnessing natural resources and for growing strategic importance. In 1933, Soviet leader Josef Stalin (1878–1953) sealed Manchukuo's borders with the USSR, with tensions erupting into the 1939 Nomonhan Incident's armed skirmishes. The area's resultant militarization by Japan, and justification for Kantô Army control, additionally assumed a "bandit problem" requiring law and order. The *Asahi Shimbun* asserts that the scientists headed into dangerous areas yet undescribed by science, and emphasized official fears of local bandit attack. Hence, each expedition member was insured for an astronomical 10,000 yen against alleged risks.[81] Yet, an "attack" occurred only once, where locals armed with one gun held the scientists at bay; threats were likely exaggerated to justify the expedition's strong military guard and augment Manchukuo's lawless image.

A mid-July *Asahi Shimbun* article calls the scientists "warriors of science," repeated in subsequent articles,[82] while many items outfitting the expedition came from military procurement:

> By courtesy of the War Office of Japan our clothes, tents, and other garments were procured in the Military Clothing Depot of Tokyo, provisions in the Military Provision Department of Tokyo, some quantity of rice and a few other food stuffs in Dairen. But a larger part of stuffs we procured in the Military Commisariat [sic] Subsistence Department of Jehol.[83]

The men ate Western-style tinned foods and used tents to avoid mingling with local populations; to further differentiate themselves from largely Chinese villagers, they wore pith helmets and khaki garments from Japan, along with sixty-seven trucks transporting their supplies and heavy caliber machine guns,[84] as they reached their destination "under unspeakable difficulties."[85] With Tokunaga and other Japanese scientists garbed like colonial British explorers from Africa transplanted to Manchuria, the paper described them as "enthusiastic" and "expressing great satisfaction with (their) new expedition uniforms."[86] Their persons thus embodied enclothed cognition and represented colonial mimesis evoking a white imperial power.

Amidst this performed mimicry of whiteness, "science was a technology of control. It provided a vocabulary with which to transform the subjugated peoples."[87] Such an expedition headed by a professor from Tokyo, the imperial center, served Japan's mission to systematize contested peripheral areas of Manchuria newly falling under Japanese occupation. The scientists, some in uniforms resembling British colonial officials or Japan's military, were accompanied by an army unit, as explorers of the empire's liminal areas perceived as dangerous and rife with "bandits." Low asserts that "Territorial expansion involved an inclusive moment when all within its boundaries would be welcome, but 'science' would be used to help differentiate and military control brought in to help manage and hierarchize differences."[88] Hence, rational, scientific ordering, and codification of Manchukuo's natural world extended to processes of increasing differentiation amongst a largely Chinese population, with significant minority groups

Figure 5.2 September 22, 1933, photograph of expedition members, in Tokunaga Shigeyasu, *Dai'ichi ji ManMô gakujutsu chôsa kenkyû-dan hôkoku* [Report of the First Scientific Expedition to Manchoukuo (and Mongolia]: Under the Leadership of Shigeyasu Tokunaga, June–October 1933, Section 1] (Tokyo: Office of the Scientific Expedition to Manchoukuo, Faculty of Science and Engineering, Waseda University, October 1934), unpaginated plate. © Image from book in author's private collection.[89]

composing the new state's alleged diversity, now run by a relatively few Japanese. The expedition's general report described Jehol as "a virgin land remaining untouched with scientific work," a manufactured "fact" which allowed Japan to inscribe whatever natural histories it desired upon the region.[90]

In 1935, Anglo-American journals began reviewing the expedition's reports—no doubt sent by Tokunaga—and revealed to foreign audiences rich flora, fauna, and natural resources under poor stewardship of the region's original Han Chinese inhabitants, described as "so simple and ignorant."[91] The general report scathingly decries Chinese ecological destruction:

> The terrestrial nature of Jehol was pitilessly destroyed by the Hans; hill and mountain were denuded of the trees; with little or no water running, the most river-beds are parched, and in consequence the larger part of the land of Jehol is dry. In this condition the species of animals are rather few in number.[92]

On March 23, the London-based British journal *Nature* reviewed the general report's 1934 publication and the first part of Section 4 (Botany), "*Plantae Novae Jeholenses, I,*" by Nakai Takenoshin, a Tokyo Imperial University botanist and Korean flora specialist, and Kitagawa Masao (1910–95), a Yokohama National University botanist in pteridology (study of ferns); it also analyzed the first part of Section 5 (Zoology), "The Freshwater Fishes of Jehol," by Mori Tamezô (1884–1962), an ichthyologist based at a preparatory school for Keijô [Seoul] Imperial University.[93] While the part on fish was praised for detailed taxonomies and twenty-one meticulous color illustrations, the botanists were particularly fruitful in discovering thirty-seven new plant species: "The November report (Section 4, Part 1) figures and describes eight species of new woody plants (by Dr. T. Nakai) and twenty-nine new herbaceous plants (by Dr. T. Nakai and M. Kitagawa)."[94] As the region's latest explorers, the Japanese scientists shed light upon species diversity in a still poorly researched area now under Japan's control.

The *Nature* article also emphasized that natural resources were crucial to the scientists: "The Japanese have lost no time in examining the resources of the new 'independent' kingdom of Manchukuo .... Never before has a scientific expedition been dispatched abroad from Japan on so big a scale."[95] Tokunaga's exhaustive expedition, described by the reviewer as "important both on patriotic and scientific grounds," covered an impressive reach of 5,000 kilometers over seventy days by truck over rough terrain.[96] Clearly, this schedule left little time for detailed investigation, while specimens and other natural objects collected were sent to Tokyo for further examination. Nonetheless, *Nature* conceded that Jehol's broad overview was required before exacting studies could pinpoint effective resource utilization.

The *Nature* article also noted the scientists encountering poor environmental conditions beyond their camp and visible signs of poverty amongst local inhabitants: "Dirty and scarce drinking water, and 'horribly poisonous insects', with the concomitant troubles of dysentery, trachoma, etc., were probably greater difficulties than the bandits, who fired upon a camp on one occasion. An endemic goitre [an enlargement of the thyroid gland] was found to be widespread in south-western Jehol."[97] For *Nature*'s editors, the Japanese expedition succeeded most in proving how prior human

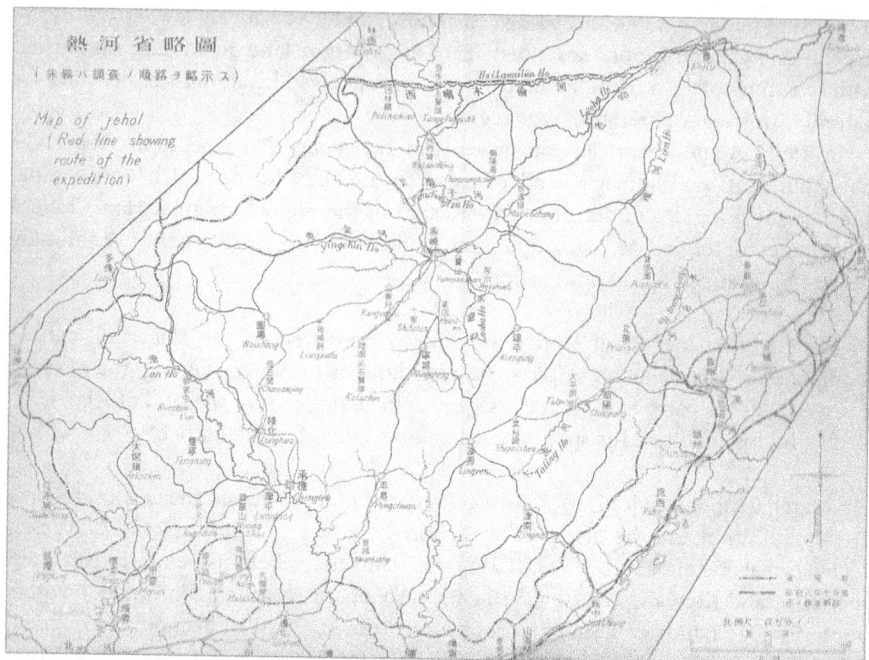

Figure 5.3 Map of Jehol [Rehe] and the route of the Tokunaga Expedition, in Tokunaga Shigeyasu, *Dai'ichi ji ManMô gakujutsu chôsa kenkyû-dan hôkoku* [Report of the First Scientific Expedition to Manchoukuo (and Mongolia): Under the Leadership of Shigeyasu Tokunaga, June–October 1933, Section 1] (Tokyo: Office of the Scientific Expedition to Manchoukuo, Faculty of Science and Engineering, Waseda University, October 1934), insert. © Image from book in author's private collection.[98]

habitation by Mongolians, Manchurians, and then Chinese failed to improve living conditions in a natural environment subjecting them to penury.

This disparaging view of Jehol's longtime inhabitants reappears in a July article in *The Geographical Journal*'s "The Monthly Record," published by *The Royal Geographical Society*, a British organization founded in 1830 and Europe's largest geographical society. It described the expedition as "the first thorough investigation of the Jehol territory in the south-western corner of Manchuria" where "disease, bad roads, and the dangers of brigandage had hitherto combined to keep Jehol almost unexplored."[99] In the early years of Manchukuo's tenure, scientific exploration served to justify Japanese semi-colonial rule. These ideological aims also appeared in the expedition's portrayal to domestic Japanese as its progress unfolded.

Nevertheless, British reviewers of these two key natural science and geography publications encountering the general report's conclusions certainly received impressions of the Chinese as poor custodians stubbornly resistant to modernization. Photographs in its last pages display the Japanese expedition's modernity and strict regimentation in traveling in trucks, performing toasts and bowing in unison, and reviewing military guards standing at attention.[100]

Yet, on the following page, images of Chinese thatch-roofed farmhouses and tile-roofed houses in Zhifeng and Chengde appear curiously uninhabited, followed by outdoor toilets in villages and rural backyards revealing lack of indoor plumbing and sanitation, while a photograph of a cave containing a farmer's preserved vegetables hints at dwellers' alleged "primitivity."[101] The report fails to explain these practices represent long-standing knowledge of local conditions to efficiently and cheaply utilize available resources: human manure was invaluable for Chinese farmers, where in-ground latrines were the fertilizer's key collection points, and caves provide naturally cool and dry depositories for safe food storage amidst hot summers and snowy winters.

Tellingly, the last page reveals six photographs of Chinese men, women, and a youth, seemingly oblivious to disabling, large goiters protruding bulbously from their throats. Captions assume this condition's epidemic-like existence, portrayed as "disease": "In the southern districts of Jehol, more than half of the townfolk [sic] suffer from Endemic Goiter, so throughout Jehol tens of thousand [sic] people must be the victims of the disease. The cause of the disease has not yet been discovered."[102] The research team failed to consider nutritional deficiencies' impacts upon Jehol's local Chinese populations, even though goiter prevails in areas lacking soil iodine and in diets without iodine-rich fish, seafood, or dairy products.[103] Prewar Japanese, with heavily fish-based diets including seaweed, rarely experienced the condition, and possibly viewed sufferers as impoverished or hopelessly anachronistic.

The general report clearly intended to solidify support for imperial Japanese "guidance" and justify quasi-colonial rule despite Manchukuo's putative independence. Here, imperial scientists imparted a highly publicized "technology of control" extending into Manchukuo's natural world, including its avifauna.

Celebrating the Kantô Army in the Expedition's Final Days, October 1933

Media accounts in Japan and Manchukuo closely followed the expedition's progress until completion. On October 12, 1933, the *Manchuria Daily News* ran a congratulatory article about its return from Jehol, along with "30 guards under Major Muraoka" provided by the Kantô Army "to guard the scientists from bandit raids during the extended journey," and noted its satisfactory results.[104] The paper quoted Tokunaga, who indicated the expedition was "a success from both the moral and the scientific point of view" and the party's brief disbanding in Mukden's (Shenyang) SMR Clubhouse, presumably for rest and relaxation.[105] On October 13, Manchukuo's (Chinese) President Zheng Xiaoxu (1860–1938), hosted a banquet in Hsinking (Shinkyô, or Changchun) for the scientists.[106] Echoing its sendoff, the expedition's completion was wrapped in layers of socialization amongst local elites, along with official receptions, to spread publicity amidst influential Manchukuo elites with economic and political clout.

On October 16, in a short piece in the "Personal and Topical" section, *Manchuria Daily News* noted that "Waseda Professor" Tokunaga Shigeyasu arrived the night before at Dairen, registering at the Yamato Hotel.[107] This was his port of embarkation for

Japan, along with the other participants. Dairen served as Manchuria's key commercial port since the Russo-Japanese War, and increased in importance with Manchukuo's establishment as a prime Northeast Asian nexus of commerce, trade, and shipping. On October 18, Tokunaga would take the Baikal Maru for travel back to Tokyo.[108]

Completed in Kobe on September 15, 1921, this large steamer arose in Mitsubishi shipbuilding yard Number 94 "as a 5,266 ton passenger ship for Osaka Shosen K.K. (OSK Line), Osaka."[109] It joined the OSK Line's passenger ships plying routes between Dairen, Moji, and Kobe, but was intermittently chartered as the Japanese military's troop transport to China in the 1920s and 1930s, and by the late 1930s and early 1940s, the ship added Korean and Chinese ports of call.[110] In 1942 during the Pacific War, the Baikal Maru served as the Imperial Japanese Army's hospital ship, circulating throughout Japan's invaded and conquered territories: On September 9, 1942, it left Singapore, and on September 13, traveled to Rangoon in Burma. From there, it arrived on September 16 in Medan, Sumatra, and returned on September 22 to Singapore. Then it went to Rangoon, Medan in Malaya, Belawan in Sumatra, and back again to Singapore, from where it steamed to Saigon, Indochina.[111] The Baikal Maru miraculously survived the war, and in 1950, was repurposed as a whaling ship, and from 1955 onwards, served in the salmon fishing industry, but was finally scrapped in 1968.[112] The ship's trajectory in war and peace highlights the militarized connections between domestic Japan and its overseas imperial possessions, and thus, fittingly served as Tokunaga's conveyance.

Notably, final commentary in the expedition's general report fully embeds the expedition within the Kantô Army. Tokunaga expressed his gratitude toward the escorting unit: "Lastly we tender our heartfelt thanks to Reserve-Major Muraoka and thirty soldiers for their warmest kindness extended towards the expedition in escorting it all the time against bandits."[113] Surveying land for future military endeavors near Mongolia's borders and the Soviet Union was one of the expedition's successes, and the Japanese government's long-term focus on a region allegedly beset by instability no doubt endeared the scientists to the Kantô Army.

Along with its military implications, Tokunaga's expedition was clearly a Japanese government publicity maneuver as much as a purportedly comprehensive investigation of the area's flora and fauna and natural features of the land and its historical inhabitants. However, this form of land surveying for the military was extremely costly and expended vast state resources, while scientists' individual reports might have critiqued imperial Japan's initiatives supported by the Kantô Army.

Kantô Army Carrier Pigeons: Loyal Surveillance from the Skies

For observation-based, efficient, and cost-effective land surveillance in real time, with meticulously detailed aerial imagery, birds trumped humans. In October 1933, during the Tokunaga Expedition, the Kantô Army created a large pigeonry and welcomed arrival of hundreds of carrier pigeons from Tokyo. Carrier pigeons were trainable, expendable, and loyal; plus, they could fly into areas not traversable by aeroplane. Also, few people suspected surveillance by birds and inclusion in intelligence reports.

In the early 1930s, the Kantô Army grew interested in carrier pigeons; consultations with Manchukuo-based ornithologists like Mizuno proved that birds could survey land and people's movements from above for security purposes. Pigeons could photograph border areas near China and observe Soviet troops, completely undetected. Such reports of avian "warriors" preceded wartime media propaganda of allegedly self-sacrificial Japanese soldiers with "selfless" martial qualities.

These mobilized birds also improved communications; they could travel hundreds of miles daily and send messages quicker and safer than ground or air mail. A July 1, 1933 *The Auk* article listed carrier pigeon speeds released by breeders, and indicated how one traveled 100 miles by air from Charlottesville, Virginia to Washington, DC in 1.8 hours, averaging 54.5 miles per hour, while another traveled the 650 miles by air from Attalla, Alabama to Washington, DC, averaging 38 miles per hour.[114] In Japan, pigeons also provided notable service for media organizations like *Asahi Shimbun*. On June 20, 1895, *Tokyo Asahi Shimbun* was Japan's first newspaper to use carrier pigeons to send a report about politician Inoue Kaoru (1836–1915), serving as Ambassador of Japan to Korea from October 26, 1894 to August 31, 1895, and returning to Japan from Korea.[115] In later decades, the imperial center's use of pigeons expanded into Japan's colonial periphery.

Coinciding with Tokunaga's expedition, *Manchuria Daily News* reported that, on August 29, 1933, Date Munetake (b. 1924), Count Date Okumure's brother, traveled to Manchuria to study carrier pigeons, supplementing earlier Japan-based studies.[116] The aristocratic amateur ornithologist spent several weeks at SMRC's Gongzuling experimental agricultural center. On October 18, when Tokunaga departed Dairen, *Manchuria Daily News* carried a front-page announcement: "Biggest Pigeonry in Orient Proposed by Kwantung Army,"[117] indicating "Bird Lovers in Japan Donate ¥4,000 for Pigeon Center at Kungchuling [Gongzhuling]—Also Contribute Number of Carrier-Pigeons to Augment 800 Now Possessed by Army."[118] Six days later, the paper ran a "Carrier Pigeon Army" photo, noting "a special ceremony was conducted to confirm the birds in their service" with the Kantô Army at Shinkyô (Hsinking), Manchukuo's capital.[119] Shintô priests symbolically blessed the birds' mission amidst state-supported religiosity. On October 23, at Gongzhuling, SMRC's experimental station located "on the S.M.R. main line," the Kantô Army received fifty-two carrier-pigeons and "pigeonry" sent from "the Japan Carrier-Pigeon Association, Tokyo,"[120] On behalf of his superior General Takahashi Hishikari, Major Okamura Yasutsugu "christened the winged messengers 'Patriotic Doves'" at a Shintô ceremony to commemorate their Shinkyô arrival. Gifts of birds connecting Tokyo-based amateur associations to Manchukuo's Kantô Army represented avian imperialism expressing Tokyo elites' direct support of Japanese militarism.

While such early 1930s Kantô Army initiatives appeared innovative, homing pigeons photographing locations with automatic cameras strapped to their breasts was nothing new. In 1907, German apothecary and amateur inventor Julius Neubronner (1852–1932) developed such surveillance, initially using homing pigeons to deliver prescriptions from his Kronberg pharmacy to far-flung towns,[121] while the Prussian military first used his pigeon camera during the First World War. Prior to widespread airplanes, such images revealed how the earth appeared from above.[122]

A October 10, 1914 *Scientific American* article indicates how cameras were adapted to conform to avian anatomy:

> It is not easy to construct an automatic camera with a maximum focal length of 2 inches– and a maximum weight of 2 1/2 ounces, including all appurtenances, but the problem was finally solved by a Frankfort firm. One form of the apparatus comprises two complete cameras, with their lenses directed forward and backward, so that at least one land view is obtained when both plates are exposed simultaneously in any position of the bird. The cameras are mounted on a thin aluminum cuirass which is attached to the bird by straps and rubber bands.[123]

This description provides clues to how the Kantô Army likely used pigeons for avian land surveillance. Mizuno had surveyed bird species in areas explored by Tokunaga's scientists but found few near railroads;[124] however, pigeons belonged to landscapes where humans lived. Such areas fell within SMRC's development, where construction, plus felling trees and razing brush, disturbed ordinary songbirds and raptors. As scavenger birds, pigeons were ubiquitous in big cities or areas featuring human activity, and therefore, their presence was overlooked during surveillance.

During the Pacific War, use of pigeon spies expanded from experimental Kantô Army deployment into other operational theaters when Japan's human-based military structure incorporated them. According to a US War Department technical manual on Japanese military field organization, the Signal Regiment, or Army Signal Unit, first developed from civilian "communication units of the corps of engineers."[125] However, after 1941, "an Inspectorate of Communications was set up directly subordinate to the War Department General Staff. This was tantamount to the establishment of a separate signal corps. Troops are classified as signal communication men."[126] The signal corps included animals like carrier pigeons for communications and intelligence purposes: "The pigeon unit has a headquarters of some 20 officers and men, including a train, as well as several pigeon platoons. A platoon has about 50 men and is divided into 3 sections, each equipped with 40 carrier pigeons."[127] Such rapidly deployed, unnoticeable surveillance and communication means were crucial amidst unreliable telegraph messaging, land- and sea-based mail, or aerial photography within wartime theatres where human conveyances were easily shot or bombed.

Pigeon camera deployment continued well into contemporary times. During the global Cold War (1947–91), the US Central Intelligence Agency (CIA) created a "pigeon camera," revealing how useful these avian surveillants were to the Kantô Army's Manchukuo operations from 1933 until 1945. The CIA's virtual museum notes aerial surveillance by carrier pigeons permitted intelligence activities to remain nearly imperceptible, since the bird blended in,[128] yet explains that "[d]etails of pigeon missions are still classified," and demurs from dates or operational theaters where pigeons were used.[129] Despite such vagueness, the CIA publicly indicates the birds' incredible usefulness as avian precursors to electronic drones: "Pigeon imagery was taken within hundreds of feet of the target so it was much more detailed than imagery from other collection platforms. (Aircraft took photos from tens of thousands of feet and satellites from hundreds of miles above the target.)"[130] Few might suspect avian surveillance during quotidian activities or combat.

Birds easily performed surveillance and compiled national intelligence, but more likely, ornithologists who carefully studied them conducted these activities. For 1930s Japanese researchers, image collection by pigeons or humans, along with other data, formed a corpus of readily employable information to compile numerous studies on areas of interest. Tokunaga's massive expedition reports featured such collaborative work, where imperial Japan's key ornithologists compiled a volume on birds.

Codifying Manchukuo's Natural World and the *Birds of Jehol*, April 1935

Nearly two years after the expedition, its results were published in a stunning data collection punctuated by attractive color plates of flora and fauna. Published in Tokyo by the Office of the Scientific Expedition to Manchoukuo within Waseda University's Faculty of Science and Engineering, the entire "Report of the First Scientific Expedition to Manchoukuo" comprises six sections, beginning with an introductory general report (1), followed by sections on geology (2), geography (3), botany (4), zoology (5), and anthropology (6). These are further subdivided into several multi-part divisions, with essays appearing in English and slightly abridged Japanese. Tokunaga clearly wished to share findings with an international, possibly Anglo-American, audience.

The pivotal zoology section comprises nearly half of the entire report, dwarfing all others with sixteen multi-part divisions covering regional animals, subdivided into fish, birds, mammals, insects, spiders and scorpions, reptiles and amphibians, crabs, and molluscs. The expedition caught "more than 500 [specimens], of which Vertebrata are about 130 species, among which are Mammalia 13, Aves 70, Reptilia 11, Amphibia 11, Pisces 32, and Invertebrata 370, among which Insecta about 300, Arachnida 40, Myriapoda 9, Crustacea 7, and Mollusca 15 species."[131] Of these, birds were the animal class most likely to traverse and surpass Manchukuo's borders, serving as truly transnational representatives of imperial Japan's fauna, and thus, symbolized formal and informal extension of power and influence.

Although none of Japan's top ornithologists accompanied Tokunaga, two team members, including the Tokyo Imperial University-educated mammologist and arachnologist Kishida Kyûkichi (1888–1968), and the colonial Korea-based naturalist and ichthyologist Mori Tamezô (1884–1962), supervised several collectors, with seventy total bird specimens harvested[132] and sent to Tokyo for the *Ornithological Society's* examining, cataloguing, and documenting to prepare *Birds of Jehol*.[133] In his October 1, 1935, review of the report, Stone indicated that Mori and Kishida's collection "contained seventy species and subspecies [*sic*] one of which *Passer montanus tokunagai* Kuroda and Yamashina (p. 87) is described as new, and named after the leader of the expedition."[134] Though a relatively common bird, this Eurasian Tree Sparrow's Manchurian varietal[135] represented the Expedition's pride in a new discovery, however modest. Such findings represent strongly collaborative work, where Tokunaga's crew members, supported by large groups of auxiliaries, engaged in collective knowledge production.

Set in Zoology's fifth section, the bilingual *Birds of Jehol* report appeared in April 1935, and composed part three of the *vertebrata* second division, with Takatsukasa as senior author; contributions by Hachisuka, Kuroda Nagamichi, Uchida, and Yamashina composed this roster of imperial Japan's ornithological authorities. Uchida wrote the English portions, and commissioned drawings of birds from Kobayashi, while he tasked Takatsukasa and Hachisuka to oversee data and descriptions of non-*passerine* species, whereas Kuroda and Yamashina took charge of *Passeres* [songbirds].[136] Within the 70 specimen collection, researchers found "13 orders, 27 families, 53 genera, and 70 forms, among which is a new subspecies."[137]

Division of labor by expertise mirrors other contemporaneous Japanese scientific endeavors, where a group of individuals focusing on aspects related to their specific fields or sub-fields collaborate to work on a larger study. Miriam Kingsberg Kadia's *Into the Field*, assessing prewar and wartime activities of Japanese ethnologists, posits such delegation of duties also resulted from political contingencies: "The expense, hazards, and logistical challenges of operating alone in an unfamiliar war zone moved Japanese human scientists toward a model of collaborative fieldwork involving multi-disciplinary teams tackling short-term survey projects."[138] By 1933, less than two years elapsed since the Kantô Army's invasion and crushing of dissent against military operations in border areas like Jiandao near Korea. Besides Hachisuka, imperial Japan's leading ornithologists lacked expeditionary experience, and like Yamashina, outsourced bird specimen collection to professional collectors. As House of Peers members, they likely feared risking their lives abroad in an allegedly bandit-ridden region.

Birds of Jehol's cover page highlights the scientists' authority in listing their elite social status or higher education: "Taka-Tsukasa" as "Prince," Hachisuka and Yamashina as "Marquis," Kuroda Nagamichi as "D.Sc." [Doctor of Science], and Uchida as "D. Agr." [Doctor of Agriculture].[139] At only ninety-one pages with twenty-eight color plates and a frontispiece errata insert, the report begins with the Japanese quail, or *Coturnix coturnix japonica*, a humble, yet agriculturally useful game bird.[140] Kobayashi, the Japanese aristocracy's preferred bird illustrator, provided its three-color images. In the errata, Plate X, erroneously captioned "*Falco tinnuculus japonensis* TICEHURST," should have read "*F. verspertinus amurensis* RADDE"[141]—the artist (or Uchida) substituted a Japanese falcon for an indigenous Manchurian one. Was such avian imperialism an unintended Freudian slip, or indicative of thought colonization by imperial Japanese power structures prompting greater focus upon the imperial center's birds over the colonial periphery? Four of Japan's most distinguished ornithologists, plus Uchida who supervised the project, might have contributed a more impressive report. Yet, they were limited to the specimens received, and subjected to a state-funded expedition's demands whose leader's impressions of Jehol were disparaging.

By the mid-1930s, Yamashina was uniquely positioned to sponsor a broader survey of Manchukuo's birds due to his influence as a Peer and great wealth as a SMRC investor. Mizuno's 1934 *A Distributional List of the Manchurian Birds* likely also played a role in prompting Yamashina to hire his collector Ori'i to amass a larger collection of Manchurian birds in 1935 to rectify deficiencies in Tokunaga's expedition. While Yamashina or his colleagues never directly criticized Tokunaga's scholarship, his *Birds* article illuminates ornithology done well: a large sampling of diverse collections of species and subspecies, collaboration with other Manchukuo authorities, and citing recent international literature. Nonetheless, *Birds of Jehol* represented "official" ornithological work by a state-funded enterprise involving collaboration of scientists from the imperial center to lend legitimacy.

Ori'i's Collection of Manchurian Birds for Yamashina, 1935

Following his work for the 1933 expedition, Yamashina strove to correct the lack of a truly definitive scholarly study of Manchurian birds. Although Kuroda Nagamichi published seven articles in *Birds* from 1930 until 1934 based on seven lots of bird skins collected in Manchuria by Mizuno,[142] these smaller pieces described partial collections as they arrived to supplement existing knowledge. In contrast, Yamashina's one-hundred-page 1939 *Birds* article, "Note on the Specimens of Manchurian Birds Chiefly Made by Mr. Hyojiro [sic] Orii in 1935," prepared in English over four years, lists an impressive ninety-four sources consulted on Manchurian avifauna in English, German, French, Russian, and Japanese to substantiate previous research.[143] In addition, the last page features a map indicating Ori'i's circuitous travels over the tortuous Xing'an Mountains[144]—visually hinting at his task's myriad challenges contrasting with Tokunaga's overt trumpeting of "unspeakable difficulties." The article meticulously describes Ori'i's collection of 1500 bird specimens, along with additions of bird skins and eggs proffered by "Mr. Loukashkin of the Harbin Museum" and "some skins from the Liaotung Peninsula" sent by "Mr. S. Ueno of Dairen."[145]

Ueno's records are lost, but Anatole S. Loukashkin (1902-88) left many materials, now cached at Stanford University. Born in Liaoyang, a key spur along the Harbin to Dalny (future Dairen) China Eastern Railroad trunk line employing his father, he later graduated from the Harbin Institute of Oriental and Commercial Studies. From 1932 until 1941, he served as curator of the Manchuria Research Institute's Harbin Museum of Natural History.[146] Now known as Heilongjiang Provincial Museum,[147] it still harbors considerable natural science collections within 107,000 collections;[148] in 1906, amidst Russian colonization, its white Baroque structure studded with red turrets was originally built as the Moscow Department Store.[149] Loukashkin, from a White Russian expatriate family, differed from Japanese counterparts in gaining little from controlling or exploiting Manchukuo's local populations as an imperialist. Like Kuroda and Mizuno, Loukashkin patronized Indigenous collectors in somewhat symbiotic reciprocal relationships.

In procuring some of Yamashina's bird specimens, one of Loukashkin's reports reveals how Indigenous hunters helped explore peripheral areas to Russia. In November 1935, almost contemporaneous with Ori'i's specimen hunting forays,[150] Loukashkin, with his field assistant, taxidermist Michael Volkov, collected specimens near the Great Khingan Mountains to supplement the Institute's mammal and bird collections. During his surprisingly easy hunt for the rare Ussurian Moose (*Alces alces cameloides*), he employed Orochens like Gelminto, a celebrated hunter, "as guides and hunters," with whom he traded new rifles and "gunpowder, lead, caps, flour, salt, oil, tea, millet, tobacco, tobacco, and more tobacco, and matches for the entire group" in exchange for their services while paying for fur-bearing animals in cash.[151] His generous exchanges of gifts with reciprocity would ensure stunning success in future specimen-hunting forays. Loukashkin also notes how Gelminto arrived with kin, "consisting of Mantusan, his wife and little boy, Sinutzan, Hovan, and Durbo," and indicated that "all these people belonged to the Kukureh Clan, of which Durbo was the Shaman, a sickly looking, feeble, and crippled old man."[152] Nevertheless, he understood how each person, regardless of age or gender, played an important role in completing his primary mission to shoot a moose—ensuring the group's spiritual welfare along with tracking, hunting, slaughtering, and preparing animal skins, all while offering protection against predators; these tasks were luckily fulfilled before a heavy snowstorm.

Loukashkin also bagged hazel-hens, black-cocks, "a good series of small rodents and resident birds," and unique chicken-like capercaillies.[153] Such rarities were added to Yamashina's collections, and marked common Russian collaboration with Japanese from prewar into postwar times, despite Japan's northern neighbor's complex political situation. In the 1960s, Mizuno wrote about Soviet birds for *Wild Birds*,[154] while in the early 1970s, Yamashina traveled to the USSR

> to try my efforts for restoring an amiable friendship between the ornithologists of [sic] Soviet Union and Japan, which has long been interrupted in the past and to promote the interexchange of scientific knowledge and research works for the conservation of wildlife, also to push forward the closer cooperation between the bird experts of our two nations.[155]

Loukashkin's additions, plus Ueno's, to Ori'i's 1,500 specimens ensured in-depth bird samplage from Manchukuo's diverse regions, featuring rarer species, since their skins often came from remote, mountainous areas.

Yamashina's article markedly includes an "Explanation of the Localities" for specimens collected by these three men, which span areas in Jehol, the Liaodong Peninsula near Dairen, and northern Manchuria's Xing'an Mountains, whose locations read like imperial survey expeditions of remote border areas.[156] Local ornithologists like Mizuno believed that Manchuria's steppelands were virtually denuded and void of bird life—certainly untrue and a viewpoint possibly engendered by colonial ideologies encountered via SMRC or the Kantô Army. Forested stretches near Japanese settler areas north of Harbin, like Tieli, and areas around Shenyang likely also yielded important bird species, so these localities' exclusion is surprising, excepting Qiqihar for Loukashkin.[157] Possibly influenced by Mizuno, Yamashina seemingly believed his "Note" catalogued the most comprehensive Manchurian bird species descriptions to date.

The majority of "Note" handles systematics and taxonomies, complementing Yamashina's earlier work, as well as Kuroda's and Mizuno's, and led him to change and revise previous conclusions about subspecies. As his key contribution, he describes five new sub-species and six new species: "The occurrence of *Nucifraga caryocatactes interstinctus, Muscicapula elisae, Myophonus caeruleus caeruleus, Hodgsonius phoenicuroidles phoenicuroides,* and *Pucrasia ranthospila* in Jehol district was determined for the first time by this collection."[158] According to Yamashina's calculations, and careful acknowledgment of earlier recent studies by German (Dresden-based carrion crow and sparrow expert Wilhelm Meise [1901–2002], 1935; Ernst Hartert and Joachim Steinbacher [1911–2005] in Berlin, 1932 and 1936), French (Jehol-based missionary P. G. Seys, 1933), Russian (Loukashkin, 1934 and 1936), Japanese (Mizuno, 1934), and Chinese (Tsen-Hwang Shaw of Qinghua University in Peiping [Beijing], 1936) ornithologists,[159] he conclusively argues that "[t]he total of 380 species are thus authentically determined by specimens as occurring in Manchuria."[160] Then, proof of existence for endemic bird species was decided by a dead individual's presence, sent to a central research institute, like Yamashina's on his Shibuya estate.[161] Yamashina portrays his work as the latest, most comprehensive, and definitive study, but still generously credits Ori'i for a yeoman's effort amassing specimens "belonging to more than 200 forms."[162]

Yamashina's 1935 work for Tokunaga and his 1939 study of Manchurian birds represent a personal resolution of conflicts between Japanese ornithologists' public and private roles. As monied aristocrats drawing upon tenant farmer' rents on ancestral lands plus investments in SMRC or other enterprises, these privileged aristocrats decided how to spend their leisure as long as they fulfilled duties in the House of Peers and delegated administration to loyal retainers. Nevertheless, pressure to conform to state dictates was high, which increased with advent of the late 1930s total war system, and accelerated after the Pacific War. As aristocrats enmeshed in an imperial system expecting their public service, their privileged position nevertheless came with increasing state obligations.

Conclusion

During the 1930s, many of Japan's scientific endeavors supported by ornithologists were composed of "avian imperialism," where birds served as a means to exert imperial control over colonized regions, or where Japanese asserted informal influence, like Manchukuo. Researching avian surveillance with carrier pigeons, analyzing data from expeditions, and creating surveys of birds in contested areas were all ways for these scientists to support the nation's imperial aims, and thus, embedded them firmly within the state apparatus.

Nowhere were these activities by Japanese researchers more expansive than in Manchukuo, a client state heavily under Japanese military control within the Kantô Army's wide net of influence—deeply impacting the country's scientific endeavors, including ornithological ones. This was markedly apparent in the numerous troops attached to the Tokunaga Expedition, with the military outnumbering scientists and their crew. With only one mild shooting incident without injuries, military assistance was primarily symbolic, and more insulated the scientists from distressing encounters with locals than protect them from unwarranted attacks. Nevertheless, Japanese militarism long suffused scientific research in Manchuria since Japan's colonization following the Russo-Japanese War. After Manchukuo's founding, the Kantô Army closely worked with SMRC in its experimental research stations like Gongzhuling, which hosted an impressive pigeonry. Manchukuo served as a "laboratory" of "modernity" while innovative endeavors within its borders presaged greater nationalistic roles for Japanese scientists within domestic Japan as the 1930s progressed.

6

Wartime Tokyo and Defeat (1937–45)—Mobilizing Imperial Japan's Ornithologists and Birds for War

Introduction

Wartime Tokyo served as an important locus for scientific endeavors, including the activities of the politically influential Duke Takatsukasa Nobusuke and his colleagues at a time when birds, ornithology, and ornithologists were all mobilized for war. In various, sometimes seemingly contradictory positions, Takatsukasa exemplifies these scientists' multidisciplinary roles in a wartime mobilization where their expertise and research could aid in ensuring military success or support the home front. While serving in the House of Peers, this ornithologist not only acted as the Meiji Shrine's head priest from 1944 into the post-war period, but also headed the Hunters' Association, Japanese Association of Zoological Gardens and Aquariums, and *Ornithological Society* (1922–46), which he headed after the 1922 death of his mentor, Iijima Isao, the "father" of Japanese parasitology, who founded the *Ornithological Society* in 1912. During wartime, Takatsukasa and his counterparts, including Yamashina Yoshimaro and Hachisuka Masauji, served in the *Shigen kagaku kenkyûjo* [Research Institute for Natural Resources], along with Uchida, who specialized in parasitological ornithology and had published numerous studies on bird-banding[1] and bird-borne *mallophaga* [biting lice].[2] To feed a starving population and minimize Japan's devastating songbird mist-netting for food, Yamashina also worked on breeding poultry in this capacity—an activity which segued into his conservation initiatives in the postwar period. During total war, the experiences of these ornithologists' experiences show how science and natural resources, including birds, were deployed to impart victory to imperial Japan.

Hence, militarized activities of Japan's leading ornithologists clearly supported imperial Japan's war effort, and likely intersected with controversial overseas biological experimentation units in China and Manchukuo. Zoologists, and especially ornithologists, maintained longstanding connections to parasitological research, as seen in academic lineages traced to Iijima, who had taught zoology at Tokyo Imperial University, where he mentored generations of ornithologists including Takatsukasa, Uchida Seinosuke, and Kuroda Nagamichi. Nevertheless, international connections with the Anglo-American world were sustained until the Pacific War's advent, initiated by imperial Japan's simultaneous attack on Pearl Harbor in Hawaii and ports in the American colony of the Philippines. These networks regenerated quickly in the postwar period, where such connections were crucial to reviving Japanese science's global reach.

Avian Imperialism and the Asia-Pacific War (1937–45)

These scientists' activities following eruption of the Pacific War, a conflict initiated by imperial Japan, can be gleaned from a 1948 report of their publications by Austin created for the Allied Occupation, who notes that "Japanese naturalists made signal contributions during the war years to their most unwarlike of sciences."[3] Austin expediently frames their endeavors within a broader US narrative blaming the Second World War on a handful of Japanese militarists who led their country astray, with all others as passive victims or forced into compliance by an authoritarian state hijacked by this group, of whom General and former Prime Minister Tôjo Hideki (1884–1948) became the greatest scapegoat, while Emperor Hirohito, commander-in-chief of imperial Japan's armed forces, was exonerated and kept from trial.[4] Recasting the Emperor as a disinterested marine biologist also helped obscure his command over military activities connected to biological weapons research and other undertakings, while it enabled successful postwar rehabilitation. Additionally, this approach permitted Americans to assimilate Japanese scientists like Kuroda Nagahisa into postwar research with US military applications—explored in Chapter 7.

The Americans' course of action was entirely consistent with Japanese transwar imperial continuities, since, as contemporary scholars like Yoshimi Yoshiaki assert, the imperial system's remnants quickly reappeared within an imperial democracy persisting until the present—characterized as a political system developed in the 1930s as constitutional governance under an imperial monarch and elected representatives supportive of centralized authoritarianism.[5] This particular narrative allowed Japan's postwar rebranding as a peaceful American ally and counter to Communist China during early rivalries between the United States and Soviet Union beginning in the late 1940s.[6] Austin's January 1948 report appeared during intensification of the International Military Tribunal for the Far East (IMTFE), known as the "Tokyo Trial." The Tribunal lasted from April 29, 1946, until November 12, 1948,[7] with many scientists escaping prosecution for activities potentially abetting war crimes.

Austin's report notes that Japanese researchers aided in its compilation, where he consulted Hachisuka for the section on birds, Takashima Haruo (1907–62) on mammals, and Kuroda Nagahisa on periodicals.[8] These men thus framed how Americans perceived their endeavors and partially controlled their undertakings' portrayal, plus chose which ones appeared in the report serving mainly as an annotated bibliography of published research. Enough information appears in the bibliography that it could deflect US attention from the nature of these scientists' activities whose findings possibly remained unpublished or whose reports were kept-in-house due to sensitive military applications. Despite some annotation for publications Austin believed important, much of the source list was left unanalyzed without further details for each article.

While these scientists, of whom some boasted expertise in parasitology and mammology, were not directly involved with biological weapons research by the notorious Manchukuo-based Ishii Unit [*Ishii butai*], known as "Unit 731," their particular research undoubtedly aided the Unit's scientists. Chapter 5 examined how Japan's leading ornithologists, despite lacking travel to Manchukuo, reviewed the 1933 Tokunaga Expedition's findings and examined its specimen collections, and thus completed their 1935 report on *The Birds of Jehol*. Nevertheless, historian Sheldon Harris concedes that "so many members of Japan's scientific establishment, along with virtually every military commander of note and members of the imperial family, either participated in chemical or biological warfare research, or supported these projects with men, money, and material, that it is difficult today to apportion exact blame or responsibility."[9] Japan's imperial family, collateral branches, and aristocrats, were deeply invested in use of science for the war effort. They also substantially invested in the South Manchurian Railway Company (SMRC) military-capital extractive complex subsuming Manchuria's transportation networks, extracting natural resources in mines and farms, and utilizing for its security the Kantô Army military force in Japanese-occupied northeast China. Yamashina was a key investor in SMRC and its corresponding Manchukuo endeavors.[10] His ornithological work also intersected with regional development aims applying resources toward the war effort.

On January 1, 1939, Yamashina published *The Eating Habits of Manchurian Birds* in Shinkyô [Changchun], Manchukuo's political capital, under Manchukuo Forestry and Agricultural Bureau auspices.[11] Established in 1936,[12] the Bureau administered resource extraction by Japanese settlers from domestic Japan, often from Nagano Prefecture's forested, mountainous areas experiencing economic distress since the early 1930s.[13] Settlements awaiting them were pre-populated by efforts of boys and young men recruited from Japan to join militarized Japanese volunteer brigades formed after Manchukuo's establishment.[14] In Binjiang Province, now part of Heilongjiang, Bureau training facilities were located in Tetsuryô [contemporary Tieli], slightly over two hundred kilometers northeast of Harbin.[15] The settlements attracted scores of young Japanese women, or *tairiku yôme* ["continental brides"], who married and served as cultivators and bearers of children whose husbands patrolled the Manchukuo-Soviet border.[16] After 1937, Japanese-occupied Manchuria became a key base for military operations in China proper and source for natural resources as the war intensified and expanded.

Yamashina's interest in Manchurian birds arose from specimens collected by Ori'i Hyôjiri in 1935, which led to a lengthy 1939 article in *Birds*, and informed the 1939 Manchukuo Forestry and Agricultural Bureau report, which resulted from investigations of the wild birds' stomach contents and Ori'i's descriptions of hunting in regional forests. At first glance, there seems little connection between forestry and aviculture, though knowledge about avian eating habits could pinpoint which areas might attract various types of birds and reveal important patterns about their dispersal. However, as a SMRC shareholder, Yamashina would also have been interested in the avicultural productivity of Japanese settlements and areas near rail-lines in the process of development in north Manchuria. Also, in-depth knowledge of political conditions potentially impacting settlements was useful for areas near the Soviet border, erupting in conflict at Nomonhan that May, and whose battles would rage until September.

The Japanese settlers' agricultural efforts included duck production for meat and feathers, while propaganda materials issued by SMRC like *Manshû gurafu* [Manchuria Graph] featured peaceful villages with bucolic duck ponds populated by smiling women and children.[17] The depiction of ducks makes sense for practical reasons. In contrast to chickens, ducks can live in harsher environments, including wetlands consisting of marshes and bracken areas. In the winter, after culling of most mallards for meat, ducks can be kept in barns where hens could be used for egg-laying. Ducks can even co-exist with pigs, whose girth and massive appetites for waste food can provide warmth trapped within barns. Ducks also are omnivorous and eat insects, including boiled silk worms cultivated using mulberry leaves in the summer, and they can also thrive on pellets of dry feed in winter.[18] Besides their meat and eggs, ducks could also produce feathers for pillows and bedding, an essential commodity in bitter Manchurian winters whose temperatures can easily dip below -30 degrees Celsius (or -22 Fahrenheit).

More darkly, feathers were also a delivery method for biological agents prepared by the Ishii Unit, whose facility was located not too far from these Japanese settlement areas. Notably, a September 30, 1944 "Japan-Bacteriological Warfare Summary" assembled by American wartime intelligence sources reported "Japanese dropping feathers from planes" and certainly indicated an increased US fear of biological weapons attack.[19] While it is unclear whether or not feathers produced in this area were harvested for this exact purpose, or whether birds, in this case ducks, were envisioned as a viral dispersal agent, it is certainly true that the research of Yamashina and other scientists could have been secretly used by the Unit to understand how avian or mammal viruses introduced into the area might spread through foraging by birds or vermin. Yet, most ornithologists directly supported imperial endeavors during wartime in more overt ways.

From the late 1930s into early 1940s, during the intensification of the Second Sino-Japanese War, Yamashina and his cohort were clearly drawn toward understanding environments of domestic Japanese birds and those spread throughout the empire and its newly acquired possessions. Notably, in a larger 1939 study on birds of Hainan Island undertaken during the so-called China Conflict, Hachisuka indicated that Yamashina "has also taken a personal interest in Hainan ornithology[,]"[20] and assisted him greatly on this topic by welcoming Hachisuka into his museum and extending help of his staff. In fact, Hachisuka emphasized that the Imperial Navy's successful 1939 wartime seizure of Hainan Island and corresponding Japanese attention to its natural resources—in which he believed birds should be included—prompted him to write "Contributions on the Birds of Hainan":

> The occupation of the island of Hainan early this year by His Imperial Majesty's Navy has aroused Japan's incredible interest in the island. The natural resources are since being rapidly investigated, and it seems an appropriate time for me to write a

complete account of the avifauna of Hainan, a subject which has interested me for many years. Some ten years ago I have described one or two birds from that island and have long felt the need of compiling a systematic list with up-to-date scientific names and synonymy.[21]

As one of Japan's leading ornithologists, Hachisuka thus felt compelled to contribute his talents at such an importune time for the expanding empire following its military conquests. He was also competing with the Americans, apparent in his introduction to the study, where he indicated that "Mr. Hugh Birckhead (1913–1943) made the last contribution from America to Hainan ornithology by recognizing a race of shrike, *Lanius schach hainanus*, from the Katsumata collection, in his paper on 'The Birds of the Sage West China Expedition' in 1937."[22] As a "talented painter and sculptor of birds," and an ornithologist, Birckhead hailed from a distinguished American family of Newport, Rhode Island.[23] He was also an intimate friend of S. Dillon Ripley (1913–2001), assistant curator of the US National Museum's (Smithsonian Institution) Division of Birds.[24] After Birckhead's 1943 combat death in France,[25] a distraught Ripley volunteered to serve as agent for the Office of Strategic Services in South and Southeast Asia during the Second World War.[26] Taking place from 1934 to 1935, the Sage Expedition was organized by the wealthy Yale University graduate Dean Sage, Jr. (1909–63) and his wife Alida (Ann) M. Sage (1915–44), with the assistance of the celebrated hunter and naturalist William G. Sheldon (1912–87), and T. Donald Carter (1893–1972), the AMNH's mammal curator. They traveled to areas in west China, including northwestern Szechuan province, one of China's most ecologically rich areas with abundant species diversity, to primarily collect mammals like the Giant Panda in eastern Szechuan for the AMNH; there, they also collected 426 bird specimens of 86 distinct species.[27]

Notably, Hachisuka indicates that Birckhead once described specimens procured from China's Hainan Island on February 14, 1902 by Japanese collector Katsumata Zensaku (1874–1940), who worked for the British Yokohama-based merchant Alan Owston (1853–1915) to procure specimens for Lord Rothschild, whose ornithological collections were partially acquired by the AMNH in the early 1930s.[28] From 1896 onwards, Katsumata lived in Hainan to develop its sericulture industry, and introduced pineapple cultivation until 1937, when the war forced him back to Japan; yet, on February 10, 1939, he promptly returned to the island as a native informant for the Japanese military.[29] Here, Katsumata served as an imperial agent in a position where geographical knowledge gleaned in ornithological collection was applied to the war effort.

To Hachisuka, it was imperative to unearth crucial Japanese involvement in ornithological endeavors undertaken earlier in a neighboring China, including Hainan Island, now partially subsumed into the empire. To this end, Hachisuka even visited Katsumata in his "native Fuji village" in 1937, and attempted to interest him to return to Hainan Island to further collect specimens for him—an activity stymied by the war.[30] Thus, he made do with 330 species of specimens collected by Katsumata from 1902 to 1909 that were presently in collections stashed in the private museums of his colleagues: Yamashina, Momoyama, Kuroda Nagamichi, and Takatsukasa.[31] Yamashina acquired his specimens through their transfer from Tokyo Imperial University, which recognized his ornithological institute's high expertise and preservation abilities.[32]

In 1941, to further build on the Sage Expedition's findings illuminated by Birckhead in 1937, Hachisuka published the article "Notes on Hainan Mammals" in *Bulletin of the Biogeographical Society of Japan*.[33] Here, he described Hainan's primary mammal populations, based on observations and specimens collected during a 1940 expedition, while he also listed Chinese animal names provided by Katsumata, a decades-long island resident.[34] Thus, Hachisuka reclaimed a definitive accounting of Hainan Island's avifauna from numerous Anglo-Americans describing them from 1870 until 1937—including illustrious ornithologists like British diplomat Robert Swinhoe (1836–77), the British Natural History Museum's curator of birds W. R. Ogilvie-Grant (1863–1924), and Hachisuka's own mentors Lord Rothschild and Ernst Hartert, who had curated Rothschild's ornithological collections at Tring, and others who described smaller numbers of subspecies and synonyms. Through such scientific work, Hachisuka performed wartime imperial masculinity in one-upmanship against Anglo-American rivals soon emerging as Japan's enemies in late 1941.

Besides broader systematics work, the late 1930s and early 1940s publications by Hachisuka, Yamashina, Takatsukasa, and Kuroda Nagamichi prioritize areas invaded by imperial Japan, allowing extension of Japanese imperial control and rapid scientific investigation to codify avifauna newly subsumed into the burgeoning empire. Titles of their publications reveal the contemporaneous incorporation of bird species from newly occupied territories into the greater Japanese empire—a process beginning in the mid-1930s with ornithology's militarization in Manchukuo. They, along with Uchida, had served as co-authors for a bird section in the 1935 report for the 1933 Tokunaga expedition to Jehol (Rehe) in Manchuria. Hachisuka further collaborated with Kuroda Nagamichi and Takatsukasa to publish "A List of the Birds of Micronesia under Japanese Mandatory Rule," which came out in 1942 as part of the *Ornithological Society*'s larger *Handlist of Japanese Birds* published each decade.[35] Yet, his marriage to Chiyeko Nagamine, a Japanese-American woman from an enemy country, prompted Hachisuka's role in joint ornithological work to cease after this date.

From 1943 to 1944, without Hachisuka, Takatsukasa, Uchida, Kuroda Nagamichi, and Yamashina published the four-part series of their *Monograph of the Birds of Greater East Asia* in a four-part folio, with color plates of paintings by Kobayashi Shigezu (1887–1975), Sakamoto K., Deguchi M., and Mizuno M. This series codified and depicted in detail the larger Japanese empire's bird life in an endeavor where claiming of bird species mirrored claiming of territory. Austin noted that more parts in this series were planned, but unfortunately, its printing company was destroyed by Allied bombings just after part four's publication.[36] The first two parts were published by Yûkôsha, a firm existing until 1944 and publishing various texts on literature, religion, philosophy, and geography,[37] while the last two parts were published by Heibonsha, a large still-extant Tokyo-based company established in 1914 by Shimonaka Yasaburô (1878–1961), specializing in encyclopedias and dictionaries.[38] These grand colonial taxonomic efforts by imperial Japan's leading ornithologists were ultimately stymied by Allied air raids.

Other wartime ornithological applications included breeding edible game birds and poultry in studies favoring a sacrificial wartime mobilization of imperial Japan's birds for food. From 1941 until 1943, Yamashina published on hybrid sterility in ducks, pheasants, and doves, with articles on chromosomal studies regularly appearing until 1945, featuring beneficial applications for food production at a time of protein-scarcity. In 1942, Kuroda Nagamichi also published *A Bibliography of the Duck Tribe* elucidating varietals and taxonomies shedding light on breeding potentialities.[39] Lastly, in 1944, Takatsukasa published the now-exceedingly rare English-language pamphlet, *Studies on the Galli of Nippon*, whose descriptions often came from his earlier *Birds of Nippon*, and which, according to Austin, included "a short historical sketch and the classification, description, and distribution of all the gallinaceous birds of the former Japanese Empire."[40] Here, the empire's chickens could serve wartime needs in Takatsukasa's book indicating their bountiful varieties throughout Asia.

While most of wartime Japan's leading ornithologists enthusiastically embraced the increased opportunities to study birds in their now burgeoning empire, Hachisuka also engaged in activities possibly viewed as "treasonous" by his colleagues.[41] This may also explain why such a prolific ornithological author as Hachisuka was not included as a contributor to the 1943-4 *Monograph of the Birds of Greater East Asia*. Hachisuka had spent considerable time in the United States from the mid-to-late 1930s, and returned to Japan in 1938 to marry Nagamine, a wealthy Japanese-American. He was banned from fraternizing with individuals from Allied forces during the Pacific War, though Nagamine was protected from internment through her husband's aristocratic status.

However, in 1942, Hachisuka boldly braved arrest from the Special Higher Police by passing the Ornithological Society's latest *Handlist of Japanese Birds* to a group of interned Australian diplomats sequestered in a Tokyo hotel with the intention that it would eventually reach American shores. Handlists came out every decade to commemorate the Ornithological Society's 1912 founding, with the first one issued in 1922, followed by the next in 1932, and the last appearing in 1942 during the war. The now controversial *Handlist* then circulated over the high seas until it got into the hands of the eccentric and somewhat reclusive American ornithologist James Greenway (1903-89) at Harvard University's MCZ. Hachisuka was well aware of Greenway's work, later inspiring him to write a book on the extinct Dodo, and knew of his friendship with Delacour. Greenway had joined Delacour's 1929 expedition, and thereafter, headed to Indochina. The story of Hachisuka's handlist reveals enduring Anglo-European networks for Japanese ornithologists that persisted even into the Pacific War's early years.

Such deep connections with the Anglo-American world also unfolded in traces surrounding ownership of an impressive set of books written by Yamashina. From the mid-thirties to 1941, Yamashina published the beautifully bound two-volume set *Japanese Birds and Their Ecology*, with text in Japanese with Latin bird species names.[42] Both volumes' covers were bound in sky blue cloth decorated with Art Deco emblems with a round crest of stylized birds and shrubbery. The first volume published in 1933-4 included systematic accounts from *Corvidae* (crow family) to *Dicruridae* (drongo family of Old-World tropical birds), while the second published in 1941 contained descriptions of *Muscicapidae* (Old World flycatchers) to *Ardeidae* (herons). The two volumes also included color plates signed by "Sugako," (Sakai Sugako) or Yamashina's wife, known for accompanying him on ornithological trips to sketch birds. Her gorgeous life-like images asserted her presence as a spectacular ornithological artist on par with the aristocratic artist Kobayashi, indicating her name in the images' lower right-hand corners in block English letters, which contrasted with Kobayashi's traditionally Japanese signature. Apparently, the second volume is extremely rare; after 1941, few books or any goods left Japan, except perhaps by diplomatic pouch or individual travelers. Therefore, any complete set of the books probably remained in Japan during the war, and surviving volumes were ferreted away in a country or seaside estate, while others trickled back into booksellers to provide their owners with much-needed cash after defeat.

During the postwar period, sets of *Japanese Birds and Their Ecology* came to two important figures in the Anglo-American world, indicating just how much Yamashina as a Japanese ornithologist had built a notable impression overseas and successfully cultivated his connections in prewar times. The Amherst College Richard L. Soffer (Class of 54) Ornithology Collection, donated by alumnus and retired microbiologist Richard L. Soffer in 2003 and 2007, contains a full set of Yamashina's books, whose rare 1941 volume features a "Bookplate of Alan Francis Brooke," otherwise known as Lord, or Viscount, Alanbrooke (1883–1963).[43]

Known as "Colonel Shrapnel," Alanbrooke served wartime British Prime Minister Winston Churchill (1874–1965) as a highly talented military advisor, with whom he was often at odds, especially in strategies regarding Japan, noting in a February 14, 1944, diary entry that "after much hard work, I began to make him [Churchill] see that we must have a plan for the defeat of Japan and then fit in the details."[44] As Chief of the Imperial General Staff from 1939 until 1944, when he became Field Marshal, Alanbrooke greatly admired General Douglas MacArthur's bold approach and detailed postwar planning, who he first met in November 1945 during a stopover in Japan to tour British colonial possessions of the "Far East."[45] Their admiration was mutual, as indicated in his wife Jean's April 1964 letter responding to a condolence note from Lady Alanbrooke following MacArthur's death: "I appreciate deeply all you wrote of my husband and I want you to know the great admiration and respect in which he held your husband. Their meeting in Japan was truly for the general a most satisfying and rewarding experience and he often spoke of it."[46]

Besides his illustrious military career, Alanbrooke was an avid ornithologist; bird-watching helped distract him from war's all-consuming responsibilities, while nature's beauty provided the warrior a refuge, and the unrelenting hunt for bird books and new species kept him involved in "an engrossing interest besides one's profession."[47] It is hard to determine exactly *when* Lord Alanbrooke received this magnificent copy, but if Yamashina gifted it, his signature would have appeared. He likely acquired it when meeting MacArthur in Tokyo, where he undoubtedly perused black market book sellers—he habitually visited bookstores with ornithological materials in wartime London and anywhere he went.[48] More likely, Alanbrooke patronized Isseidô Booksellers in the Kanda area of Tokyo's Jimbo district, a business founded in 1903 to stubbornly survive Allied bombings with an unusual reinforced concrete structure completed in 1931.[49] Also possible is that Alanbrooke purchased it during the May

1960 ICBP meeting in Tokyo (discussed in the next chapter), as vice president of the Royal Society for the Protection of Birds between 1949 and 1961.

However, more telling of these transwar continuities is that, in the late 1940s, Yamashina indeed signed the set of *Japanese Birds and Their Ecology* which he gave to Austin, with a dedication of "To my friend O. L. Austin, Jr. With the author's compliments Yamashina Yoshimaro"; he additionally included his name written in Japanese characters.[50] This bi-lingual signing communicated Yamashina's comfortable postwar resituating of himself in both Anglo-American and Japanese ornithological worlds. Austin affixed to the book's frontispiece his Japanese square-shaped signature chop, which showed an Ikky Bird in the middle (his official American signature of the name "Austin" stylized to look like a sea bird), and traditional Japanese characters for "king," "longevity," and "pacification/garrison" at the top that read Ôsuchin ["Austin" in Japanese pronunciation], while "O. L. Austin Jr." appeared at the bottom in English. Whoever chose the Japanese characters for Austin's name poked fun at his official Occupation position, since they also meant "the king overstaying his garrison" or "ruler [over] long-term pacification."

Until the early 1940s, Japanese ornithologists were clearly in dialogue with the Anglo-American world by surpassing earlier scientific work on birds while imperial Japan still appeared successful in conducting the war. As Japan's military claimed new territories under their occupation, these scientists labored to codify avifauna now subsumed under Japanese imperial control. Flush with victory, Japanese media disseminated such endeavors to an avid domestic public. Concurrently, Japanese diplomats keenly explained aspects of Japanese national identity to rivaling Anglo-Americans—notably including birds. Attracting tourists to the victorious empire heading the Greater East Asian Co-prosperity Sphere would also help solidify support for its competent control over former western colonies and Chinese treaty ports.

A Wartime Bird Book for Anglo-American Tourists: *Japanese Birds* (1941)

Takatsukasa's perhaps most interesting wartime contribution was the English-language tourist book *Japanese Birds* published in January 1941, barely a year before the Pacific War's outbreak.[51] Printed on high-quality parchment paper in paperback, this book was the thirty-fifth volume in the Tourist Library of the Japanese Government Railways' (Ministry of Railroads) Board of Tourist Industry (BTI), an organization established in 1930 to oversee and provide financial assistance to the Japan Tourist Bureau (JTB), founded in 1912 to primarily serve foreign visitors to Japan, and a precursor to the later postwar Japan Travel Bureau (JTB). The first volume in the Tourist Library, or *The Tea Cult of Japan* by Y. Fukukita, was followed by short handbooks on topics including Noh plays, cherry blossoms, gardens, Shintô, hot springs, food, music, family life, education, Kabuki, dance, history, folk toys, *sumo* wrestling, and other subjects, all by Japan's leading authorities. These booklets, composing the BTI's goal of a hundred volumes, served to educate English-reading tourists, primarily

Anglo-Americans, on imperial Japan's cultural, geographical, and social merits: "By studying the entire series the foreign student of Japan will, we hope, gain a general knowledge of the country and its people."[52] During the Asia-Pacific War, articles in the last issues of BTI's quarterly magazine *Kokusai Kwankô* [International Tourism] published from the late 1930s onwards emphasized Japan's unique beauty, and saw tourism as means to avoid conflict with the United States:

> In order to forestall an American entry into the war, which would be problematic for Japan, Japan had to take full advantage of all available forms of propaganda. By promoting more tours from the US to Japan, Japan would be able to win over the hearts and minds of common people, without whom the government would be unable to fight a winning war. It was also an ideal strategy to use with regard to a neutral country that might yet serve as an intermediary.[53]

Prior to the war, tourism could disseminate positive messages about the Japanese nation; political scientist David Leheny reveals "[b]etween 1900 and 1938, members of the Japanese government struggled over tourism, and the Ministry of Railroads in particular formulated strategies to increase inbound tourism, largely by creating an image of Japan that would be appreciably exotic yet sufficiently comfortable and familiar to appeal to European and American tourists."[54] Providing well-to-do cosmopolitan tourists with a bird guide was certainly one draw, plus topics believed to spark foreign curiosity about Japan. By the early twentieth century, Japan was an immensely popular tourist destination, with wealthy Europeans and Americans providing the most revenue.[55] The National Diet proposed a 1929 bill favoring national support for the hotel industry, where the House of Peers argued that "international tourism" led to "the creation of 'international understanding' and the improvement of Japan's global image," while the House of Representatives asserted that tourists offered desirable foreign exchange earnings.[56] Takatsukasa, in his position as a member of the House of Peers, keenly followed these issues, and during a critical time for Japan's international image in the Anglo-American world amidst the burgeoning China conflict, he contributed his own ornithological expertise.

The cover of *Japanese Birds* shows a multi-colored woodblock print of Japanese Little Egrets (*Egretta garzetta garzetta*), or *kosagi*, in breeding plumage. This guide reveals that Japan's avian life drew so many English-speaking visitors it was believed merited. Bourgeois Anglo-Americans widely enjoyed bird-watching and amateur ornithology, and presumably engaged in this hobby while abroad. Takatsukasa's introduction begins with descriptions of the Japanese empire's expansiveness, which allowed its plants and animals to benefit from "the charm of variety."[57] He envisioned the slim volume as a welcome guide aiding visitors' enjoyment of Japan's avian life: "The following pages do not purport to be anything like an exhaustive monograph on Japanese birds. The reader will find a brief description of common species, such as the casual visitor to Japan will frequently see on the beaten track, mostly in Japan Proper."[58]

Figure 6.1 Front cover of Prince Nobusuke Takatukasa [sic], *Japanese Birds* (Tokyo: Board of Tourist Industry, Japanese Government Railways, 1941). © Author's personal collection.[59]

Despite its brevity, the handbook is peppered with commentaries on birds from classical premodern Japanese texts like *Chronicles of Japan* (720), *Tales of Ise* (794–1185), and the popular wartime "Tale of the Peach Boy," where, with a band of animals, including a pheasant, a young boy saves his surrogate grandparents and entire land from marauding *oni*, or foreign devils.[60] Takatsukasa noted that during the Edo period, "men of taste" traveled to Tokyo's outskirts in elegant excursions to listen to quail calls.[61] Certain birds sparked significant pleasure amongst connoisseurs: "The first twitter of the *uguisu* [Japanese bush warbler], spring's harbinger, arouses keen delight amongst Japanese bird-lovers. Men of taste ... used to take the trouble to go after [sic] to seek this source of esthetic pleasure, just as Englishmen make much of the nightingale's song."[62] Thus, Takatsukasa firmly sets Japan's birds into a long classical canon of Japanese civilization comparable to the Anglo-American West, by evoking iconic texts signaling a particular Japanese identity commonly referenced during the war, while connecting bird appreciation to stylish Tokugawa-era aesthetes.

An intrinsic commentary comparing behaviors of Japanese and Chinese birds hints at Takatsukasa's mounting frustration with Japan's growing military quagmire in China proper. As a House of Peers politician, Takatsukasa well-understood domestic Japan's dire budgetary situation. Months before the handbook's publication, on September 27, 1940, imperial Japan joined the Tripartite Pact, or Berlin Pact, known as the "Axis Alliance," with Nazi Germany and Fascist Italy joining a military alliance tacitly against the United States. In the handbook's "Birds of Rural Regions," appears a photograph of a Japanese blue flycatcher (*Cyanoptila cyanomelana cyanomelana*), or "one of the three best songbirds of Japan,"[63] which feeds a Chinese hawk cuckoo (*Hierococcyx fugax hyperthrus*), in a posture where the oversized, raucous, and voracious "Chinese" bird accepts handouts from the elegant, composed "Japanese" bird.[64] Takatsukasa extends a pointed commentary: "All these species of cuckoos are well-known for their retention

Figure 6.2 Japanese blue flycatcher feeding a Chinese hawk-cuckoo, Prince Nobusuke Takatukasa [sic], *Japanese Birds* (Tokyo: Board of Tourist Industry, Japanese Government Railways, 1941), 73. © Author's personal collection.[65]

of the reptilian characteristic of not sitting on their eggs. They make other birds do the task of brooding for them."⁶⁶ Unsurprisingly, similar complaints arose in upper-echelon quarters about imperial Japan's costly mission to pacify China.

The Government Railways' desire to profit during a frugal time also played a major role in BTI's foreign tourist promotion through handbooks. In July 1940, the Japanese government initiated the frugality campaign "Regulations Restricting the Manufacture and Sale of Luxury Goods,"⁶⁷ sparking the phrase "luxury is the enemy" in popular media to incite consumers to save for the nation. However, BTI's well-meaning, though misguided, intention to avoid war through cultural diplomacy promoting Anglo-American tourism, failed miserably once Japan preemptively attacked the United States. Following the empire's 1941 Pearl Harbor bombing, Japan's government eliminated BTI, and in 1943, merged it with the Greater East Asian Travel Company, the Ministry of Railroads' travel representative in Japan's conquered territories, renaming it the Greater East Asian Travel Public Corporation.⁶⁸

Despite Takatsukasa's key role in this handbook supporting BTI's optimistic aims, rising appreciation for wild birds by Japanese and foreign tourists dampened following the Pacific War's advent. After 1941, he oriented ornithological activities toward assisting the war effort, in actions countering earlier promotion of wild birds, stewardship of wildlife through responsible hunting, and curatorship of zoological life. When the tide turned to favor the Allies, all Japanese were compelled into sacrifice amidst this dire historical moment. Imperial Japan's ornithologists, like Takatsukasa, were mobilized to effectively and efficiently utilize the nation's natural resources. This included the empire's animal subjects—wild and caged.

Mobilizing Wild Birds as Food Resources during Wartime, 1937–45

As the head of Japan's hunters' association, Takatsukasa well-understood how the war impacted edible bird populations of all kinds. Japan's ornithological resources came under his purview as a distant member of the imperial family, in whose stewardship they remained during four years at the *Research Institute for Natural Resources*, where Uchida, as a former Ministry of Agriculture and Forestry official, also worked. The war accelerated a harvesting process already decimating Japan's domestic passerine bird populations, while also affecting waterfowl. Postwar shortages two years after defeat rapidly increased such activities until Japan's general food supply stabilized by the late 1940s.

Though Japan's allegedly "traditional cuisine," known as *washoku* since the late nineteenth century, heavily emphasizes fish and rice-based meals, with pickled vegetables and fermented bean paste soup,⁶⁹ historian Hans Martin Krämer asserts that Japanese ate meat for centuries as an integral part of their diets.⁷⁰ Though some Buddhist sects prohibited meat-eating for strict adherents, early Tokugawa cookbooks indicated extensive animals consumed in "recipes that contain deer, racoon dogs, wild boar, rabbits, fish otters, bears, and dogs."⁷¹ From prehistoric times into the present, meat was so common in the Japanese diet that preparation instructions circulated widely. According to folklore scholar Suga Yutaka, in the Edo period, Japanese consumption of avian meat included songbirds, wild and domesticated fowl, and birds of prey, although crows were rarely eaten.[72]

Trends on harvesting wild birds from wartime into the postwar are detailed in Austin's 1947 SCAP report "Mist Netting for Birds in Japan," whose data collection was aided by scientific expertise by Takatsukasa and Uchida who had headed management of avian natural resources during wartime. Such quarry were mostly passerine songbirds, caught in pockets of traditional two- or three-shelved mist nets after enticement by a trained bird serving as a song decoy, in a practice allegedly perfected in Ishikawa prefecture during the early Meiji period. From 1937 until 1947, according to the Ministry of Agriculture and Forestry, around 3,500,000 thrushes (or 300 metric tons) were harvested annually, while all types of sparrows caught amounted to 4,500,000 birds annually (65 metric tons).[73] Though total birds netted remained relatively steady due to increasing protein demands after 1937 during the war, numbers of licensed A-class hunters (using nets, lines, or traps instead of firearms) precipitously decreased until 1942, likely from men's drafting into the military, after which the war destroyed data, making reliable statistics unavailable for 1943 until 1945.[74] Contrary to waterfowl, and despite enormous numbers netted each year, songbirds represented a fraction of the Japanese diet's animal protein, since they were primarily a delicacy sold fresh in local markets or pickled in barrels to anticipate certain holidays.[75]

In contrast, a subsequent SCAP report published in 1949 on "Waterfowl in Japan," compiled by Austin and again assisted by Takatsukasa and Uchida, reveals illuminative data on wartime and postwar hunting for wild waterfowl with nets, lines, and shotguns. Interestingly, the American ornithologist now worked for the Wildlife Bureau of the NRS within SCAP's General Headquarters in a capacity resembling his Japanese colleagues' employment in wartime positions for the *Research Institute for Natural Resources*. According to statistics from the Ministry of Agriculture and Forestry, which administered hunting licenses and tracked data on hunting of ducks and mergansers, from 1924 to 1938, the annual take of geese declined, but gradually rose from 1938 to 1942.[76] The cull of ducks and mergansers reported by licensed hunters grew slightly from 1924 to 1935, and until reaching slightly over 800,000 caught in 1939, remained at similar levels, and dropped to about half that number following defeat.[77] Interestingly, numbers of licensed hunters with firearms increased from 1939 until 1945, which according to Austin, indicated the government's expansion of shotgun owners in the general population for wartime self-defense purposes.[78] Yet, in 1946, a staggering number of geese were still reported killed by licensed hunters—7,096,000—which probably reflected partial data for 1945 and 1947, but also revealed the striking necessity to seek food by any means necessary in postwar Japan.[79]

Harvesting of any wild birds reached unsustainable levels amidst the war and its aftermath. Putting these numbers into perspective, Austin noted how, "Approximately 10 percent of all the bird meat consumed in Japan at present comes from wild waterfowl."[80] In 1946, the Ministry of Agriculture and Forestry's statistics indicated that "11,300,000 pounds of domestic poultry were slaughtered," while "the kill of wild waterfowl alone during the same period was reported as 1,500,000 pounds."[81]

All reported wild bird harvesting, which Austin estimated as "probably less than one-half the actual kill," was "4,500,000 pounds during the 1946–47 hunting season."[82] Certainly, Austin's role as an occupation official was to gather data to scientifically implement laws regulating postwar Japan's social, political, and economic landscape, where the NRS served as a clearing house for such statistics.

More importantly, the reports that Austin wrote with Japanese assistance give historians an indication of how even wild birds of all varieties became casualties of accelerating wartime resource mobilization as food sources dwindled into the immediate postwar. This utilization of natural resources in a systematized fashion was very much a part of how birds and ornithologists were mobilized for war, in a process that became codified in a centralized research institute in 1941.

The *Research Institute for Natural Resources*, 1941–5

With the European outbreak of the Second World War, tensions with the Anglo-American world rose following imperial Japan's occupation of Vichy-controlled French Indochina in September 1940, and escalated further following American embargos of scrap metal, freezing of Japanese assets, and an oil embargo in summer 1941. At this critical time, Japan's government planned to create the *Research Institute for Natural Resources* to harness science and natural resources for the war effort while supporting and funding projects with measurable outcomes.

During the Pacific War, many scientists, including ornithologists like Takatsukasa, Hachisuka, Uchida, Kuroda Nagamichi, Yamashina, and Kiyosu Yukiyasu (1901–75), contributed expertise to this wartime research institute located in Tokyo. Not surprisingly, the *Institute*'s research activities paralleled the war's progress. From late 1941 until mid-1942, prior to the Japanese naval defeat in the June Battle of Midway, Japanese troops invaded the Philippines, with military forces then turning southwards with few impediments to conquer new areas in Southeast Asia. Consequently, laudatory newspaper reportage of the *Institute*'s expeditions in northern China, Indochina, and Malaysia appeared in *Asahi Shimbun* when these areas came under Japanese occupation, with their natural resources soon investigated for allocation toward the war effort.

Historian Benjamin Uchiyama asserts that Japan's media constructed a "national public sphere" that transformed into a "fascist public sphere" to "consolidate public support for the war."[83] Mass mobilization permeated domestic Japan's quotidian life, while reports of overseas imperial undertakings, including scientific management of newly colonized areas, consolidated fascist control at home by exhibiting superior Japanese methods of categorizing the natural and human worlds. The *Asahi Shimbun*'s fairly regular reporting of the *Institute*'s activities from December 1941 until October 1943 allows a glimpse into its functions. By late 1943, only ads for the *Institute*'s bulletin appeared until December 1945. Articles in this leading Japanese paper reveal how the *Institute*'s activities were disseminated to a patriotic public yearning for inspirational news of superior imperial management and military interventions abroad. This partially helped justify sacrifices endured domestically.

The *Asahi Shimbun* reported the first meeting of the *League of Academic Associations for the Study of Natural Resources* held on January 16, 1941, at the exclusive aristocratic society known as the *Peers Club*,[84] a venue revealing imperial elites' intimate connections to Japan's colonial exploitation of the Continent. Presided over by the meeting's chair, biochemist and botanist Shibata Keita (1877–1949), with Duke Takatsukasa named as the *League*'s Director and Viscount Tôki Akira (1892–1979) as its Chairman, the *League* was intended to unify aims harnessing the Greater East Asian Co-Prosperity Sphere's natural resources through promoting research investigations. Known as the "lord of bread," Viscount Tôki studied fermentation science at Tokyo Imperial University, started a bread manufacturing company, and led a Tokyo-based rice wine business before becoming a House of Peers politician.[85] His career and that of high-born colleagues reveal how closely aristocratic class status, science, and politics were intertwined during Japan's imperial period and beyond.

On July 10, 1941, to announce the *Institute*'s inauguration, the *Asahi Shimbun* ran the article "Seeking Natural Resources—Launching a Scientific Research Institute," and noted Shibata's direction of eight staff members, twenty-five assistants, and two secretaries.[86] Shibata's education and career paralleled trajectories of scientific pioneers like Iijima Isao; he studied from 1910 until 1911 in Leipzig, Germany, and served as professor in Tokyo Imperial University's Department of Botany from 1912 until 1938, concurrently working as the Iwata Institute for Plant Biochemistry's director until 1949.[87] The celebrated American plant biologist Andrew Benson (1917–2015), who discovered the carbon cycle, deemed Shibata "the most outstanding plant physiologist in Japan."[88] Boasting distinguished qualifications, Shibata determined natural plant resources and agricultural products best-suited to the wartime empire for harvesting, cultivation, and management. His leadership saw formation of eighteen research groups, including on pisciculture, applied microbiology, groundwater, ectoparasites, and vermin.[89] On December 9, 1941, the day after Pearl Harbor,[90] Emperor Hirohito's imperial decree announced opening of the *Research Institute for Natural Resources* under Ministry of Education auspices, in the elegant Aoyama area's Takagi-chô in Tokyo's then-Akasaka district.[91] The newspaper boasted that the *Institute* would mobilize a "scientific Japan" and send exploration teams to investigate the Greater East Asia Co-Prosperity Sphere.[92]

From April until June 1942, the *Institute* sponsored an expedition to north China under Japanese military protection to investigate five districts in Shanxi and collect mainland specimens for comparative purposes.[93] Japanese troops successfully captured railways, county seats, and roadways, but rural areas vacillated between Chinese Nationalist and Communist forces resisting Japan's army.[94] The 1933 Tokunaga Expedition's Kantô Army security detail was largely symbolic, but Shanxi was situated in a strategically important area with rich natural resources crucial to imperial Japan's wartime success, including one of the world's largest coal deposits near Datong.[95]

According to art historian Tracy G. Miller, Shanxi boasts at least "70 percent of China's monumental timber architecture," with many tenth- to twelfth-century buildings dating from the Song period (960–1279).[96] Now under Japanese control, China's province Shanxi composed ancient settled areas along the Yellow River basin, located equidistant from Xi'an and Peiping [contemporary Beijing], both key cities.

The Japanese military and Tokyo-based *Institute* scientists believed themselves a modernizing force upon a traditional Chinese area.

In 1942, numerous *Asahi Shimbun* articles assured the public that scientists strove to develop newly pacified territories under imperial Japan's control: "Covering Young Toshihide—Miscellany on the Shanxi Province Research Team" (March 21);[97] "Shanxi Province Academic Investigation Research Team Project (1)/Flexibility Towards Development/Tôki Akira" (April 19);[98] and "Establishing an East Asian Ethnicities Research Institute—The Basis for Founding a Co-Prosperity Sphere" (May 20).[99] Detailed in the press, such projects ran concurrently with the Shanxi Expedition, embedding it firmly within Japan's larger imperial endeavors and deploying of science to lead the Co-Prosperity Sphere. Ornithologists also directed expertise toward these ends.

Born into Japan's nobility with ancient Chinese roots, the Zoology Team director Count Kiyosu graduated in 1924 with a zoology degree from Tokyo Imperial University, and worked for the prestigious Tokugawa Biological Institute, established by the last Shôgun's scion, until 1927 when he began Kyoto University graduate studies in avian physiology under Kawamura Tamiji (1883–1964), Iijima's former student. In 1932, Kiyosu began working for the Bird and Animal Research Office in the Ministry of Agriculture and Forestry's Livestock Bureau.[100] In 1940, with Shimomura Kenji (1903–67), a colleague and leading ornithological photographer and cinematographer, he co-authored *Photo Collection on the Ecology of Wild Birds*,[101] which supplanted *Photo Collection on the Ecology of Bird Varieties* (1930),[102] written by Uchida and featuring Shimomura's photographs, and received positive acclaim by *The Auk* editor Stone that same year.[103] Hence, Kiyosu was well-positioned to describe ornithological ecology in a developing northern Chinese area superimposed over the older cultural wellspring of his ancestors.

On June 22, 1942, Kiyosu wrote an article on "Bird types in the Hengshuizhen and Yuanqu Areas/Zoology Team," where he listed and described birds his team observed near Taiyuan, Dayu, Linfen, Henglingguan, and the Yellow River.[104] Within gates of Chinese cities on the loess plain like Taiyuan, Dayu, and Linfen, he found abundant Rooks [*Corvus frugilegus*] frequenting raucous mobs in tree-lined streets, along with Common Swifts [*Apus apus*] colonizing eaves of walled homes and turrets. In agricultural areas, the Crested Lark [*Galerida cristata*] and Lesser Short-Toed Lark [*Alaudala rufescens*], emerged in prominence.[105] To the right of Kiyosu's article appeared the special report "Towards the Development of the Natural Resources of Shanxi—Researchers' Notes from the First Scientific Expedition,"[106] and two articles on yellow and red soil transformations and coal veins next to a Shanxi map, with tinier passages on arable and forested areas, plant life, and climate.

Judging from the article's prominence and large-sized title, besides Shanxi's natural resource exploitation, bird ecology greatly interested readers and revealed Japanese environmental concerns as the region's new stewards. In February 1943, when Kiyosu finally analyzed the Expedition's findings and closely examined bird specimens collected in Shanxi, he collaborated with Yamashina to describe "A New Race of a Sparrow from Shansi [sic], China" in the *Bulletin of the Biogeographical Society of Japan*, which they termed *Passer montanus shansiensis*.[107] Here, ornithologists functioned as colonizers by describing avifauna and discovering new species like analogous Western European and Anglo-American colonial enterprises.

On April 2, 1943, the *Asahi Shimbun* publicized the "Exhibition of Items Collected by the Shanxi Survey Group," featuring 500 plant, soil, geological, and archeological specimens exhibited in a weeklong exhibition at the National Museum in Ueno Park along with a film featuring the Expedition.[108] The article boasted, "Here, through the arrangement of our academic circles, the so-called natural resources of Shanxi, a continental storehouse of treasures, will be made entirely clear scientifically."[109] Elucidation of an exotic, far-off place like Shanxi, whose culture once served as Japan's template, proceeded through curation of artifacts and film to reveal Japanese development's potential for a once-moribund space. Just as the Tokunaga Expedition's report revealed poor Chinese stewardship of Jehol, so had the Shanxi Expedition, whose abundant material collections categorized a plethora of riches. Historian Noriko Aso views national museums in late nineteenth and early twentieth-century Japan as "sites specifically designed to call imperial publics into being,"[110] and asserts that they also disciplined Japanese subjects in supporting imperial aims.

Such ambitions correlated with Japan's plans for administrating conquered territories and harvesting resources for the war effort, but also illuminated a scientific approach to empire in efficiently allocating resources from areas of abundance to those less-endowed. From March 1943 to June 1945, the *Institute* published findings in eight issues of the *Natural Resources Research Institute Bulletin*, published by Kasumigaseki-shobô, a bookseller ending the publication's circulation when May 1945 Allied bombings rendered the *Institute* defunct.[111] The war's relentless cannibalization of manpower and resources also halted further projects, such as a planned series of a "Bibliography of Natural Science in the Great[er] East(ern) Co-(p)rosperity Zone," where only the Philippines' portion was published, and an aborted series on the "Geographical Maps of the Great[er] Co-(p)rosperity Zone."[112]

Austin's March 1948 SCAP report carefully detailed these scientists' *Institute* activities through their publications,[113] where Hachisuka assisted on the bird section, Takashima Haruo (1907–62) aided the mammal portion's completion, and Kuroda Nagahisa provided its periodical list.[114] Partially prepared by Austin's Japanese colleagues, the report exhaustively listed extensive wartime materials published by Japanese zoologists and ornithologists, including those by the *Institute*'s "higher vertebrate staff, [that] nominally at least, consisted of Okada [Okada Yaichirô (1892–1976), an ichthyologist and herpetologist], Taka-Tsukasa [sic], Kuroda, Hachisuka, Kiyosu, and Yamashina."[115] Notably, no *Institute*-supported publications or others, appeared for Kuroda Nagahisa, who worked for Japanese Army during the war.

Austin indicated the *Institute* served as a wartime "repository and clearinghouse for all the scientific material sent to Japan from the invaded territories, and many specimens filtered through its hands as the Japanese advanced through the South Pacific."[116] Much of the *Institute*'s research therefore centered on mapping and understanding geographical topography for conquered areas, possibly for agricultural production and colonization, while additionally examining these lands' cultivation value as sources of food for troops and civilians during wartime when rationing increasingly limited the Japanese diet. Hence, this assemblage of data and specimens was part of imperial

Japan's extension of power through scientific codification of lands now under wartime control and concurrent absorption into a burgeoning empire.

Yet, as the war's tide turned to favor the Allies, reports began to focus more heavily on scarce domestic natural resources, with food in particular. Thus, Austin indicates how, from 1943 to early 1944, the *Institute* studied alternative means of nourishment for Japanese subjects, like "the food value of seaweeds, and the calcium content of marine shells"[117]—a topic that he was also tasked with by SCAP's Natural Resources Section in the postwar period. Notably, in November 1943, Kiyosu published "On Birds That Can Become Food Resources."[118] By the end of 1943, the Japanese government sought protein sources to feed its domestic population beset by food shortages and rationing, and examined exploitation of songbirds, fowl, and ducks. According to Austin, the report lists species of birds throughout Japan's northernmost territories in the Kurils to Okinawa whose harvesting was permitted: "15 sparrows, 1 bulbul, 3 thrushes, 13 ducks, 3 doves, 27 shorebirds, 3 rails, and 6 gallinaceous species."[119] Kiyosu also included details on each of these species' "habitat, voice, nesting, food, migration, general distribution, and the totals reported captured, with their market prices for the past 16 years."[120] However, as Austin discovered during research for his writing of a 1947 SCAP report on mist-netting, harvesting of wild birds for food was nothing new, with a centuries-old history. It accelerated during the war, but represented an already established process built upon the delicacy trade of commercial songbird netting.

Besides ornithological research like Kiyosu's, though not directly related to the *Institute*, Austin's report also revealed a wartime bibliography of scientists researching vermin: mice, rats, and bats—all of whom potentially transmitted illnesses to humans as zoonotic disease vectors, while certain mice and rat varieties could serve as potential laboratory animals. In particular, 1942 seemed a quite active year for such research. Shimizu Mitsuo (unknown) published findings on bone growth of certain rats (*Ratus ratus ratus* and *Ratus norvegicus*) and voles (*Mircrotus montebelli*), while he and Shimoizumi Jûkichi (1901–75) and Nagura Otohiko (unknown) compared bone growth of the Japanese long-winged bat (*Miniopterus schreibersii japonise* Thomas) with the house wren.[121] Tanaka Ryô (unknown) wrote about rats in Formosa, and with Endô Tadashi (unknown), described seven different types of rats collected during his November 1940 trip to China's Hainan Island.[122] In 1944, he published a comparative study of its rats and mice, following a 1942 expedition to the island sponsored by Taihoku [Taipei, Taiwan] Imperial University.[123] Revealingly, from 1942 to 1943, Tanimoto Kentarô (unknown) published a series of studies on mammals' relationships to bubonic plague's spread in Manchuria, in particular emphasizing the Manchuria kangaroo rat [*Sipus sagilta sowerbyi* Thomas (*Sipodidae*)], and a 1943 report on the habits and preferred living spaces for various rat species in western Manchuria, "a permanent pest zone."[124] In 1942, Tateishi Shinkichi (1894–1977) published "Advice to Rat Catchers," on "[a] delineation of methods of catching rodents for scientific study" where "[v]arious types of traps are described and pictured."[125] From 1941 until 1944, Tokuda Mitoshi (1906–75) published taxonomies and distribution studies on East Asian rats and mice, inhabiting Japan and imperial colonies like Taiwan, Korea, Manchuria, and Mongolia.[126]

Tokuda's and other scientists' research on rodents, including those connected to the *Institute*'s vermin group, possibly intersected with the notorious Manchukuo-based Japanese research Unit 731, located in Pingfang, a Harbin suburb, headed by Imperial Army medical officer and microbiologist Ishii Shirô (1892–1959), whose laboratory bred flea-ridden, plague-infected rats for bioweapons research to weaken resistance against Japan's military in Chinese cities.[127] According to interviews by journalist Hal Gold, in Japan's Saitama prefecture during the war, farmers raised rats in "wood-framed rat cages ... built to house six rats" and brought them to "a center which raised animals for research," whereupon "they were transported to Tokyo and the [Tokyo] Army Medical College, where they were put upon the return flights to the continent [to Pingfang]."[128] While Japan's leading ornithologists or mammologists were not directly involved with the Unit, research findings developed in Tokyo and based on specimens procured by collectors elsewhere could have informed its studies. Though specific connections to Unit 731 are unclear, such research intersections reveal the preponderance of transnational connections of Japanese scientists, including ornithologists, who mobilized their talents for the state in prewar times and beyond.

The *Institute* organized scientists and projects to efficiently and scientifically utilize the empire's natural resources, including birds and other animals, to optimize military victory and support home front provisioning. By spring 1945, it failed to escape wartime ravages, and succumbed to flames of encroaching American bombing raids in the evening of May 25, burning down in the worst Allied air raid in history with Tokyo illuminated by mile-high raging fires rendering the sky bright as daylight.

Preparing Japanese Civilians for Wartime Sacrifice: Takatsukasa and the "Great Zoo Massacre," 1943

As war intensified with 1942 Allied victories, the empire's focus on mobilizing natural resources soon included prized collections of caged animals, now subjected to their human counterparts' spiritual mobilizations, and sacrificed for wartime propaganda purposes, often by the same individuals tasked to maintain their physical welfare and symbolic importance. These included Takatsukasa, whose scientific purview extended beyond birds and included stewardship of imperial Japan's animal collections housed at Ueno Zoo. Located in imperial Japan's political capital, the zoo boasted a long and storied history involving state formation enterprises, including during wartime.

In 1881, government bureaucrat and naturalist Tanaka Yoshio (1838–1916), a student of physician and biologist Itô Keisuke (1803–1902), took over the National Museum of Natural History's menagerie and reorganized it as the Ueno Zoo, which opened in 1882. His initiative joined imperial Japan's state-making apparatus where exhibition politics merged with state aims when the zoo disciplined spectators into a regimented public. Aso indicates that "Ueno was a location particularly well-suited to making a statement with regard to the cultural legitimacy of the state."[129] This centralized display locus for foreign and domestic animals showcased myriad living things, with charismatic megafauna like elephants, tigers, and exotic animals gleaned from throughout the empire and its diplomatic exchanges. Despite periodic modernizations and updates of displays to resemble shop windows in attracting fickle spectators, the zoo served as a prominent symbol of a prosperous, wide-reaching empire with deep scientific knowledge to showcase diverse thriving life-forms.

In *The Nature of the Beasts*, historian Ian Miller details Japan's wartime "Great Zoo Massacre" at Ueno Zoo in summer 1943, a description which cites Robert Darnton's now-classic 1985 historical work on a politically motivated killing of cats in eighteenth-century Paris.[130] Governor General Odachi Shigeo (1892–1955), the Tokyo high official leading this initiative, need not have killed megafauna in the zoo's collection, including its elephants, which posed little threat to the general population—unlike bears and big cats—but instead, zealously strove to prove his resolve (and the state's) amidst a total war atmosphere.[131] To achieve this goal, the elephants first had to quietly be poisoned with strychnine to prevent causing the public undue alarm, as gunfire would have frightened individuals amidst the wartime climate.[132] The much-beloved elephants might have been transferred to another location; a regional zoo even offered to take them to an area possibly spared by Allied bombings and with more abundant food. After the elephants stubbornly refused poison-laden provisions, they slowly starved to death, in a process extended due to an empathetic keeper who secretly fed them uncontaminated food.[133]

On September 4, 1943, during a long-organized ceremony to assuage the dead zoo animals' souls and further discipline the public to accept worsening wartime strictures, the elephants still lingered in a slow and painful dying process.[134] A solemn parade of Buddhist clerics, officials, and curious children preceded the "Memorial Service for Animal Martyrs."[135] Eyewitness accounts in *Yomiuri Shimbun* recounted that flower bouquets were laid upon the prepared altar by Governor General Odachi and Takatsukasa.[136] Miller argues that this chilling ceremony was used as a symbolic means to prepare the Japanese public for defeat's possibility in a growing "culture of total sacrifice"[137] that symbolized "some of the central dynamics of life on the Japanese home front in the context of imperial collapse."[138]

Following a retinue of Buddhist monks led by Abbot Omori Tadashi, chief abbot of the Asakusa-based Sensôji Temple who performed the ritual, was Governor General Odachi with Takatsukasa, who boasted direct familial connections to the Emperor as a distant blood relative and father-in-law to Hirohito's daughter.[139] Besides Takatsukasa's connections to the Throne and imperial Japan's commander-in-chief of the military, his position as leader of the Japanese Association of Zoological Gardens and Aquariums, and head of the Hunters' Association, lent him the authority to accompany this ceremony—as did his House of Peers membership.

Odachi's career trajectory reveals his knowledge of publicity and propaganda in bureaucratic administration; he previously served as occupied Singapore's mayor beginning in 1942,[140] as a Home Ministry bureaucrat until 1939, and before that as Secretary of Internal Affairs and Communications for Manchukuo's Management and Coordination Agency under Hoshino Naoki (1892–1978).[141] His accompaniment by Takatsukasa cited the Emperor's absent figure, to whom Takatsukasa was related, and perhaps, lent legitimacy to the moral correctness of wartime extirpation of beloved zoo animals by a man who oversaw all of Japan's zoos. The zoo's centrality in the empire's capital sent a strong message to a broader Japanese public: its cages lay barren, intimating that the public should desist from viewing confined wildlife when more pressing matters concerned the human population. Through Takatsukasa's actions highlighting appropriateness of these exigent measures, spectators were instructed that viewing animals, some acquired as imperial gifts, was no longer an acceptable pastime during wartime, with the state requiring the population's full-time efforts devoted toward total war. Several animals' extirpations literally represented the empire's required self-sacrifice of all lifeforms, not only humans.

Besides this bleak propaganda message, there were other possible reasons for Takatsukasa's involvement. The Shintô belief system retains a longstanding fear of *onryô*, or vengeful spirits. In the past, shrines were built to appease warriors killed in battle or those who died wronged in exile, like the scholar Sugawara no Michizane (845–903), with slain enemies sometimes granted shrines and worshiped. Thus, these spirits were transformed into protective deities rather than remaining as spiteful specters who could wreak havoc upon the living: even Tokyo's Yasukuni Shrine where souls of Japan's imperial war dead are enshrined can be viewed as representing these beliefs.[142] In particular, wrongful or violent deaths could generate potentially harmful *onryô*. Few likely feared that ghosts of the elephants Tonky or Wanri would spiritually trample Ueno Zoo's environs and blight the area, but negative emotions of shock and horror expressed by children and other onlookers may have required appeasement and sweeping out of public consciousness in a spectacular *harai* [exorcism].

Since Takatsukasa's family cared for the Meiji Shrine, and in 1944, he would become its head priest, quite possibly the animals' disembodied souls found their way to the deceased Meiji Emperor's spiritual resting place in a beautiful, bucolic wooded park, and thus, their possibly dangerous nature might be neutralized. However, since the zoo's animals perished in unnatural deaths due to their wartime sacrifice for the nation, the Yasukuni Shrine was likely their souls' final resting place in a religious structure already absorbing spirits of numerous animal soldiers including horses, carrier pigeons, and dogs serving in a military capacity in past imperial wars.[143] Numerous animals displayed at Ueno Zoo were also gifts of tribute to imperial Japan, war trophies, or participants in past military service. Hence, the zoo symbolized a long-term history of mobilizing animals for military purposes.

However, the horrific, and largely unnecessary 1943 massacre of zoo animals failed to prepare Japanese civilians for the war's devastation of Tokyo when Allied air raids started to level the city in late 1944. In the second of two major firebombing attacks appearing two days apart in late May, the US XXI Bomber Command under General Curtis Lemay (1906–90) of the United States Army Air Force utilized 502 B-29 Superfortress bombers to destroy 44 square kilometers of central Tokyo with incendiary bombs.[144] Just like in Berlin and Dresden, these incendiary bombs fueled by napalm created firestorms, whose burning caused swirling winds with vortices swallowing everything in their path and sucking the oxygen out of living beings. Following this holocaust, wartime mobilization of the Japanese empire's human and animal populations would shift to pure survival, and imperial Japan's superior scientists could do little to save the domestic population or the nation's infrastructure.

When the War Came to Tokyo: Wartime Contingencies and Beyond

Two other casualties from the May 23 and 25, 1945, air raids were Takatsukasa's and Kuroda's estates in Tokyo, which included their libraries, aviaries, and specimen collections.[145] Regrettably, Takatsukasa lost everything: his house, collections, and even land where his estate was built. As the Meiji Shrine's head priest, he expediently moved himself and his family directly into its attached precepts and outbuildings. In a 1960 obituary for Takatsukasa in *The Auk*, Delacour, now at New York City's AMNH since his 1940 wartime exile from France after Nazi-requisitioning of his estate, fondly remembered his friend's genteel attitude toward loss: "he never became bitter nor discouraged. As the High Priest of the Meiji Shrine, he lived his last years in a beautiful park where wild birds abound despite its location in Tokyo and he enjoyed them greatly."[146] The war's devastation pushed Takatsukasa into the city's last unspoiled area, where wooded paths of the Shrine's lands bordered Yoyogi Park, soon populated by the American occupiers as Washington Heights. However, Takatsukasa was not alone in loss of hundreds of specimens and years of research materials in the air raids' ensuing flames.

Kuroda Nagamichi and his son Nagahisa also lost nearly all their specimen collections. Luckily, Kuroda kept his twenty most valuable specimens in his *kura*, a small stone warehouse, on his estate, which survived the flames as a traditional unburnable structure built in a fire-prone city; these treasures included his beloved pair of exceedingly rare *Pseudotadorna* (Sheldrake ducks),[147] unique varieties of *Astrapia recondita* (Lesser Ribbon-Tailed Bird of Paradise) found in Southeast New Guinea's Morobe District, which he and Uchida had described in 1943, and *Erythrura trichroa pelewensis* (Palau Blue-faced Parrotfinch), described by him in 1922.[148] All, except for seven skins of thirty-eight specimens from western New Guinea, were destroyed in the mid-1945 conflagrations.[149]

Not only material damages plagued these Japanese scientists. During the war's final year, Hachisuka, whose *Handlist of Japanese Birds* had traveled the wartime high seas after he smuggled it illegally into the hands of soon-to-be deported Australian diplomats in 1942,[150] unknowingly suffered a betrayal from at least one Westerner he had trusted and once considered a close friend. Similar to the majority of ornithological handlists that bore at most a two-to-three decade shelf life, Hachisuka's allegedly conclusive volumes on the birds of the Philippines were eventually due a revision. Yet, this came in 1945 from his close friend Delacour and his correspondent Ernst Mayr, who collectively worked on a new edition of *Birds of the Philippines*, by dividing up tasks where each scientist worked on half of all the islands' avian families for systematic revisions. After the war on November 15, Delacour and Mayr published the reasons for their revisions separately from their new volume. Mayr recalled their furious pace of work at a point when the Allied war with imperial Japan dragged on just prior to the dropping of two American atomic bombs in early August:[151]

> In our work on the "Birds of the Philippines," Jean and I exchanged each morning the manuscripts of the parts we had done the previous day and night, and this kept

both of us working at top speed. We had started the project on March 13, 1945, and the completed manuscript was mailed to the publisher three months later, June 13, 1945.[152]

Their *Notes on the Taxonomy of the Birds of the Philippines* thus emerged at a time of American victory, when Japanese imperial designs were expelled out of East and Southeast Asia, and Philippine rule was restored back to the United States. Only in 1946 would the Philippines finally be granted independence, while the islands continued to host American military bases on their soil for decades to come.

Despite initial enthusiasm for Hachisuka's early-to-mid-1930s study, after imperial Japan's defeat, Delacour and Mayr challenged many of the new taxa he allegedly discovered, determining his findings as merely "variants," and thus, diminished his studies' import. They scathingly noted "none of the new generic names proposed by the Marquess Hachisuka is valid in our opinion."[153] Hachisuka's fortune as a once rising-star in prewar Anglo-American ornithological circles rapidly fell during the war's denouement but experienced an all-too-brief revival in the immediate postwar period. In his 1953 obituary for Hachisuka in *The Auk*, Delacour mentioned that after 1937, he never saw Hachisuka again, and asserted that "The war period was difficult for him because of his western ways and friendships."[154]

Nevertheless, after the war's end, Hachisuka was happy to reengage with ornithologists in the Anglo-American scientific community. In the June 1947 BOC members list, he appears as "Hachisuka, The Marquess; Mita Shiba, Tokio, Japan"[155] on the same page as Delacour, whose address is noted as New York City's Stanhope Hotel, where he lived with his mother. Hachisuka was joined several pages later by "Kuroda, The Marquis Nagamichi; Fukuyoshicho, Akasaka, Tokio, Japan,"[156] "Momiyama, Toku Taro; 1146 Sasazak, Yoyohata-mati, Tokio, Japan,"[157] "Taka-tsukasa, Prince Nobusuke; 1732 Sanchome, Kami-meguro, Meguro-Ku, Tokio, Japan,"[158] and "Yamashina, The Marquis; 49 Minami Hiradei, Shikuya-ku, Tokio, Japan."[159] Out of a total 149 members, 5 were Japanese,[160] in a list overwhelmingly composed of Anglo-American men from Great Britain, the United States, or British colonies; one Belgian man; and a few British women. They were the only non-white members in a club whose purpose was "the promotion of social intercourse between Members of the British Ornithologists' Union and to facilitate the publication of scientific information connected with ornithology."[161] This renewed inclusion into ranks of the British ornithological world was a Pyrrhic consolation, since during the Occupation, Hachisuka was prohibited from travel outside Japan to attend the BOC's monthly meetings in the Rembrandt Hotel within London's exclusive South Kensington area.[162]

Those who knew Hachisuka from war's end until his untimely 1953 death remembered his difficult personal life and occasional angry outbursts aside from his pioneering research. Hachisuka certainly had much to lament. Not only were his social activities subject to media scrutiny, he also lost much of his family's land in the early Occupation's land redistribution to former tenants, which prevented him from receiving a once-steady income. In a 2015 interview, Hachisuka's disciple, ornithologist and bird migration specialist, Nakamura Tsukasa, though nearly ninety, still remembered his mentor's strong anger toward Occupation authorities, whose seizure of provincial land dissolved holdings belonging to his clan since the late sixteenth century.[163] As the Hachisuka family's sixteenth descendant of a clan of Tokushima feudal lords dating back to the fourteenth century, Hachisukai came from a wealthy *daimyô* lineage once annually receiving nearly 300,000 *koku* [4.5 western bushels—a measure intended to feed one man for a year] of rice from their domains. Nakamura believed that the shock of lost wealth caused Hachisuka's health to rapidly decline, with his illustrious family lands originating from gifts by the great unifiers Oda Nobunaga (1534–82) and Toyotomi Hideyoshi (1537–98). He was frightened when Hachisuka leapt from his chair to rant, raving against the postwar land redistribution, especially in Hokkaidô, while his Tokushima holdings were likely also affected.[164]

Added to these problems, Hachisuka was plagued by marital difficulties. His union with Chiyeko Nagamine produced only one child—a daughter Masako born in 1941—and soon after their 1939 marriage, became estranged, despite a potentially more relaxed partnership countering social expectations for aristocrats. The war's intrusion undoubtedly contributed to relationship troubles, plus imperial Japan's growing tensions with the United States, which erupted into war after December 1941. Since April 1944, when Allied bombings regularly harmed Tokyo, Hachisuka and his wife lived in separate dwellings.[165] To avoid bickering, fire bombs, and social opprobrium while devoting his studies to the Dodo and other extinct birds, he moved permanently into his comfortable Mediterranean-style villa in Atami several hours from Tokyo—where he set up his female secretary Usui Mitsue, who some alleged was his mistress. Hachisuka's postwar contacts in the ornithological world noted that he traveled everywhere to ornithological meetings with Usui, the only woman depicted in the commemorative 1947 photograph of the revived *Ornithological Society*.[166] Regardless of rumors, her services were crucial to the scientist, since Hachisuka's hearing diminished to where normal conversation was difficult.

In April 1948, Hachisuka filed for divorce, alleging his wife's adultery, but surprisingly, this petition was unsuccessful.[167] In November 1950, Nagamine responded to this humiliation by accompanying her daughter to Los Angeles, purportedly for schooling, and in 1951, contested the divorce, alleging Hachisuka's cohabitation with another woman (Usui) for nearly eight years after their separation; she also requested a 40 million yen alimony payment to end the marriage.[168] Despite past opprobrium between her parents, Masako Hachisuka was later proud of her father's scientific achievements.[169] After Nagamine left for the States, in late 1950, Hachisuka sold his Tokyo estate to the Australian embassy, which presumably paid for divorce proceedings and the considerable sum demanded by his estranged wife as compensation.[170] Notably, though Hachisuka's stately home avoided destruction, the Australian Embassy promptly requisitioned it; only his Spanish-style Atami villa escaped the capital's fiery air raids and provided him with a peaceful refuge from mounting troubles until his untimely death.

In 1950, Hachisuka also loaned 30 million yen at 20 percent interest per year to his neighbor Yokoi Eisuke, viewed by other aristocrats as a brash young nouveau-riche upstart with yakuza connections who earned his fortune in the textile industry and outfitted troops at American military bases during the Occupation era. Hachisuka heard from Yamashina that Prince Nashimoto Morimasa (1874–1951), in whose household he was raised as a youth, sought to sell his vacation home, which Hachisuka casually mentioned to his friend Yokoi. Based on this tip, in 1946, Yokoi purchased the villa next door from the distinguished Nashimoto-no-miya family, whose royal blood failed to insulate them from postwar wealth taxes and a perceived low-class parvenu. Yokoi eventually paid back 10 million yen, but after Hachisuka's death, he refused to repay the remaining portion to Nagamine, to whom her husband's death had transferred the debt.[171]

Nagamine likely wished to help her parents back in the United States, who likely lost their property and wealth during the American government's 1942 to 1945 compulsory internment of Japanese-Americans during the Second World War. In 1948, the Japanese-American Evacuation Claims Act allowed some individuals to recoup fractions of their losses, indicating the huge economic blow to this community: "Twenty-six thousand, five hundred and sixty-eight claims totaling $148 million were filed under the Act; the total amount distributed by the government was approximately $37 million."[172] The divorce proceedings, whose news followed Nagamine to Los Angeles, certainly brought her further shame. In April 1951, news reports in the Tokyo *Yomiuri shimbun*,[173] and in the Los Angeles *Pacific Citizen*, mentioned the divorce, indicating Hachisuka's allegations that his wife had "committed adultery on numerous occasions" with a French diplomat and aviation officer in 1943, and American lieutenant in 1949, for which he offered letters as evidence.[174] Nagamine's attempts at dissolving the marriage stalled for years, concluding only with Hachisuka's sudden death, which dissipated his estate and scattered ornithological treasures. For Hachisuka, the war's negative psychological and material effects lingered far beyond its end.

Others faired far better. Fortunately, before the Pacific War, Yamashina absorbed into his museum Hachisuka's libraries and specimen collections and those of Tokyo Imperial University, all evacuated to his villa in the mountainous Nagano Prefecture's Karuizawa resort town for safety prior to the Allied bombings.[175] Yet, his Nampeidai estate in Shibuya burnt to the ground, while Allied air raids miraculously spared his museum laboratory harboring much of his specimen collection. During the Occupation, the American scientist H. Elliott McClure (1910–1998) described their wartime fate: "The estate adjoining the museum had been shattered by bombs, but firebombs landing on the institute had failed to ignite it. Its second floor was filled with cabinets housing the collections of other ornithologists as well as those of Dr. Yamashina."[176] While Yamashina lost his home, outbuildings, and aviaries, his adjoining personal museum was intact, where he and his wife took up residence when in Tokyo.

Most of the time, the couple lived what Yamashina called a "semi-rural postwar life" at their unharmed seaside home in Yokosuka's Kuruwa beach area, seventy kilometers distant from Tokyo.[177] In his autobiography, Yamashina recounted:

Because the garden at the Kuruwa villa was large, we turned it into a field and made our own food. My hand that manipulates the microscope gripped a hoe in unaccustomed agricultural labour, but because the people of neighbouring rural families all kindly did it for me, this really helped. Also, because raising birds is my forte, we raised chickens and took their eggs, and even sold them.[178]

Nakamura remembered that, though his first mentor Yamashina raised chickens (at Kuruwa) for research purposes, he generously distributed eggs to his Shibuya neighbors, while Hachisuka told Nakamura to sell them when he became his *shosei* [live-in disciple].[179] Valuable edible gifts of eggs and chickens improved relations with Yamashina's less-fortunate neighbors, where "amidst the widespread burnt plain, the single building of my research institute stood out."[180]

The grief-stricken couple nevertheless experienced a precarious moment when the American military placed a requisitioning tag on the Shibuya laboratory's front door.[181] Yamashina's scholarly reputation and ornithological pursuits saved him from destitution, when Colonel Hickey, a Civil Affairs Officer from Yamanashi Prefecture, came unannounced one day, and after inspecting the specimen room and library, suddenly ripped off the requisition slip and cancelled the order.[182] Deep within Occupation military ranks existed bird lovers like Major General Wolfe, a raptor expert befriended by Hachisuka during his 1929 Philippines expedition, now leading the Yokohama military base headquarters, who propitiously swooped in to save the day.[183] This was followed by his weekly Saturday visits to examine eagle and hawk specimens in Yamashina's collection and peruse books.[184] Wolfe even collaborated with Kuroda Nagahisa to "read through the part of the birds of prey giving [him] valuable advice" for his 1949 draft of *The Birds of Japan* (1953), co-authored with Austin.[185] Nevertheless, in the late 1940s, high taxes forced Yamashina to sell land from his ruined estate to save his laboratory and specimen collection.[186] He then busied himself to transform his Tokyo laboratory's holdings into an internationally recognized research center attracting the world's ornithologists.

Both Yamashina and Hachisuka admitted American occupation officials into their homes, mirroring the late Meiji practice of performing Western rituals to communicate a level of civilization to outsiders, or in the postwar case, of actively reforming an image of former "enemy" into potential ally. Most lodgings of high-ranking Occupation personnel consisted of the same housing populated by Japanese aristocrats, of whom a fair number were scientists, and quite likely enjoyed ornithology. Yamashina and Hachisuka's Tokyo estates boasted Western portions, closely resembling homes in affluent Anglo-American suburbs, and traditionally built Japanese sections. In the prewar period, Japanese families circulated between both estate portions with ease, in a symbolic bifurcation of domestic and public lives, while after defeat, many retreated into Japanese portions of their estates if they had escaped major damage.

Within this environment arrived an American interloper with the Occupation forces, initially imposing his colonial gaze on those he encountered, while later interacting as a colleague with other Japanese scientists. In work for SCAP, Austin arrived in Tokyo with a relatively close connection to General Douglas MacArthur, and shot photos generating nearly a thousand photographic slides of a reconstructing Japan that memorialized his experiences and revealed his interests.[187] First by entertaining in the Daichi Hotel, an initial lodging place for high-ranking Occupation officials, and then in a requisitioned aristocratic home with his dependents, Austin strove to acquire social connections building his Japanese counterparts' trust, whose collaboration he needed for success of his NRS mission. Despite unequal power relations between occupier and occupied, the renewed growth of colleagueship arose out of these encounters within the home's domestic spaces and precipitated revival of earlier embedded networks.[188]

Conclusion

How can a wartime history of Japanese ornithology be written when so many sources were lost? A solution might be to examine victors' documents—the US Occupation's quasi-colonial amassing of knowledge and data resembled wartime Japanese research institutes. Such an approach naturally requires sensitivity toward the Anglo-American conqueror's careful curation of information deemed useful to the political, social, and economic reconstruction of postwar Japan. As a counterpoint, one could consult extant wartime texts produced by Japanese ornithologists to sample how they codified knowledge to a domestic and international audience amidst an imperial framework. Some are carefully cached in libraries, while others command a small fortune on the open market, especially if they are rare—prewar and wartime ornithological books seem to command a particularly steep price.

Interviews are another method, which rely on experiential memories of informants who lived through the period. Yet, as anthropologist Marc Augé asserts, "Oblivion is a necessity both to society and to the individual. One must know how to forget in order to taste the full flavor of the present, of the moment, and of expectation, but memory itself needs forgetfulness: one must forget the recent past in order to find the ancient past again."[189] Inevitably, lacuna appear that threaten to fragment historical narratives or lead researchers toward what historian Vicky Ruiz once so eloquently called "creative non-fiction."[190] Thus, only a wholistic approach to a variety of interdisciplinary primary sources, including non-traditional historical sources like photographs, paintings, drawings, prints, and even traditional Japanese signature stamps, can supplement a historian's mainstays like diaries, letters, journal articles, newspaper columns, published studies, interviews, and so on. Combining these sources can reveal gaps, lacuna, inconsistencies, imperfectly concealed secrets, and even blatant lies collectively allowing historians to piece together the rich and complicated lives of a select group of politically influential Japanese aristocrats who also happened to be deeply passionate about birds.

However, the Pacific War undeniably subsumed the talents of all of imperial Japan's scientists, including ornithologists, into the wartime apparatus, so that, by 1941, they actively published and worked to support an imperial Japanese agenda. Some, like Takatsukasa and Yamashina, were even related by blood to the Japanese emperor, who served as the military's symbolic commander-in-chief, and felt obligated to provide their own contributions. As indicated by Hachisuka, most felt inspired rather than compelled to devote their expertise toward describing the burgeoning empire's birds as imperial Japan grew flush with victory in the Pacific War's early years. As the conflict deepened into its second year, Japan's leading ornithologists also wrote about cytological studies on fowl, which helped farmers breed improved varieties for food as the war began to cannibalize the empire's resources. As Allied bombings reached

Japan's main islands by 1944, many of these scientists' homes and laboratories in the imperial capital were destroyed, but as wealthy elites, they relied upon country homes surrounding Tokyo, where some cached their collections. Nevertheless, the war's end and Allied Occupation provided them with opportunities to redefine their roles by necessity, and for some, to dispel dark memories of wartime activities by working with the occupiers to gain special privileges and regain access to influence political workings of their ruined country.

7

Tokyo under the Allied Occupation (1945–52)—*Yankees with a Mission amongst Threadbare Aristocrats*

Introduction

From the 1920s until 1945, imperial Japan's scientists worked to aggrandize personal and state aims in an imperial context. Zoologists, of which a sizable portion considered themselves ornithologists, were highly connected to political centers as House of Peers members and aristocratic men. Yet, the mid-August 1945 defeat of imperial Japan by Allied Forces led by the United States displaced the locus of political power into the putative conquerors' hands until the Occupation's 1952 completion. Following the Japanese empire's swift collapse, the Anglo-American occupying powers applied new ideologies to fill Japan's political vacuum, including their versions of democracy and social interactions to supplant those of a defeated population. This included new paradigms for bird conservation.

This chapter[1] shifts focus to the occupying Americans by exploring relationships of Japanese scientists with the American ornithologist, Oliver L. Austin Jr., who set up and worked for SCAP's[2] NRS Wildlife Branch from 1946 to 1950 during Japan's postwar Allied Occupation.[3] He also compiled a report detailing Japanese zoologists' Second World War-era work, which appeared as a bibliography listing many of the ornithologists' studies.[4] Austin knew his mission could only succeed with close relationships with Japanese ornithologists and zoologists like Yamashina, Hachisuka, and Takatsukasa, who could push through legislation that he proposed to protect songbirds as a "democratic right" for all Japanese. He also organized specimen-gathering expeditions, assisted by prominent collectors like Ori'i and Nakamura, and was often accompanied by Hachisuka and Kuroda Nagahisa, with whom he co-authored "The Birds of Japan, Their Status and Distribution" in 1953.[5] This chapter discusses Austin's collaborations with Japanese scientists to revive Japan's bird and wildlife populations, and his broader position as an Occupation official representing US imperial interests in interactions with former elites from a vanquished empire in Korea and Japan.

As scientists from comparable social strata, Austin and his collaborators spoke the common language of ornithology in Japanese and upper-class Yankee or British accents to expediently transcend barriers of nation, race, and politics within embedded networks during a military occupation. Austin's relationships with Japanese scientists of similar elite backgrounds and education helped smooth cooperation with US authorities. However, the Japanese, despite high social status, also needed the American's help to access scarce postwar necessities like clothing, medicines, and medical care, while offering their expertise as a reciprocal measure. Thus, Japanese input toward the US scientist's work was considerable, as both sides developed mutually symbiotic relationships and networking communities fostering scientific inquiry and conservation endeavors while resolving wartime antagonisms. The relationships initiated between conqueror and conquered often enjoyed prewar antecedents, which allowed Japan's leading ornithologists a later re-emergence onto the world stage beyond the Occupation where Austin's role slowly disappeared as they reclaimed positions as prominent conservationists of global bird life.

Pheasants over Raptors: The *Ornithological Society*'s Postwar Rebirth

A large traditional painting of a pheasant, likely by a Tokugawa-era Rimpa School artist, hung on a scroll in the office of Dr. Matsuda Takeshi, president of Kyoto University of Foreign Studies, and a renowned US-Japan relations expert, whose research has highlighted "soft power."[6] In summer 2014, I was tasked with initiating a student exchange program between my American university and his.[7] Matsuda noticed my admiration for this sublime artwork, indicating that the green pheasant (*Phasianus versicolor*) was Japan's national bird. His choice of adornment for an office where he often met foreign visitors and Japanese dignitaries was no accident; yet, it took archival research to uncover how the pheasant, a popular upland hunting quarry, became a symbol of postwar Japan.

A year later while researching this book, I soon discovered that the *Ornithological Society* chose this particular bird in 1947 to emphasize Japan's peaceful global relationships. On April 10, 1947, newly designated as Japan's first official "Bird Day," Austin mentioned in a speech geared toward a children's audience that this group picked Japan's national bird "because they know more about the Japanese birds than anyone else in Japan .... they decided that first of all, the national bird should be one that is well known, and that lives in Japan and nowhere else."[8] No doubt, the group favored the green pheasant due to input by Kuroda Nagamichi, author of an acclaimed 1926 pheasant study who held the *Ornithological Society*'s prewar meetings at his estate, coupled with Austin's favorable late October 1946 SCAP assessment of growing green pheasant populations compared to other birds in Tokyo's outskirts.[9] Instead of an aggressive raptor for a national symbol, *Ornithological Society* members picked a tasty favored hunting quarry, whose colorful plumage delights the eyes and tempts palates with succulent meat, and also allegedly sacrifices itself for its young.[10] Nonetheless, despite beneficial attributes, the pheasant still intimates prewar shadows in appearing in Japan's eighth-century mythohistories and Peach Boy Momotarô's retinue of loyal animal friends, lauded for bravery in a folk tale used to motivate children's patriotism from the Meiji into Asia-Pacific War eras.[11]

The *Ornithological Society* consisted of a revived group of scientists including Japan's most prominent zoologists and ornithologists, all elite men with a strong national political impact. Its early postwar meetings were initially held in the requisitioned home of former colonial administrator and shipping magnate Ariyoshi Yoshiya (1901–84), then housing Austin and his family. The national bird arose from their discussions, where the pheasant symbolized new paradigms of democracy developing in peaceful interactions by former wartime foes. Common ornithological interest by Japan's elites now translated into new scientific concepts and collaborations with international political ramifications.

In the postwar period, attempts were made to shed ornithology of its socially stratified reputation as an elite pastime or aristocratic scientific pursuit through general emphases on "democratization" in all aspects of Japanese society. Historian Morris Low asserts that "Instead of portraying knowledge as the domain of a select elite, scientists sought to work towards the democratization of knowledge."[12] This process radically took root in the late 1940s. Austin, as an occupation official, framed enjoyment of songbirds and bird-watching as a "democratic right" for all, while Yamashina saw these aims intersecting with his prewar goals for bird conservation. For Yamashina and his cohort, bird conservation now was an expedient means to expand upon broader, popular democratization initiatives amongst the Japanese people, while re-entering the world stage with studies of avian species as symbolic of freedom and borderless flight unhindered by international squabbles. Notably, the pheasant and other Japanese birds like cranes became enduring symbols of Japan's soft power, a late-1980s term developed by American political scientist Joseph Nye,[13] and popularized in the early millennium as an international relations concept often linked to US influence.[14] Yet, as a Japanese political scientist, Matsuda also sees such soft power as linked to dependence upon the United States.[15]

When the first postwar issue of *Birds* appeared in November 1947, it conspicuously carried Austin's preface "To the New *Tori*," noting his SCAP position.[16] He frames the publication's renaissance in quasi-paternalistic terms, where American-style "constructive criticism" requires fostering to allow democratic thought to penetrate "all Japanese science":

> Of greatest value to ornithology both in Japan and abroad would be the establishment of a regular review section, to note, summarize, and criticize all publications on the birds of Japan or of interest to Japanese ornithologists. Honest, constructive criticism is one of the basic tenets of democratic, scientific thought, and is badly needed for the improvement of all Japanese science … It is my earnest hope that the Ornithological Society of Japan and *Tori* will continue to grow and prosper under the new democratic precepts adopted by Japan, and take their rightful place among the learned societies and scientific periodicals of the world.[17]

Austin infers that the alleged *lack* of scientific traditions of review and debate led imperial Japan down a road toward fascism, with American-style critical discourse as remedy to infuse postwar Japan with a spirit of proper global citizenry and democracy. As asserted by historian Shinobu Seizaburô, imperial Japan indeed enjoyed a thriving, emergent democracy during the Taishô period,[18] with universal manhood suffrage initiated in 1925, but it was slowly derailed by government concerns over leftist representation and rising militarism from the early 1930s. The 1931 Manchurian Incident, Second Sino-Japanese War, and Pacific War each ushered in greater repression against political and academic freedom,[19] with scientists increasingly pressured and cajoled to contribute expertise toward the war effort.[20] Many proudly offered their scientific knowledge to contribute to military victory. After the August 1945 dissolution of empire rendered Japan's imperial project defunct, Japanese scientists, including ornithologists, needed to deeply rethink their research foci. Zoologists now sought to reintegrate their research into a more peaceful, internationalist paradigm in the postwar Anglo-American world order. Therefore, they appropriated the Occupation's rhetoric of democracy to their endeavors, and pursued a renewed global outlook, reconnecting with Anglo-American communities whose contact they once had enjoyed. Austin, as a highly placed occupation official close to General MacArthur, was thus an expedient figure whose connections were used to good advantage.

Birds lent themselves well as symbols of democracy in long migrations over national borders and seeming peaceful coexistence with other species wherever they landed and made their homes. First, a vast cognitive shift was required for how postwar Japan's general population viewed birds—not just as pretty objects depicted in art or as creatures caught by falcons, nets, or firearms for hunting and science, but as organisms enjoyed from afar by all. The war had accelerated traditional use of mist-netting to capture wild songbirds and liming of ducks[21] for food during a protein-scarce time. In peacetime, spyglasses, binoculars, or cameras of distanced scientists and committed bird watchers could now capture these individuals, allowing them to live free in prose, checked off on personal handlists, or shot in photographs rather than as prepared skins spending their eternity in specimen collections. This also mirrored paradigms where a once militaristic Japan no longer used force or arms to take colonial possessions or acquire specimens, but now peacefully enmeshed itself within global networks where animals joined an international patrimony transcending borders.

Deposed Japanese Aristocrats and SCAP Democratization— Conquerors Become the Conquered

The Allied Occupation's early democratization efforts rested upon a tripod of progressive, late 1940s initiatives: radical agrarian reform termed "land reform"[22] transferring cultivated areas into tenants' hands for nominal prices, abolishment of aristocratic status, and institution of a new American-written postwar Constitution inspired by New Deal principles also allowing women the vote. On October 14, 1947, SCAP eliminated the imperial family's eleven collateral branches, and disbanded special aristocratic privileges.[23] The new Japanese Constitution, promulgated on November 3, 1946, and effectuated on May 3, 1947, enacted these changes into law; Article 14 codifies equality and nondiscrimination where "Peers and peerage shall not be recognized."[24] Now, all "citizens" rather than "subjects" were to relate to another as commoners, while the Emperor was made to renounce his divinity, and publicly proclaimed his personhood to Japan's people.²⁵ Yet, when faced with prospects of a revived and vocal left-wing movement possibly destabilizing US democratization efforts and allying toward Moscow or Beijing, the Americans' focus on fashioning a liberal democracy soon evolved into reliance on former powerful elites. Historian John Dower asserts

> [i]nitially, the Americans imposed a root-and-branch agenda of "demilitarization and democratization" that was in every sense a remarkable display of arrogant idealism—both self-righteous and genuinely visionary. Then, well before their departure, they reversed course and began rearming their erstwhile enemy as a subordinate Cold War partner in cooperation with less liberal elements of society.²⁶

This characterization of the Occupation and its shift in aims aptly describes the heady political and social atmosphere experienced by the aristocratic Japanese scientists, who now faced a world newly repudiating their position. Their aristocratic status disappeared, but their social and political capital was still useful for the occupiers.

Rapid postwar refashioning of Japan's society and politics to American ends prompted a cognitive shift toward a new Japanese worldview privileging US global influence. Hence, the postwar represents minimal rupture of a long structure of avian imperialism where ornithologists worked to uphold personal careers and national aims throughout the transwar period (1920s to 1950s). Austin serves as a lens to understand how these aristocrats were viewed by an influential American occupation official, and how they rearticulated their positions in a transformed society. Japanese aristocrats now saw themselves and their world as radically altered where they were no longer the dominant class. Therefore, it is worth examining how the occupier perceived them and their environment through Austin's perspective. A certain worldview and the particular imperial roles as scientists that they embodied as colonizers now came home to them in Tokyo as the putative "colonized"—a condition persisting for them until at least the Allied Occupation's completion in 1952. Since these former aristocrats originally lacked knowledge of how long the United States planned to occupy their country, they were obliged to find new places as "conquered conquerors" who could no longer embody prewar imperial masculinities.

In *Tell Me the Story of How I Conquered You*, Latin American literature scholar José Rabasa examines how knowledge was transmitted by Indigenous Mesoamerican *tlacuiloque*, or scribes, decades after their sixteenth-century conquest by Spain into a corpus of depictions and descriptions of Indigenous life.²⁷ Supervised by Dominican Friars, they were tasked to draw and write in a European format on European paper, with blank spaces for later annotations.²⁸ Hence, the work that they produced was a translation on many levels: linguistic, cultural, and material. Yet, it provoked much unease in their religious overseers. In his analysis of folio 46r from Codex Telleriano-Remensis (ca. 1563), where Spanish Catholic authorities supervised *tlacuiloque* to describe their own conquest, Rabasa proposes the notion of "elsewheres," places of belonging forever outside colonizers' realms of imagining and everyday existence, along with their inability to fully partake in Indigenous knowledge.²⁹ He asserts that "*Elsewheres* are not merely spatial locations," and describes them as esoteric social spaces: "They consist of forms of affect, knowledge, and perception underlying what a given individual in a given culture can *say* and *show* about the world."[30] While the scribes carefully followed supervisors' instructions, they still depicted a world largely inaccessible to the conquerors.[31] The Spaniards' original plan for this codification of knowledge using Indigenous scribes was for the creation of "an album of Mesoamerican writing systems to facilitate the indoctrination of the Nahuas," but instead, they uncomfortably found themselves inserted into the text as depicted objects alongside others rather than the main subject.[32]

Interestingly, Rabasa's characterization of relationships between Spanish friars and *tlacuiloque* describing the Spaniards' conquest of the Americas somewhat resembles Austin and his Japanese counterparts. While the American scientist and SCAP official believed in his centrality in their world—strikingly communicated in an iconic 1947 photograph where he sits at the front-center of the revived *Ornithological Society* posed amidst stark limestone ruins of Yamashina's estate—theirs only temporarily absorbed his presence.[33] Deeper analysis of Austin's notable collaboration with Kuroda Nagahisa on an important study of Japanese birds reveals this aspect. In 1949, soon before his US return, the American scientist commissioned Kuroda to write the lengthy study "The Birds of Japan, Their Status and Distribution," published under both their names in 1953 in *Bulletin of the Museum of Comparative Zoology at Harvard College*.[34] With a younger Japanese associate and subordinate, Austin added his own nation's latest, and allegedly, most complete contribution to codifying bird life in Japan.

Prior to his boss's US departure, Kuroda prepared a Japanese-language manuscript on Japan's birds, but following Austin's instructions, obligingly finished translating it into English on December 31, 1949, to which Austin added his interpretations and corrections in June 1950; Kuroda then sent back his revised manuscript to Austin on January 6, 1951.[35] This manuscript, initially intended by Kuroda to supersede British amateur ornithologist Henry Seebohm's (1832-95) "The Birds of the Japanese Empire" (1890) and German ornithologist and botanist Hermann Jahn's (1911-87) *Zur Oekologie und Biologie der Vögel Japans* (1942), would bring to light Japanese ornithologists' significant transwar research to the Anglo-American world. In his 1954 assessment of Kuroda and Austin's study in *The Quarterly Review of Biology*, the German ornithologist Erwin Streseman (1889-1972) called it the "fruit of an American-Japanese collaboration," but conceded that Kuroda was "won over to the plan of a joint work" by Austin after having completed a manuscript on the subject.[36] Their study, or rather, Kuroda's, included full-descriptions of 415 bird species, including "ecology, reproduction, and migrations," with "morphological suggestions that will be welcome to the systematist."[37] Here, Austin attempted to take credit for work laboriously completed by a Japanese quasi-subaltern, whose final version he edited to make it conform to American standards—similar to the broader purpose of the US-led Occupation to instill a tutelary democracy.

Intriguingly, over half a century later in 2004, Kuroda self-published a limited "memorial publication" of eighty-eight copies of the comprehensive draft, originally called "The Ornithologists and Avifauna in [sic] Early History of Japanese Field Ornithology."[38] He dedicated it "in memory of the four years collaboration, 1947–1950, with Dr. Oliver L. Austin, Jr. at the Wildlife Branch, Fisheries Division, Natural

Resources Section, GHQ,"[39] and thus inserted his work into a years-long hierarchical "collaboration" under the American scientist described by title and position rather than as colleague or friend. Also, Kuroda describes his draft as "reference material" sent to Austin, minimizing its import, but also indicating that the American scientist's revised version of it was far from comprehensive and subjected to extensive editing.[40]

Five years before his death, Kuroda's publication of the original draft, a study he was clearly proud of, served as a polite rectification of the soft "violence" performed by the conqueror's cuts, omissions, and dismissals refashioning the work into echoing his own voice. The republished text's first section with Kuroda's handwritten annotations provides a more complex, personalized impression of ornithology prior to the American occupiers, along with a detailed systematic inventory of Japanese bird life peppered with personal anecdotes and earlier scientists' revelations. More so than the comprehensive handlist that Austin desired, Kuroda wrote a full history, inserting into the text his personality and that of cohorts, along with indices of lost aristocratic privileges, like the seigneurial domains where they leisurely observed and caught birds.

On its first page, a notable asterisk hand-drawn by the author appears on the top right-hand corner right after the first chapter's title "A History of Field Ornithology in Japan." At the page bottom, in a hand-written footnote, Kuroda provides the following annotation: "This historical description has been omitted in Austin & Kuroda, 1953."[41] This chapter, detailing early natural history contributions and then field ornithology in Japan's four main islands of Hokkaidô, Honshû, Shikoku, and Kyûshû, runs for a total of fifty-three pages, which comprises considerable omissions of text. Portions of Kuroda's prose reveal an uncomfortable position where once fabulously wealthy aristocrats now served as near-lackies for their overlords in an arrangement few knew was temporary. The chapter's last paragraph describes a late March 1948 expedition to Fukuoka and Kagoshima where Kuroda accompanied Austin. By referring to himself in the third person as "Kuroda, Jr." and "he," Kuroda attempted to erase his own voice or personal viewpoints prior to Austin's revisions, though when mentioning Austin and himself, he still used "we."[42] Jobs such as Kuroda's, where Japanese worked directly with Occupation officials and received steady income at a precarious economic time, were highly sought after, but still entailed adjustments for the defeated. Notably, the first chapter's last section on Kyûshû included descriptions like "[t]he influence of Kuroda was remarkable in Chikuzen, Fukuoka Prefecture the former domain of his family"[43] which revealed immense past wealth, former landholdings, and social capital enmeshed within. Here Rabasa's notion of "elsewheres" intrudes, as putative *lieux d'ailleurs*[44] that Austin never penetrated though a Yankee upper-class US occupation official.[45]

Befriending Austin: Colleagueship with a Benevolent Conqueror

Despite clear power differentials between putative conquerors and conquered, in postwar Tokyo, ornithologist Oliver L. Austin's leadership of SCAP's NRS Wildlife Branch serves as an intriguing lens into the reconstruction of Japanese conservation activities from the perspective of an outsider and former enemy of imperial Japan.[46] His experiences as a scientist working on wildlife policies in US-occupied Korea (1945–6) and Japan (1946–50) illuminate the war's impact on individuals and their political and social environments. Austin collaborated closely with elite Japanese colleagues, despite their ruined laboratories, burnt collections, inadequate shelter, and cached despair. For them and the American official, science and conservation provided a common language for establishing sometimes intimate connections amidst a ruined landscape.

The postwar peace allowed for new collaborations between scientists from Japan and the United States, transcending earlier wartime animosities. These human relationships were rebuilt amidst drastically changed environments impacted by war. The US-led Allied occupation radically restructured Japan's society, including conservation and wildlife management, and aided in leveling class distinctions. This encompassed land redistribution, ending tenant income for Japanese landed upper-classes, but helped support a rising middle class. In October 1946, landlords owning over certain acreages were required to sell land to the government, which then sold it at reduced prices to existing tenants who benefited until 1948 from rising inflation; this further harmed landlords, yet resulted in wealth redistribution to initiate a rising postwar middle class.[47]

Greater scientific networks of Americans and Japanese regenerated, while Anglo-American ornithological contacts rekindled quickly. Earlier prewar transnational "embedded networks"—a term borrowed from electrical terminology[48]—of a hierarchical nature slowly recoupled in Japan. What anthropologist and occupation official John W. Bennett (1915–2005) calls "'colleagueship'" (citing sociologist Everett C. Hughes [1897–1983]), or "'the establishing of intellectual links across political and cultural boundaries in the modern world,'" offers a useful model to understand revival of these oftentimes transwar relationships between Americans and Japanese. Here, military hierarchies, and those between conquerors and the defeated, were smoothed by a modicum of intellectual egalitarianism and desires for genuine professional interchange.[49] From 1948 to 1951, Bennett worked for the Public Opinion and Social Research (PO&SR) division of SCAP's Civil Information and Education (CIE) section, and became chief of PO&SR in 1949. Bennett indicated several important organizational goals:

> The PO&SR's mission was to plan and execute studies—often with the assistance of Japanese universities and government institutes—leading to social reform, or evaluating their role and progress. Studies of the agricultural land reform, prostitution, local political development, family structure and many other things were accomplished.[50]

Despite encountering initial obsequiousness and deference, Bennett noted that collaboration with Japanese proceeded amicably, though he felt communication styles differed due to cultural specificities, and noted how "discussions tended to be slow, involved, and replete with delicate suggestions instead of forthright disagreement or recommendation."[51] These measured interactions were likely responses of once influential elites carefully gauging occupiers' reactions prior to extending their own opinions or voicing concerns.

Lengthy interactions requiring time to develop also resembled Austin's experience, like his hosting of the *Ornithological Society*'s first meeting at his requisitioned house on February 19, 1947, where "At first they were all a bit ill at ease. But I put out lots of beer, coca-cola, cigars and cigarets [sic], doughnuts and crackers, and that soon broke the ice."[52] While refreshments helped establish conviviality amongst the twenty-eight Japanese ornithologists present, Austin's November 1946 Dai'ichi Hotel invitation preceded this meeting, plus close professional interactions with members including Hachisuka, Takatsukasa, and Kuroda since September 1946. When assembled in more informal company in winter 1947, "For the first time in my experience, they all relaxed, and showed signs of enjoying themselves, gathering in little knots and gossiping, and talking shoptalk, like similar groups would do at home."[53]

However, contrasting with most occupation officials like Bennett who never spoke Japanese and adjusted to an opaque linguistic atmosphere, Austin benefited from six months of intensive daily Japanese-language training school (including Japanese cultural and social norms) at Stanford University until 1945, and continued his studies at the now Monterey Institute from September to November 1945.[54] Austin's at-least intermediate level of Japanese in speaking and reading helped in improving encounters with Japanese counterparts, along with work translating ornithological texts (often containing English-language abstracts), while they also conversed in similar levels of English, while reading the language. Both Hachisuka and Takatsukasa boasted remarkable English fluency due to studies or extensive travel abroad, while most Japanese aristocrats received a modest English education at the *Peers School*.

Austin and his Japanese counterparts interacted using each other's languages to begin collaborating amicably in the intermediate postwar period by laying aside their nations' wartime violence. Ornithology and a new conservation focus provided them with mutual goals, and slowly permitted them to connect intimately as colleagues and friends in informal social gatherings and on specimen-collecting trips. Colleagueship, along with revival of earlier embedded networks, presents clues to why such connections failed in Korea, but succeeded in Japan. Nevertheless, scientific research and environmental policies were indelibly tied to the political realities of Japan's empire and formerly victimized colony, Korea—occupied by yet another foreign power—a fact that Austin clearly overlooked.

An American Ornithologist's Perspective of War's Impact on Avian Populations in Korea and Japan

Birds, in transnational migrations, care little about respecting territorial boundaries or political squabbles of human neighbors. However, in what atmospheric chemist Paul Croutzen calls the "Age of the Anthropocene,"[55] animals, plants, and other species can scarcely avoid human activity's pervasive detrimental impacts upon environments. When national enemies engage in warfare, birds are literally caught in crossfire, succumbing to soldiers' bullets or nets set by hungry home-front populations. Historian William Tsutsui also indicates harm by severe deforestation to faunal and avian habitats caused by disproportionate demand for ship-building timber, though the Pacific War spared over-fishing due to manpower shortages.[56] Yet, wartime often intensified earlier patterns of severe human-inflicted ecological stress in Korea and Japan.

During wartime and its aftermath, Austin circulated amidst such ravaged spaces. From 1942 until 1944, he served in the Western Pacific as an onboard US Navy communications officer, reaching Lieutenant Commander rank, and took the opportunity to hunt with a "riot gun" or "old sawed off twelve-gauge automatic" to collect over 2000 bird and bat specimens between missions on Pacific islands despoiled by combat.[57] He promptly sent these samples packed in bacon boxes by parcel post to ornithology curator "Jimmy" Peters at Harvard's MCZ, where Austin enjoyed personal connections due to his doctoral education at the university.[58] Following military service in the Pacific, where the war raged on, he was assigned to military government school at Princeton University and Civil Affairs Training (CAT) school at Stanford University to prepare him for future duty amidst American occupation forces in East Asia upon victory. This included training in understanding both Japan's language and culture, from Japanese-American *Nissei* [second-generation] and former Christian missionary tutors.[59]

Following this Japanese Studies crash course, Austin was shocked to discover his assignation instead to Korea as second-in-command of the Suwon Agricultural Experimental Station outside Seoul from late October 1945 until June 1946.[60] He was tasked to assist US Army Major Vic Shumber and work under the station's Korean director, Dr. Keh Ung Sang, who spent time running "a big silk weaving concern" as "president of the old Mitsui company" in Seoul instead of staffing his US-appointed position.[61] Though knowledgeable about agronomy, Keh took the opportunity afforded by American military forces to advance his own interests. During the US occupation of the Korean Peninsula's southern portion, to purge power structures of former Japanese colonial authorities, Americans relied on rehabilitated Korean appointees, who viewed their occupiers as yet another exploiting foreign power like the Japanese, so some concentrated on furthering personal material gain rather than aiding US interests.[62] The four Japanese engineers heading the station's sections on agronomy, animal husbandry, soil research, and administrative offices were left in place, but the station's Korean workers in subordinate posts now refused to cooperate.[63] In a March 2, 1946, letter to his father, Oliver L. Austin, Sr. (1871–1957)—a wealthy medical doctor who was also an amateur ornithologist running bird-banding initiatives in Wellfleet, Massachusetts at his Cape Cod beach home—Austin vented his frustration and used a highly derogatory term to refer to Korean underlings; undoubtedly, this mirrored imprecations he heard from US military counterparts: "to Hell [sic] with trying to make the damn gooks [sic] get their experiment station going. They're going to do just as much as they want to do, and no more, and all the urging and exhorting and explaining and ordering is vain, useless and futile. You're just banging your head against a stone wall."[64] Here, he absorbed earlier Japanese colonial ideas about Koreans' alleged laziness and sullen recalcitrance against Japanese authorities, along with Anglo-American ideas of racial superiority reminiscent of "The White Man's Burden,"[65] while he increasingly turned toward Japanese around him, who as the defeated, obliged him with helpful respect. With Major Shumber's blessing, Austin decided to dedicate work to full-time ornithology due the station's allegedly incorrigible corruption and neglect[66]—motivated by personal gain not unlike his putative Korean superior.

As Austin fulfilled ornithological goals while expecting the station to run under minimal supervision, he took on as research assistant Kurosawa Shôju, Suwon Experimental Agricultural Station's former administrative head, who continued on in his position after most of Korea's Japanese overlords left during the country's decolonization. Kurosawa also helped Austin translate Japanese-language copies of *Birds* and *Wild Birds* gleaned from neglected Keijô (Seoul) University archives repurposed for American military storage after 1945.[67] In these journals, and other publications, he first encountered in print scientific contributions by future Tokyo-based Japanese collaborators as he labored with his assistant to translate Japanese terms into English. This allowed him to understand gaps in scholarship to systematically catalog Korean birds.

To obtain a collection of nearly 2000 birds representative of Korea (excepting the Dagelet and Quelpart Islands),[68] Austin ventured forth with his Browning A5 semi-automatic shotgun to hunt specimens in areas surrounding royal tombs, temples, and mountain forests, relatively free from human encroachment due to religious taboos and popular superstitions.[69] He also set mist-nets near the agricultural station to trap smaller birds, discovering once, to his dismay, a hungry Korean intruder had destroyed his precious net by cutting out a portion to get at its quarry. This prompted Austin to write colorful outbursts peppered with racist slurs in an April 9, 1946 letter to his father now indicating candid feelings about Koreans, who he vituperously expressed were "the lyingest [sic], stealingist [sic], crookedest, most improvident, selfish, wasteful, deceitful, filthy bunch of sons of whores on the face of the earth. I loathe them through and through."[70] Clearly, the American scientist deeply misunderstood ordinary Koreans' straightened postwar conditions, where protein from a netted bird likely prolonged survival for yet another day. Wherever Austin went, he found astounding lack of species diversity for any wild fauna, decimated by starving populations, but despite this challenge, took the opportunity to use Italian mist nets[71] and a shotgun to hunt and prepare bird specimens for his study "Birds of Korea"; this was eventually published in 1948 after extensive revisions and collaboration with Japanese ornithologists who he befriended upon his later Tokyo posting.[72] To process his extensive collections, he also decided to hire as assistant his servant Kim Sung Jang, who he called his "houseboy,"[73] and trained him to prepare bird skins for shipping to the MCZ. Despite assuming a semi-colonial relationship in serving another occupier, Kim appreciated valuable meat remaining from skinning birds to bring home to his family, which consisted of his aging mother and pregnant wife. However, when Kim and his wife fell ill, they received precious antibiotics from Austin, though the ornithologist sadly lacked medical training to address Kim's wife's postpartum infection and that of her infant girl, who died soon after birth.[74]

Due to Austin's ignorance of Korea's layered history in dealing with foreign occupiers and his own racist attitudes no doubt reinforced by his Army superior, he adopted a similar colonial mentality common to other US military members involved in its postwar occupation, and blamed the Korean national character for stymying of his work and the dearth of birds and animals, both of which were affected by decades of colonial land exploitation and Japanese resource extraction for the Asia-Pacific War. On March 23, 1946, he complained about environmental conditions in the US-occupied southern portion of the Korean peninsula, while also criticizing American authorities, who he believed attended to only trivial matters:

> The Koreans themselves are wasting their country. The higher-ups are madly playing politics, each crookedly trying to line his own pocket, and to Hell [sic] with his brother, much less say his neighbor. The lower down are cutting the forests flat, denuding every bit of the land itself with no thought of the future. What do the Americans do about it? They issue an edict that all traffic from April 1 will drive on the right hand [sic] side of the road instead of the left.[75]

When Austin arrived in postwar Korea, the country already suffered over seventy years of negative impacts from foreign intervention, colonial rule, and the natural environment's unsustainable exploitation. After the 1876 Treaty of Kanghwa, resulting from Japan's use of US-inspired gun-boat diplomacy, rapid socioeconomic change accompanied this forced "opening" of Korea for trade. Like Meiji Japan, but on a smaller scale, Korea was slowly incorporated into Northeast-Asian transnational trade networks in telecommunications and transportation infrastructures, which intensified after Japan annexed the country to its empire in 1910. Such development detrimentally effected the environment as urbanization and rural exploitation for burgeoning Japanese domestic rice consumption swallowed up land. With infrastructure building and changing land-use patterns (codified by Japanese colonial rule), rural villagers increasingly encroached upon mountainous areas near Seoul and other growing cities to rake up forest undergrowth for burnable fuel and green manure. In his same April 9, 1946, letter to his father, Austin bitterly complained as a frustrated conservationist about this practice, which he added to litanies of accusations in his profanity-laced rant against the Koreans who he was paid to assist.[76]

Although clearly performed as a survival measure, Korean villagers' efforts to gain burnable fuel led to progressive deforestation and hillside denuding, causing desertification and terrain change. The Japanese empire's high taxes and wartime agricultural requisitions also prompted hungry Korean farmers to hunt wildlife to protect crops and supplement diets—leading to stark drops in bird, rabbit, and deer populations. In early twentieth-century Japan, similar effects resulted from tenant farmers' desperate measures to combat famine. In colonial Korea, Conrad Totman also mentions impacts from a wood-boring beetle epidemic on pine trees, with forest conservation and pest removal once managed by village associations.[77] Such earlier Yi Dynasty (1392–1910) conservation measures handled by villages apparently disappeared during the Japanese annexation period (1910–45), and domestic Japan's rice and resource requirements during wartime further accelerated destructive environmental exploitation.[78] For Japan also, earlier patterns of ecological devastation accompanying industrialization preceded total war exigencies from the late 1930s until 1945 and led to similar negative stressors on wildlife and bird populations.

In letters home, Austin also initially used slurs to refer to Japanese, which he likely acquired during his US Navy wartime military service. However, as the months passed in Suwon, the scientist increasingly praised them for efficacious prewar forestry management and wildlife conservation, revealing his privileged position, feelings of appreciation for Kurosawa's assistance, and dismissive attitude toward colonial Korea's conditions of inequality. His descriptions clearly appropriated an earlier imperial Japanese colonial gaze, while revealing the persistence of wartime American prejudices against Japanese by using the once commonly used racist term "Jap." Though Austin rarely used this term in private correspondence after his subsequent Japan posting, a passage from nearly six months into his Korea assignment reveals rising admiration for Japanese methods and transformations in his perspective:

> From what we've seen, the Japs [sic] did a pretty good job of putting this land back on its feet in the forty years they were in power. They did it with an iron fist, the only kind of treatment that these blankety-blanks (I am running out of suitable vilification) understand. But forty years of conservation and planning and education weren't enough to teach the Korean the first principles of how such an over-populated land must be managed to support its peoples, and they've gone right back to their principles and practices of fifty years ago, the moment the rigid control and planned economy was lifted. They've despoiled and ruined more this past winter than can be replaced in half a century.[79]

Interestingly, like his Japanese counterparts during Japan's imperial period, through ornithological activities, Austin imposed an avian imperialism upon colonized peoples now hosting the American occupiers.

Once in Tokyo, Austin notably spared Japanese of the vituperative tirades launched against Koreans in letters home to his father, and even praised the men he encountered. What accounted for this difference in viewpoint? After all, Korea was part of the Japanese empire, and Tokyo was its exact political, economic, and cultural epicenter. Why had Austin adopted Japanese colonial mentalities in Korea, but then treated defeated Japanese in Tokyo with greater reciprocity and respect? Clues can be found in Austin's relationship with his assistant Kurosawa, where he likely first encountered residual Japanese attitudes toward their colonized subjects, now complicated by imperial Japan's defeat and expediencies to curry favor with American occupiers for protection and preferential treatment. The American scientist's renewal of positive working relationships with Kurosawa in postwar Tokyo within his earlier embedded network with the Japanese administrative head of the agricultural research station, and later flourishing of collaboration into genuine colleagueship, points toward the sociological mechanisms at work between individuals involved in contemporaneous scientific research.

Renewed Ornithological Collaboration in Early Postwar Japan, 1946–7

Following what Austin perceived as an unsatisfying stint in Korea, the ornithologist was thankful for his subsequent Tokyo posting in a plum position, designated as the civilian equivalent to Lieutenant Colonel to head the NRS's Wildlife Branch of the Fisheries Division. For this arrangement, he relied on assistance from extensive social networks of well-positioned individuals. Austin's friend Claude M. Adams (1907–74), who once worked as a California cannery inspector, and now helped re-establish Japanese fishing and whaling industries for the NRS, finagled him an interview for a Japan-based position.[80] Adams's boss was geologist Lieutenant Colonel Hubert G. Schenck (1897–1960), who conducted by phone what likely resembled more a conversation than interview with Austin.[81] Adams, who had commissioned with the US Army as a captain during the Second World War, had met Austin, who had served with the US Navy at a similar rank, when they studied at Stanford University's CAT School and received Japanese language instruction at the Monterrey Language School during the last years of the war. This fateful phone conversation finally allowed Austin to work in Japan, but his bid for employment was certainly improved because his family and the Schencks had established amicable relations prior to his Tokyo sojourn.[82]

According to Schenck, the NRS was established on October 2, 1946 to inform General Douglas MacArthur on "policies and activities pertaining to agriculture, fisheries, forestry, and mining (including geology and hydrology) in Japan" and consisted of scientists from "the Army, the Navy, and other governmental agencies, from academic and research institutions, and from private industry."[83] Their role included coordinating surveys and making reports to recommend proper measures to revive Japan's economy, examine land reform, and manage natural resources to pay for the Occupation and return Japan to environmental sustainability. These were priorities toward which Japanese scientists in various fields also lent their expertise; arguably, such collaborators were vital to the NRS mission. The NRS served as a putative institute for policymaking consultation for the Allied Occupation's top authorities.[84] NRS activities resembled those undertaken by Japanese in colonies like Korea to effectuate imperial control, and could be likened to the past Tokyo-based *Research Institute for Natural Resources* where natural resources were mobilized for war. However, for a democratic system and a New Deal-inspired reorganization introduced by the American authorities to actually work, they required assistance from Japanese elites influencing lawmaking in the National Diet. These men served as classic subalterns, a term referring to the valuable roles of "go-betweens"[85] amongst colonizers and colonized, also occasionally termed "comprador class." The origins of this practice harkened back to the Portuguese in early sixteenth-century Africa and Spaniards in mid-sixteenth-century Latin America, who initiated cooption of local elites for their colonies, whereupon other colonial powers soon mimicked them. Luckily for Austin, many Japanese elites also maintained deep interests in ornithology.

Austin's task for the NRS Fisheries Division's Wildlife Branch originally consisted of finding new sources of animal protein for a still-starving population[86]—possibly through whale meat or fowl—while addressing concerns about wildlife management and conservation. To effectuate personal ornithological aims while fulfilling his military employer's needs, Austin was assisted by Japan's most powerful political and imperial elites, including Diet members like Yamashina or those with behind-the-scenes influence and imperial family connections, including Takatsukasa, known as "*chô no kôshaku*" [the "Bird Prince" (actually Duke)], who helped push SCAP's desired laws through the Diet. Yet, the American scientist was also irascible at times, as indicated in his *The Auk* obituary: "There was certainly something about Oliver's strong personality and opinion—you really liked him or didn't."[87] Certainly, Austin's work on systematics and categorizing birds in Korea familiarized him with these Japanese aristocrats' ornithological writings, whose own scientific expeditions to Korea after annexation enmeshed their scientific endeavors within privileges garnered by Japan's imperialism, much like those accorded to Austin by SCAP.

To his relief, Austin's earlier nine-month Korea tour earned him three months Stateside leave prior to his Japan posting and granted him a coveted air ticket back to Tokyo. On September 4, 1946, after fifty flight-hours, Austin landed at Haneda Airport, a new airfield constructed atop Kuroda Nagamichi's famed waterfowl sanctuary, whose portions still exist as the Tokyo Port Wild Bird Park harboring 226 species of birds.[88] The plane's route to Tokyo from Fairfield, California, via Pearl Harbor, Johnston Island, Kwajalein, and its last Iwo Jima stopping point before reaching Japan's mainland mirrored a reverse itinerary for earlier US Second World War island-hopping campaigns. Austin's final destination was the city's Shinbashi-area Dai-ichi Hotel, serving as billeting quarters for GHQ or SCAP field-grade officers, but, luckily, his new lodgings also allowed Japanese access for his new friends. First on his agenda was finding his former assistant Kurosawa from Suwon, Korea; an efficient search by Tokyo police and post office officials found the young man, who reported for work the next day at the Wildlife Branch on the Mitsubishi Shoji Building's third floor.[89] Paid by Japan's government on Austin's behalf, he served as translator and interpreter, and accompanied the SCAP scientist on several expeditions, helping him understand the history of duck liming and songbird mist netting[90] in rural Japan, and eventually became his valued colleague.

Amidst helpful conquered people, and envisioning fine prospects for scientific work and collaboration with Japanese scientists, whose work he read in Japanese and English with Kurosawa's assistance in Korea, Austin's spirits lifted considerably. In a September 8 letter to his father, Austin outlined the coming year's plan:

> I'm starting off with an investigation of Japanese game laws, and methods of enforcement, which will lead to a general study of Japanese wild-life conservation measures. Then, we've got to find out what they've done and are doing with game management, and I'm going to concentrate after that on two major subjects, the upland game birds and the waterfowl.[91]

Already familiar with their names from *Birds*, Austin soon endeavored to meet the Japanese scientists who would serve as his collaborators—aristocrats who spoke English and shared privileged social backgrounds with the upper-class Yankee, including the ornithologists Hachisuka, Yamashina, and Takatsukasa, who published internationally recognized studies on birds of Japan, China, the Philippines, India, and Australia.[92] Such studies of birds in East Asia and the Pacific echoed a "collecting imperialism"[93] also practiced by ornithologists in the British Empire and US territorial possessions.

One of the first Japanese scientists meeting Austin, Hachisuka proved his indispensability in ornithological matters to the US Occupation official, and quickly befriended him. Undoubtedly, a potentially well-paid position with SCAP also motivated him. In Austin's September 19 letter to his mentor Jimmy Peters, he described a budding friendship with Hachisuka—"I like Hachi very much, and I think he likes me"—but mentioned that his new colleague suffered from "the added burden of personal troubles which have added to his worries."[94] These included Hachisuka's loss of land wealth and squabbles with his estranged Japanese-American wife, who alleged his secretary Usui Mitsue (unknown) was actually his mistress. Yet, her secretarial presence in taking notes during meetings and other occasions was necessary, since Hachisuka's hearing was rapidly failing. Austin quickly learned more about Hachisuka's personal situation during forays into Japan's countryside to investigate postwar bird populations.

The American official then strove to acquaint an even more influential Japanese ornithologist. In a September 20, 1946, letter to his father, Austin described a productive initial meeting with Yamashina at his intact Shibuya laboratory and specimen museum—extending into the next day until Yamashina needed to report to the Diet.[95] Presumably, Yamashina would mention his amicable visit with the American to House of Peers political colleagues, and garner support for conservation efforts that he personally favored since the mid-thirties. Austin's chatty correspondence that same day with his father, laced with profanity and rarer use of the racial slur "Jap," helps illuminate his strategy of utilizing connections and negotiating skills from these Japanese scientists, who he increasingly appreciated in balancing increasingly complicated American military politics:

> My first big job is going to be to get the Japanese game laws revised, in which I'll have the combined help of all the Jap [sic] ornithologists. The laws are a mess, and are primarily responsible for the terrifically bad conditions of Japanese wildlife [sic]. The open season runs from 1 October to 30 April, and the only birds protected are those with religious significance such as cranes and storks. Thrushes, finches, and most small birds are considered game birds. At that, I think it's going to be easier for me to get the Japs [sic] to alter their own laws of their own volition on my suggestion rather than to force them to do it by getting SCAP to issue a directive to that effect, for I'll be bucking the high American brass that would just as soon go hunting right through the nesting season, and to Hell [sic] with the Japanese birds. It's going to take some damn careful and tactical and political handling on my part, but it's well worth doing, so here goes.[96]

In this epistolary rant, Austin still referred to Japanese counterparts with the wartime racial slur "Japs," but began to recognize political imperatives in collaborating with them to solve mutual problems like continued wildlife over-harvesting, sustained by protein-starved Japanese and US soldiers believing in their right as vanquishers to shoot any animals they encountered. In an American perspective, smaller birds like thrushes and finches were considered song-birds for enjoyment by bird-watchers, and not for food, as for hungry Japanese in immediate postwar Japan. Austin believed that he would experience a protracted struggle, since the long game bird hunting season in Japan impinged upon sustainable conservation, while certain US military officers viewed duck hunting as the quasi-imperial privilege of conquerors.[97] Therefore, he hoped to collaborate with elite Japanese allies to make conservation a cooperative endeavor gaining well-connected political proponents in Japan's Diet to change laws from within. However, to effectuate his plans, the American scientist first needed to understand conditions affecting the defeated nation's bird species, which required compiling existing knowledge on birds in Japan and enlisting his new Japanese allies for fieldwork.

On September 23, Austin wrote a nine-page letter detailing wartime effects upon Japanese bird research and its scientists to ornithologist John Todd Zimmer (1889–1957), the AOU journal *The Auk*'s influential editor, and Curator of Birds at New York City's AMNH. This letter informed Austin's 1948 SCAP report "Japanese Ornithology and Mammalogy during World War II."[98] On the 25th, he finally received SCAP permission to allow Hachisuka to accompany him on a thirteen-day expedition to observe wildlife refuges in Kôfu, Yamanaka, Numazu, Tsuchiura, Chôshi, and Ajiki.[99] Austin, despite his often racist portrayal of Koreans, learned an important lesson to co-opt local elites into his plans. For Japan, similarities in class and education with local elites made this process easier. Austin's actions composed a concerted strategy to avoid the problems encountered in Korea, and enlisted support of powerful Japanese allies by building relationships eventually leading to colleagueship.

Though Austin and Hachisuka were similar ages and both wealthy elites in their respective countries, they radically differed in situations of power. By the late 1940s, Hachisuka's hearing worsened such that, when assisting Austin in field research to assess bird populations, the latter would "shout at him all the time, which is a nuisance as well as embarrassing to both of us."[100] Here, the former Peer now assisted a past enemy hired by the occupation forces, so Hachisuka's reticence in responding was likely an attempt to control frustrations in balancing his desire for renewed connections with the Anglo-American ornithological world via Austin with his own personal pride. To ameliorate this situation, Austin begged his father, a prominent New York society physician, to send Hachisuka a pair of newly invented Sonotone hearing aids: "He's a swell little fellow, and he's going a great deal for me. In fact he's the biggest help I've found in Japan, and if I can help him in any way, I'm glad to do it."[101] Though Hachisuka was indeed short, barely measuring past five feet, and thus may have seemed physically slight to Western observers, Austin's enthusiastic descriptions betrayed a sense of superiority and paternalism toward a social better in a diminished political position. Here, imperial masculinity was redefined as expedient hierarchical colleagueship for Hachisuka.

Such relationships were common where US officials became benefactors after receiving what they needed from defeated Japanese; historian Naoko Shibusawa asserts, "By conceiving the bilateral relationship in the mutually reinforcing frameworks of gender and maturity, many Americans began seeing the Japanese not as savages, but as dependents that needed US guidance and benevolence."[102] These aspects allowed US occupation forces to circumvent racial difference and perceive them as "honorary whites" amidst a greater ideological mission. However assistance to the white occupiers temporarily wounded his pride and impinged upon his masculinity, Hachisuka also viewed Austin's Japan arrival as an opportunity to steer his country's future in a positive direction, gain allies abroad, and mentor new generations of scientists.

In an interesting transwar imperial continuity, Hachisuka recommended Nakamura Yoshio as an experienced collector to join Austin's expeditions, considering his exemplary service to Hachisuka in the Philippines, Manchukuo, and southern China. To reciprocate the favor, in 1947, Austin allowed Nakamura's son Takatsukasa to live in his requisitioned Shibuya house, where he informally learned ornithology and skinned specimens during university studies. For nearly a year, Takatsukasa served as the Austin household's *shosei* [disciple or apprentice], helping his sons Timmy and Tony prepare for school and practice their lessons. In 1948, following tensions with Elizabeth, Austin's wife who viewed Takatsukasa as more a servant than apprentice, Austin arranged for him to move into Hachisuka's Atami dwelling where he flourished. Takatsukasa then served Hachisuka as *shosei*, first organizing his specimen collections and raising chickens, and then, as his secretary, in increasingly intimate ornithological activities.[103] Late 1940s photographs hint at the possible nature of Hachisuka's ménage: shot by Takatsukasa, one reveals Hachisuka stretched out in casual shorts lounging on his second-floor porch amidst an exotic background of Mexican agave plants, while another taken unawares in a candid shot by Hachisuka reveals a young Takatsukasa in the outdoor hot springs bath holding a rare monkey.[104] The camera reveals their tenderness and mutual appreciation, and perhaps even love.

Despite complex negotiations of power relations between the American occupiers and vanquished Japanese, Austin, Hachisuka, and Nakamura Yoshio appeared friendly, and well-complimented each other's roles for Austin to effectuate his SCAP mission. Accompanied by Hachisuka and Nakamura, Austin, during numerous expeditions throughout Japanese backwoods country from late September 1946 until late 1949, discovered aggressive harvesting with efficient, but cruel methods, using lime or mist-nets and crushing the birds' skulls, later illustrated in detail with historical background for the SCAP reports "Mist Netting for Birds in Japan" (1947) and "Waterfowl of Japan" (1949).[105] From September 26 to October 7, accompanied by Hachisuka, who had recently received permission from GHQ to serve as the American's assistant, Austin traveled throughout Yamanashi and Shizuoka prefectures to investigate wildlife and birds—familiar territory for Hachisuka, whose retainer and collector Nakamura lived in Kôfu.[106] In mid-November 1946, Austin embarked on several weeks of expeditions with Japanese collaborators in Shizuoka and Gifu, including Nakamura and Ori'i, who aided him in assessing wildlife conditions and Japanese usage of mist-nets near *toyoba* [bird-calling decoy sites]; these amounted to 1,000 in Gifu alone, "with a 10% increase this year because of the food problem, and so many returning veterans looking for work."[107]

In Austin's descriptions, Japanese methods of killing birds en-masse, like pressing on their sterna, skulls, and backbones, or simply sitting on a full sack, represented for him residual feudal barbarism reminiscent of imperial Japan's wartime conduct. These accounts were undoubtedly colored by the ornithologist's own Second World War military service. From 1942 until 1944, when Austin was posted on a ship in the South Pacific as a US Navy communications officer, he certainly heard stories of Japanese atrocities exacted upon American troops. In a December 15, 1946, letter to his father, Austin sarcastically recounted how ducks encountering *mochi-ami* [sticky line nets] during *tsuri-gamo* [duck fishing], were caught: "Bird swallows the hook, and there he stays, the trawl being staked down firmly. Nice people. How the members of the Brookline Bird Club[108] would appreciate this!"[109] He also explained how more efficient methods of killing ducks had developed in the prewar period, when markets for their meat arose in urban centers:

> [A] big dealer from Tokyo showed them four other better ways …. The quickest and easiest method takes a strong thumb, and considerable knack. Shimada [a duck limer] just went down the net and killed each bird in a second or so, by simply pushing the back of the skull in with his thumb. You could hear it crack. I couldn't do it. Couldn't even bend the bone. But he shoved them in like so many egg-shells [sic], and says you have to do it just right. If you push in too far so the bird bleeds, it won't die for quite some time. The second method is to bite the back of the skull, which several of the boys were doing, and which is just as fast, just as effective, though perhaps harder on the teeth …. My, what these clever people will think of.[110]

For Austin, the Japanese bird harvesters' workaday "barbarity" in processing kills—sometimes with their teeth—clearly contrasted with more "honorable" game-bird shooting by hunters or ornithologists, carefully collecting select specimens for study as cosmopolitan gentlemen of science. One of his primary goals was to outlaw indiscriminate songbird mist-netting, while only allowing shooting of relatively abundant game birds. Arguably, this effected common people greater than wealthy gun owners on leisurely weekend pheasant hunts or specimen collecting forays. For SCAP, Austin's notes on these excursions culminated in his 1948 report on "Japanese Wildlife Sanctuaries and Public Hunting Grounds."[111]

Countering initial preconceptions, Austin discovered that wartime represented little aberration in aggressive mist-netting of song birds and duck-liming, whose practitioners boasted an over two-hundred-year history. In Austin's letters to his father, he lamented Korea's dearth of birds, but found conditions in Japan far worse, sustained by venerable traditions of hunting edible avian populations for the delicacy trade: "There are a hundred times more birds in Korea than in Japan, and you know what I thought of Korea."[112] The opening sentence of his 1947 report expounds, "[t]he abnormal scarcity of small insectivorous birds in Japan today, resulting from unregulated slaughter in an overpopulated land, is felt economically as well as aesthetically," and indicates "[t]he principle instrument by which Japan's birds have been depleted is the mist net, a unique Japanese invention about which the rest of the world knows almost nothing."[113] The scientist's castigating voice sets its tone, blaming the Japanese for squandering their bird life as natural resources. Soon afterwards, he begins a detailed analysis of mist-netting techniques, and an extensive history of their use. Invention of the *kasumi-ami* [mist net] for songbird harvesting purportedly began in the seventeenth century during the Tokugawa era under an Ishikawa domain *daimyô*.[114] For the "pickled thrush industry," skinned birds were preserved in barrels of bran mash and sent onwards as New Year's specialties into cities, where they were enjoyed as *yakitori* [roast bird skewers].[115] Harvesting continued into the late 1940s prior to the Occupation's ban.

Ducks and other fowl were also a favored quarry for centuries in Japan, while elites and prefectural governments boasted reserves. Feudal-era Shoguns ran *kamoba* [duck preserves] within seaside estates like the Hama-Rikyu detached palace in Edo

Figure 7.1 Oliver L. Austin Jr., circa 1947–1950. © Oliver L. Austin Photographic Collection, permission granted by curator Annika A. Culver. https://austin.as.fsu.edu/files/original/d19ff1a7b13e950d0a694e7e2995e98b.jpg.[116]

[now Tokyo], taken over by the Imperial Family in the Meiji era until postwar reversion prompted its opening to the public on April 1, 1946. These estates later served as places where Japan's postwar government entertained foreign and domestic dignitaries with popular duck hunts and subsequent *sukiyaki* [meat fondue] feasts; invitations to an imperial duck hunt were coveted by high-ranking SCAP members. However, Austin was incensed when he received an invitation on April 17, 1947, when the NRS had already closed the hunting season by law.¹¹⁷ The imperial family was actually exempt from such strictures, but the American official's attendance as Wildlife Bureau director would have represented contradictions to his conservational aims. Austin eventually attended with his family, and in a February 1950 letter to his father mentioned the Imperial Household's send-off gift of a *kamoshaku-ami* [spoon-shaped duck net].¹¹⁸ Japan's imperial family still maintains important symbolic connections to this heritage; Crown Prince, and now Emperor, Naruhito, took his future wife Masako hunting at an imperial *kamoba* on their first date.¹¹⁹ In areas like Yamanashi and Kôfu, prefectural governments allotted preserves for commoners to engage in duck-liming in swamps and ponds in government forests.

Birds were also harvested for their feathers. From wartime into the postwar period, Japanese hunters' associations donated feathers to the government via drives organized by the Ministry of Agriculture and Forestry to outfit disabled veterans with down pillows and blankets. In the late forties, prefectural governments forwarded such feathers to Tokyo, which sent them onwards to the Aomori-based *Hane Kôgyô* [Feather Industry] Company to manufacture "pillows for American forces," which required a mix of duck *and* songbird feathers for resiliency and softness.¹²⁰ Amidst scarcity, like wartime, mist-net exploitation of songbirds usually went for village use, while plusher years allowed conversion of bird protein and feathers into extra income.

Armed with knowledge of historical conditions impacting harvesting of songbirds and game, and to enlist Japanese assistance for conservation efforts, Austin decided to revive the prewar *Ornithological Society* to draw upon talent from prewar embedded networks of Japanese ornithologists. This organization was originally formed in 1912 by Iijima Isao, a celebrated Tokyo Imperial University zoology professor known as "the father of parasitology in Japan," who had trained at University of Leipzig in the 1880s, and was a noted ornithologist.¹²¹ His students Takatsukasa and Uchida Seinosuke served as the organization's second (1922–46) and third (1946–7) heads amidst its wartime dissolution, while Kuroda Nagamichi, a former student, served as fourth head (1947–63) after its postwar revival.¹²² According to ornithologist Tôzawa Kôichi, the *Ornithological Society*'s prewar incarnation welcomed a disproportionate number of aristocrats; however, he believed that its structure headed by a prominent scientist represented not so much feudal hierarchies, but rather, promoted parallel cooperation and support of organization members, while allowing it ample financial contributions.¹²³

On November 23, 1946, Austin invited his future Japanese colleagues to the *Ornithological Society*'s first postwar meeting in a banquet room at the Dai-ichi Hotel,¹²⁴ which remained his lodgings until a requisitioned Shibuya villa, or "Req House #665," became serviceable for dependents arriving in late December. This large estate, with Western and Japanese wings, belonged to the widow Ariyoshi, who lived in its Japanese-style wing partitioned by a garden wall,¹²⁵ and whose deceased husband was Yokohama's mayor, and Korea's former vice governor general¹²⁶—inserting former and current inhabitants squarely within imperial projects of their respective countries. Though Austin clearly took her photo, he mentions little about her in letters to his father, so it appears they rarely maintained contact.¹²⁷ The estate's western portion thereafter housed *Ornithological Society* meetings during Austin's Japan sojourn. The American scientist paid $15 ($211 in 2020 dollars)¹²⁸ for the fateful dinner,¹²⁹ a timely investment enabling future success of his work in Japan by fostering friendships and engendering colleagueship. This meeting also proposed revival of the ornithological journal *Birds*, defunct since paper shortages and air raids stopped publication in September 1944.[130]

The fifteen attendees included Japan's finest ornithologists (Yamashina and Takatsukasa), zoologists, and biologists (Uchida).[131] Most served the wartime regime with the *Research Institute for Natural Resources* whose bird and mammal studies furthered state interests. However, the group lacked Count Kiyosu Yukiyasu, who only visited Austin on April 16, 1947, following the American ornithologist's well-publicized April 10 speech initiating "Bird Day," explaining that he "was out in the country, and hasn't been able to get into town before."[132] When they finally met, Austin described him as a "[v]ery nice chap, very much of a gentleman, though his English is very poor, and we had to do most of our talking through an interpreter."[133] To mollify Austin for the possible oversight, Kiyosu extended assistance in an area where Austin maintained deep interest: "He spent about three years on one of the Izu Islands, and gave me a lot of good dope and useful information about it."[134] Also, the Count's wartime expertise was crucial for Austin's own SCAP mission to investigate options supplementing the Japanese postwar diet with protein. In November 1943, Kiyosu published the wartime study "On the Birds for Food Resources," "giving description, habits and distribution of those species considered suitable for eating, and tabulating the annual kill of each for the past fifteen years."[135] This study provided valuable historical data for inclusion in later NRS reports for SCAP.

Participants at Austin's banquet discovered that the Second World War provided an opportunity for continuing research and collaborating with far-flung authorities in often exotic locations. With the Pacific War's intensification in summer 1942, Hachisuka snuck into a Tokyo hotel to pass his "Handlist of Japanese Birds," produced in late 1941, to Australian diplomats interned there, who were subsequently deported by ship to West Africa. Thereafter, the list circulated for two years via the Philippines until it reached James Cowen Greenway, Harvard College MCZ's Assistant Curator of Birds,[136] from whence it traveled to evolutionary biologist Ernst Mayr, who promptly microfilmed and disseminated it to prominent American libraries with strong ornithology components.[137] While the Japanese Handlist rode the wartime high seas, Austin enlisted in the US Navy to fight the Germans, but received orders in July 1942 for three months of communications school training, followed by active duty in the Southwest Pacific for twenty-two months. Whenever his evacuation transport docked, he ventured ashore to collect specimens for the MCZ after the Marines finished defeating entrenched Japanese troops, and soon discovered two new bat species.[138] All fifteen participants at the fateful Dai-ichi Hotel dinner, including the American ornithologist, were deeply enmeshed within earlier discourses of imperialism or national conflict, where science profited the state or its interests, and discoveries of new species prompted national and personal pride.

In her study on scientific nationalism in imperial Japan, Hiromi Mizuno asserts that "Science is never simply about experiments in a laboratory or new discoveries of phenomena and laws of nature. It is a highly contested social field in which its legitimacy, territoriality, and definition are constantly challenged, negotiated, and asserted by various participants in the discourse of science."[139] As Austin's relationships with these Japanese scientists suggest, this "highly contested social field" continued into the postwar era, with social relationships and parallel elite hierarchies arising from colleagueship soon smoothing over acceptance of a "democratic" discourse of conservation under SCAP's authoritarian tutelage, while the Japanese also helped steer this process in a not completely one-sided transfer. Notably, for Americans, "democracy" was now a convenient buzzword to eliminate so-called prewar "feudal" remnants, and allow a bright, new postwar ideological orientation to flourish. This rhetoric prompted rapid, unprecedented change, and promoted peaceful rebuilding under liberal reforms implemented under New Dealers' idealistic guidance. These were characteristics of a hopeful atmosphere of intellectual exchange enveloping Japanese scientists around Austin.

Wildlife and Conservation in Postwar Japan—Austin's Contributions

To address issues prompted by the November 1946 Dai-ichi Hotel meeting of the newly revamped *Ornithological Society*, and further prevent endangering of avian populations more broadly, Austin recommended aggressive education campaigns and US-inspired modes of conservation by using a particular language and ideological rhetoric emphasizing American-style democracy and its favorable links to science. A 1918 Japanese law originally prohibited commercial hunting of birds, except for sport; however, it failed to address mist-net harvesting or liming, arguably more detrimental on a larger scale since these were geared toward harvesting for the market. Though he felt it was a top priority to eliminate these traditional Japanese practices, as an avid hunter, Austin did not favor a complete ban on shooting game birds. Instead, he desired stricter hunting laws to prevent wholesale slaughter without bag limits like that embarrassingly exhibited in the early occupation period by undisciplined military brass and personnel who abused their power by shooting thousands of ducks in Niigata duck preserves on traditional netting grounds.[140] As historian Christopher Aldous asserts, American military authorities and prefectural Japanese authorities often stymied Austin's plans in complex entanglements of competing power structures.[141]

However, for Austin, socializing with powerful collaborators positively influenced potentially divisive politics, with close, personal connections and similar class backgrounds alleviating potential tensions. Principles of colleagueship also worked for the ornithologist's relations with high-ranking US military members: notably, General MacArthur's son Arthur (b. 1938), accompanied by military police, attended Cub Scout den meetings in Austin's household.[142] Concurrently, Austin befriended influential Japanese elites by inviting them as guests into his home and hosting subsequent *Ornithological Society* meetings in his living room, complete with copious postwar luxuries—cigarettes, snacks, and libations.[143] Yet, some Japanese ornithologists had earlier cultivated personal relationships with American military authorities preceding the Occupation, like Hachisuka's friendship with Colonel L. R. Wolfe, an amateur raptor specialist who liked to talk birds with Austin during his Japan posting. In 1929, Wolfe enjoyed hunting with Hachisuka in the Philippines near Lake Tal, and in the mid-1940s after the war, habitually visited Yamashina at his Shibuya laboratory to peruse his remarkable raptor specimen collection. Wolfe later reviewed Kuroda's descriptions of birds of prey and extended advice for Kuroda's 1949 manuscript prepared for Austin.[144] Because Austin allowed him to reconnect with old friends and make new ones amongst like-minded bird enthusiasts, Wolfe championed him amidst opprobrium from trigger-happy military brass like Brigadier General Joseph ("Jumping Joe") May Swing (1894–1984) and Lieutenant General Walton Harris Walker (1889–1950).[145] Austin pointedly viewed these men as "to-the-victor-belong-the-spoils advocates"[146] opposing hunting restrictions and disregarding their actions' effects upon local populations, potentially hindering Occupation efforts to enlist Japanese support for broader initiatives. When General Muller called Austin into his office to indicate bag limits for geese were "an injustice to the Occupation," MacArthur, grateful for the Austin family's warm welcoming of Arthur into their home, himself praised the scientist and his conservational work.[147] Certainly, Japanese scientists' influence like Hachisuka also assisted Austin's creation of proper connections amidst an oft-stymying US military hierarchy. Ultimately, with helpful allies from both the occupied and occupying forces, Austin succeeded in pushing through his game and hunting laws within Japanese and American political channels. His accomplishments were thus attained by symbiotic relationships with Japanese aristocrats like Yamashina and Takatsukasa, who could influence the Diet, and MacArthur's personal approval, receiving one of the occupation period's only two personal commendations from him.

Embarking on these goals also required collaboration between American and Japanese educational authorities to inspire a new conservational consciousness amongst the public to ensure future protection of birds and wildlife. In a December 3, 1946, meeting with Japan's Ministry of Education officials, Austin recommended four NRS courses of action: establish an effective Department of Wildlife Conservation under the Ministry of Agriculture and Forestry; revise game laws to prohibit sale of wild birds, eliminate mist-netting, protect all birds (besides "pheasants, quail, grouse, the pigeons and doves, the rails and gallinules, and the ducks and geese"), and shorten the hunting season to sixty days with a January 15 end date; establish multiple inviolate wildlife sanctuaries; and initiate an intensive educational campaign.[148] He also proposed revision of game laws to "prohibit the taking of birds by any method other than with firearms,"[149] which essentially outlawed the highly effective traditional Japanese practice of duck-liming, and severely curtailed Japanese access to hunting due to the defeated population's now-limited access to firearms and ammunition.[150] Yet, this also meant that the new game laws would not prohibit military occupation personnel from hunting fowl. The authorities present included Shimizu Saburô (1918–2004), Director of Scientific Education; Wada Hiroö (1903–67), Minister of Agriculture and Forestry; Takatsukasa, president of the *Ornithological Society* and Japanese Hunters' Association; and Uchida, Head of the Bird and Animal Research Institute. Those on the American side included Austin and Dr. Vivian Edmiston Todd (1912–82) of the Education division—more commonly known as the Civil Information and Education (CIE) section—who later introduced him to Elizabeth Gray Vining (1902–99), an American

Quaker educator who also served as English tutor to Japanese Crown Prince Akihito, whose name she Americanized as "Jimmy."[151] Vining later arranged for Austin's son Anthony (Tony) to serve as the Crown Prince's conversational English partner to learn to informally interact with American children.[152] All these individuals served as influential allies, acting within principles of a seemingly democratic colleagueship, while some assisted with Austin's last recommendation, increasing education for Japan's bird protection.

On February 25, 1947, Austin gleefully recounted to his father the New Year's momentous events leading to the government's implementation of all his recommendations outlined in December:

> Did I tell you the Japanese have capitulated to my requests and are adopting all of my recommendations for the new game laws, even to the elimination of mist-netting [sic]? Quite a victory, and how it came about, I'm darned if I know. The opposition just suddenly wilted. Have a hunch Taka-tsukasa's [sic] fine hand is responsible somewhere in the background.[153]

Friends in high places—including General MacArthur and a former imperial prince (Takatsukasa)—helped convince US military authorities and Japanese politicians of wildlife and conservation measures to replace less stringent Meiji and Taishô era laws. The Ministry of Agriculture and Forestry promoted the Bird and Mammal Department to a higher rank, which accorded greater budget and ampler staff to deal with conservation issues. The highly respected zoologist Dr. Uchida left retirement to head it, with plans for his assistant, ornithologist Kuroda Nagahisa, to succeed him after two to three years. In 1953, Kuroda also published a study on Japan's birds with Austin. Newly armed with these accomplishments, the American ornithologist proudly wrote to his father: "I think I've really done something."[154]

Austin reveled in flurries of media interviews following the initiation of new game laws—news reports revealed the US occupation's purported successes in bringing democratic values to Japanese in all aspects. Influential American journalist Sydney B. Whipple (1888–1975)—incidentally former lead United Press reporter for the 1932 Lindbergh case—painstakingly detailed Austin's educational aims for the Japanese public in a long, in-depth February 27, 1947, *Nippon Times Magazine* article: "Let's Protect Japan's Birds: Serious Economic Loss to a Nation Foreseen Unless Destructive Custom of Netting Wild Fowl Abandoned."[155] The article's recommended extirpation of what Whipple described as a "destructive custom" echoed Occupation initiatives intended to purge postwar Japan of its aggressive imperial past. This newspaper also outlined the occupiers' goals to Anglo-American and English-language readers, and therein promoted US interests. Whipple thus devoted a large portion of text to Austin's accomplishments as a SCAP official, while additionally outlining rationales for extirpating harmful wholesale songbird mist-netting, and duck-liming, now deemed cruel. In careful interviews with Austin, the journalist detailed large-scale commercial killing of songbirds, like 4.5 million thrushes captured each season, which served as pickled delicacies for wealthy elites, and supposedly amounted to high profits for netters, but only added mouthfuls of protein to diets of seventy million ordinary

Japanese. Whipple portrayed such practices as barbarous, and decidedly undemocratic while indicative of rapaciously unchecked capitalism in language betraying New Deal sentiments just prior to the Reverse Course.[156] He also criticized the Japanese as "backwards" in wildlife policies, and in a paternalistic paean, praised the North American Wildlife Conservation Model as a guide for a not-yet-enlightened postwar Japan:[157]

> So it is imperative that Nature's balance be restored. How can this be done? By law and by education. By demonstrating that Japan is not only destroying part of her beauty but part of her natural wealth. Other nations have passed through critical periods like this and, where they have been enlightened, have corrected the trouble. America, about the turn of the century, was awakened to the danger by the great scientist William Dutcher [1846–1920], and effected [sic] a cure through the dynamic apostle of conservation, Theodore Roosevelt. The Audubon Society, formed to protect bird life from the extravagances of hunters, had its share in educating the public to the necessity of game laws.[158]

This model cited by Whipple, honoring American ornithologist Dutcher, originated from late nineteenth-century conservation movements, where proponents believed wildlife should be held in public trust, and called for market hunting to be abolished; in the 1930s, naturalist Aldo Leopold (1887–1948) proposed wildlife management should proceed scientifically, and supporters argued sustainable, managed hunting was a nation's citizens' democratic right—echoing avid hunter and conservationist Theodore Roosevelt.[159] Whipple noted historical American trajectories toward conservation, and asserted how other countries could enlighten themselves and progress, indicating lessons for Japan.

As outlined in Whipple's didactive article, Austin initiated an educational campaign for bird conservation, which he first discussed with officials from both nations in December, and aimed to promote "Bird Day" to Japan's public in April 1947. In starting this campaign, Austin created a subtle form of outside US pressure, or *gaiatsu*, toward Japan to ensure continued widespread public support of laws he promoted. On April 10, to honor Japan's first official Bird Day, Austin presented a simple English-language speech at Tokyo's Hibiya Park Public Hall, geared toward Japanese children and other language-learners, which he opened with: "Children can do a great deal to help the birds in Japan. And the birds need your help."[160] He notably stressed democratic aspects of enjoying birds as a common right for all:

> We do not regard wild birds as the property of any one individual person. In a democracy we regard all the wild birds as the property of everybody, regardless of whose land they are on. They are public property in which everybody has an equal right, and in which nobody has more right than anyone else.[161]

Completing his speech, Austin unveiled the *Ornithological Society*'s choice of the green pheasant, or *kiji*, for Japan's new national bird to inspire Japanese citizens' pride in domestic bird life.

By spring 1947, Austin accomplished many goals by capitalizing on a mutual love of ornithology amongst Japanese scientific elites, which helped him revive embedded networks and stimulate useful colleagueship. However, his Wildlife Branch work was far from finished. Austin continued surveying birds throughout Japan with colleagues, and completed findings for several SCAP reports published in the late 1940s. Future legislative battles concerned Russian and Japanese pelagic seal hunting near the Pribilofs, whose complex politics carried the Wildlife Branch's mission into the Cold War.

Following Austin's speech and its democratic focus, the Japanese public, now including women, exercised their democratic voting rights after May 3, 1947, under Japan's new postwar Constitution.[162] Crafted by idealistic American New Dealers in February 1946,[163] and following October deliberation in the House of Peers (which Yamashina attended), it became law on November 3 with the Emperor's approval. Concurrently, American awareness of Japanese war crimes increased in the late 1940s, with the 1948 hanging of General Tôjô Hideki (1884–1948), blamed for instigating war with the United States. Targeting and sacrificing a small group of militarists for symbolic execution allowed Americans and Japanese to re-establish amicable ties and mollify wartime viciousness, a process transferring into the scientific community, who attempted to lay aside wartime pasts. Even expatriate children were inculcated in this philosophy laying the groundwork for American-style democracy. In a school bus heading for the American School in Japan located in Tokyo's Meguro district, Austin's son Tony (1935–2019) remembered that two or three times a week, it drove past another bus at an intersection carrying Japanese prisoners to their trial. The school once brought Tony's class to the Tokyo Trials, where they entered a room and saw prisoners waiting in the dock.[164] While Japan's government addressed judgments of the US-led proceedings of the IMTFE, ordinary people also began tentative steps toward effectuating new roles of democratic citizenship following imperial subjecthood.

Amidst these changes, and composing the occupation's rhetoric of democratic tutelage, scientists from both countries discussed shedding feudal remnants and adopting new democratic idealism, which extended into conservation and bird protection. Japanese elites opportunistically latched onto such rhetoric to hasten Japan's reintegration into the global postwar order while reinitiating earlier transnational collaborations and potentially conflicting rivalries. After Austin left Tokyo in February 1950, Japanese ornithologists published multiple books on ecology and bird conservation to "reclaim" discourses of wildlife protection from the Americans. These included Niibe Tominosuke's (1882–1947) 1951 book, notably published on Bird Day (April 10) initiated by Austin, and his *The Ecology of Birds in the Field*, with a forward by Uchida;[165] and an influential book published ten days later by Yamashina, *The Ecology and Protection of Japanese Birds*.[166] Since the mid-thirties, Yamashina attempted to implement stiffer Japanese conservation laws, also aims of the *Wild Bird Society* to which he belonged since 1934. Yamashina's English-language study *Birds in Japan: A Field Guide*, geared toward a general Anglo-American tourist audience (echoing Takatsukasa's 1941 tourist handbook), also appeared in 1961[167] to supersede Austin and Kuroda's scholarly 1953 volume and to help popularize Japanese ornithology abroad. Yamashina received awards in the late seventies further heralding his international recognition: in 1977, he received the Jean Delacour Prize in ornithology, and in 1978, the World Wildlife Fund granted him the Golden Ark Award for international efforts in conservation.[168]

With loss of Japan's empire and subsequent Cold War rivalries, Japanese ornithologists adopted a transnational scientific discourse of conservation paralleling political developments fostering democracy in Asia that served to further their interests as a revived nation on the scientific forefront. Temporarily, through principles of colleagueship, they collaborated with a well-connected American interloper representing Allied occupation forces over their defeated country. As a US official, Austin temporarily migrated to Japan amidst structures formed to dismantle Japan's empire and to undermine Japanese scientists' exalted aristocratic positions amidst democratic rhetoric. He worked with Japanese counterparts in seemingly symbiotic roles through colleagueship, but their experiences and illustrious family histories placed them in a condition that Rabasa terms as "elsewheres"—impenetrable to the American scientist regardless of the free-flowing liquor and generous hospitality that he extended. Hiromi Mizuno's analysis provides further explanation for these interactions: "Scientific nationalism is paradoxical because the universality of science and particularity of nationalism often conflict with each other. And because of this paradoxical nature, the examination of scientific nationalism reveals dynamics, tensions, and politics that are a crucial part of modern nationalism."[169] Once deeply embedded in empire, the postwar decolonizing world impacted ornithology as a social practice in an era of imperial dissolution following a previous global landscape of fixed sovereignties.

Conclusion

In the late 1940s, earlier structures of scientific exchange reemerged, while the 1950s marked revival of transnational connections enjoyed by Japanese scientists in the Anglo-American scientific world from the 1920s into the 1930s, briefly interrupted by the Pacific War. In the 1950s, old connections were reactivated, and ornithology was framed within a context of peaceful internationalism that echoed the 1920s. Thus, the war briefly ruptured the trans-imperial continuities of their exchanges. During this decade, science was strongly connected to "democracy" with Japanese scientists characterizing their work as dedicated to peaceful endeavors connected to the ruling classes and political power.

In Austin's postwar SCAP-enabled encounter with Japanese scientists, this symbiotic, albeit unequal, exchange proved profitable for all to varying degrees. Back in the United States, bird-banders also benefited from Austin's early 1950s introduction of Japanese mist-nets, which boasted greater efficacy and convenience than Italian mist-nets.[170] Ironically, Austin succeeded in banning mist-netting in Japan precisely because of the nets' deadly efficiency. The US scientist's research in Japan culminated in "The Birds of Japan: Their Status and Distribution," which he co-authored with Kuroda Nagahisa and published in October 1953, after his US return. In his "senior author's foreword," Austin critically assessed his allegedly historic role in Japan:

The new constitution, the various land, educational, and labor reforms, were perhaps the most important innovations, but nowhere was the contrast between the two ideologies more marked than in the field of wildlife conservation. Here, two radically different concepts of man's relation to wildlife were brought together, those of the very old world and those of the comparatively new, and two radically different methods of attaining more or less the same ends. The degree of success attained by our attempted reconciliation of the two cannot be appraised as yet, and will have to wait for the judgment of time.[171]

Austin's rhetoric sounds Orientalist in retrospect, but reflects the colonial mentality of Americans working for SCAP who believed that they served an important mission to bring democratic tutelage to postwar Japan, including in joint scientific endeavors.

The 2012 publication of a volume commemorating the *Ornithological Society*'s hundredth anniversary described Austin's meeting with his co-author, the younger Kuroda, as "a fateful encounter" [*unmei no de'ai*] leading to significant collaboration.[172] Through his father Nagamichi, Kuroda could have developed an equally successful career without the American scientist's mentoring, while later publications by Yamashina and others minimize Austin's role or omit him altogether. However, Austin did provide initial funding and impetus for Japanese scientists to reconvene activities and rebuild morale after wartime defeat. He also extended them an opportunity to reconnect in meetings located in his requisitioned villa after wartime Allied air raids scattered them to locations outside the city, like Karuizawa and Atami, where many had summer or seaside homes. Most importantly, the American ornithologist's contributions to revisions of Japanese game laws were lasting, and remain in effect today. In ornithology, Austin provided initial postwar steps toward reestablishment of flourishing networks of scientific exchanges between Americans and Japanese—a process also mirrored in other fields like physics and biology where embedded networks revived and led to colleagueship across borders.[173]

After Japan rebuilt its economy following the 1950-3 Korean War, named a Japanese "Marshall Plan" or a "gift of the gods"[174] by Prime Minister Yoshida Shigeru, Japanese scientific discourse began to re-adopt a more autonomous tenor while economic success aided in rebuilding a flourishing, capitalist East Asian nation under democratic principles. After accepting American postwar assistance, scientists now felt it was time to make independent contributions, while still enmeshed in transnational networks with Western scientists and expanding colleagueship worldwide. Japanese physicists, like Hideki Yukawa, invited in 1948-9 for membership at Princeton's Institute for Advanced Study, exemplified this triumph when he received the 1949 Nobel Prize. Similarly, as early as 1951, and following Hachisuka's untimely 1953 death, Japanese ornithologists, like the Kurodas and Yamashina, emerged from shells of "aristocratic dilettantism" and exhibited positive trends toward international assertiveness.

Not surprisingly, the Korean War soon played a key role in determining Japan as an American ally in containing Communism, against which most former aristocrats were vehemently opposed. Logistically, Japan hosted US military bases as deployment points, often built upon the ashes of the imperial military forces' wartime airfields and naval yards. On an ideological level, for the Americans, the Korean War served as the

Cold War's first proxy war where US interests centered on containing communism after China "fell" to communist forces in 1949. Though an operational failure, the Korean War's 1953 stalemate ending in an uneasy armistice highlighted American resolve through military involvement, and emphasized Japan's corresponding "support" for maintaining or militarily aiding a fledgling democratic government on the Korean peninsula's southern portion. Thereafter, internationalist initiatives like conservation of animals, plants, and natural environments, and peaceful uses of nuclear technology, resonated with Japanese scientists functioning amidst a postwar democracy under American tutelage and beyond. However, some, like Kuroda Nagahisa, soon lent wartime expertise to serve the Americans in clandestine research during the emerging Cold War. Such relationships highlight the intriguing transwar imperial continuities of Japan's ornithologists in careers where studying birds was inherently political in an avian imperialism initially connoting empire, then potential military victory, and finally, hard-won democracy.

8

Tokyo and the United States (1940s–70s)—*Cold War Ornithological Collaborations between Japanese and American Scientists*

Introduction

Transwar imperial continuities within the lives and careers of Japan's leading ornithologists continued to suffuse their activities deep into the postwar and stunningly reveal key global roles played by these cosmopolitan scientists so invested in supporting national goals. Moreover, garnering high status as non-white leaders in a field dominated by Anglo-Americans often involved complex strategies of collaboration and competition, masking rivalries enmeshed in politically transcendent colleagueship. This chapter focuses upon the activities of Kuroda Nagahisa, who assisted Oliver L. Austin, Jr., and then H. Elliott McClure (1910–98), a US military-funded American entomologist who studied insect-borne illnesses carried by birds.

Such collaborations reveal how, from the late 1940s and into the 1950s, Japanese scientists like Kuroda were absorbed into the American Cold War apparatus at a time when sensitive scientific information aided in cementing loyalty for a new US ally—often relying on past wartime expertise. After the Allied Occupation ended, Japanese ornithologists began engaging in independent endeavors occasionally tethered by American military financial contributions to their research.

Japanese Ornithology in the late 1940s: "Fateful" Encounters and Cold War Remobilization with the Conquerors

Historians John Dower in *Embracing Defeat* (1999)[1] and Naoko Shibusawa in *America's Geisha Ally* (2010)[2] reveal how a defeated Japanese population successfully "disarmed" the American victors in multiple ways to soften the military occupation, and thus, a former maligned enemy transformed into a newly trustworthy ally. As exemplified in Japanese ornithologists' postwar relationships with US Occupation official Oliver L. Austin, Jr., these scientists strove to charm the occupiers, and thereby gained personal agency in reframing a postwar new order no longer predicated upon imperial masculinities. Moreover, close collaboration with the conquerors allowed them some control over their defeated country's treatment while ensuring them a role in its postwar refashioning. Thus, they engaged their vanquishers in a complex process of reciprocity caching defeat's uncomfortable reality.

These encounters ran smoothly on the surface, but fell short of the occupiers' desires for spontaneously generated camaraderie optimistically transcending hierarchical differences between conqueror and conquered. Actually, as go-betweens, Japanese served the Occupation as valuable subalterns. Through requisitioning, American authorities literally encroached into former Japanese elites' private domestic spaces, with householders enduring the newcomers' temporary presence by walling off estate portions and distancing themselves with polite reticence. In addition, SCAP-administered laws, which allowed Allied officials access to protected or sacred spaces, certainly caused hidden resentment amongst an allegedly compliant former enemy. Nevertheless, complicated American connections with Japan's leading ornithologists initiated under Austin's tenure, and continuing after his departure, starkly highlighted these aspects.

For some, like Hachisuka Masauji, the postwar was exceptionally tragic. Multiple distressing factors accumulating over a short time gravely impacted his health. Besides Hachisuka's personal troubles, his protégé Nakamura Tsukasa emphasized that the shock of losing wealth and lands originating as gifts from Oda Nobunaga (1534–82) and Toyotomi Hideyoshi (1537–98), caused rapid declines in his mentor's health.[3] Though Hachisuka's death was purportedly from a "sudden illness" or heart attack on May 14, 1953 at his Atami estate, he possibly committed suicide, like Natsumé Sôseki's (1867–1916) character Sensei in the 1914 novel Kokoro [Sincerity] and other notable Japanese figures keenly disappointed by their life trajectories, along with numerous others in the immensely challenging immediate postwar years.[4] In Japan, suicide often meant taking responsibility for "failure," while displaying resolve for now-obsolete convictions.[5]

However, another Tokyo-based American official, scientist H. Elliott McClure, who worked closely with Japan's ornithologists to further the Allied Occupation's aims, even hinted at a less "honorable" demise. Arriving after Austin in 1950, McClure, as a self-described "entomologist turned birdman"[6] and gossip collector, worked for the US Army 406th Medical General Laboratory in Yûrakuchô to eradicate Japanese encephalitis, transmitted by mosquitoes, and possibly birds, as a disease rampant amongst Japanese children. Through his research, he met Hachisuka and Takatsukasa, and received close assistance from Kuroda Nagahisa, who McClure described as "a quiet little man with great stature as a scientist."[7] Yet, McClure tempered admiration for Hachisuka's pioneering research by scathingly dismissing his personality: "His ornithological work in the Philippines was a major contribution, but he was not in good favor among other ornithologists because he was brash, argumentative, and married a wealthy American woman. I met him only a few times before he died on a cruise."[8] This contradicts Scottish ornithologist Norman Boyd Kinnear's (1882–1957) obituary, stating that Hachisuka died "after a brief illness,"[9] and Biogeographical Society of Japan's director Morinaka Sadaharu's explanation of a heart attack.[10] Considering Hachisuka's past tendency to elicit intrusive interest in his personal life, his cause of death unsurprisingly sparked gossip and occasioned debate even amongst his closest allies. Regardless of what precipitated Hachisuka's sudden demise, it was clearly unanticipated.

Notably, most successful Japanese scientists, including leading ornithologists, assimilated US Cold War interests and goals even prior to the Occupation's end. Such connections were established and facilitated through friendly encounters with Austin and his American visitors as "nodal points."[11] Austin, who served in US naval intelligence during the Pacific War, and received intelligence affairs training at Stanford University's Civil Affairs Training school before it ended, served as a bridge to reconnect Hachisuka with ornithologist, wildlife conservationist, and occasional intelligence agent S. Dillon Ripley. As an Office of Strategic Services agent during the Second World War, Ripley headed US intelligence services in Southeast Asia, and was likely a Central Intelligence Agency (CIA) operative or informant during the Cold War, despite denial by Smithsonian Institution officials.[12] In November 1946, Ripley took a short stopover in Tokyo while heading to Calcutta, where he spent six months collecting specimens for Yale University's museum and the Smithsonian Institution,[13] where Hachisuka had sent specimens in prewar times. Upon his arrival amidst Tokyo's ruins, Ripley contacted the NRS Wildlife Bureau in the Mitsubishi-Shoji Building, whereupon its director Austin "[c]alled him, and went for a drink and dinner with him at the Imperial hotel [sic]."[14]

Some record of their discussions in this elegant venue designed by American architect Frank Lloyd Wright (1867–1959) exists in a letter that Austin wrote to his mentor "Jimmy" Peters. Peters had taught both Austin, who earned a Harvard PhD in ornithology in 1931, and Ripley, who earned his in 1943. Austin noted that, in his November 19 conversation with Ripley, "[H]e has been looking over the Pacific Islands as chairman of the committee on using the bases for scientific research stations for living war memorials."[15] The document reveals that Ripley considered possibilities for ornithological research on remote islands with once strategically useful Second World War–era infrastructure now retained for commemorative value and more.

Austin, during his wartime US Navy service from 1943 to 1944, came ashore on multiple western Pacific island bases, including New Britain, the Solomon Islands, New Caledonia, Vanuatu, and elsewhere to collect over 2,000 specimens and gain knowledge of the islands' terrain, flora, and fauna, along with ornithological life. The MCZ's archives reveal astounding numbers of specimens collected in far-flung locations: 620 records include battlefields like Noumea, Guadalcanal, Bougainville, Emirau Island, Bismarck Archipelago, Tulagi, Treasury, Green Island, and Florida Island.[16] These specimens were carefully hand-catalogued in ledgers cached in the museum's Ernst Mayr Library, listing their catalogue number, original number, name, sex, locality, nature of specimen [generally skins], [where] collected from, and when collected.[17] These bird skins attest to Austin's perceived invincibility as he alighted upon islands not-yet-fully pacified from Japanese enemy troops; utilizing his naval officer privileges while tasked with communications, for Austin, the war became an opportunity for scientific exploration.[18] Particularly proud that his wartime contributions to MCZ collections would astonish the ornithological world, he noted that "I collected the first birds to come out of Bougainville for a great many years."[19] Such arrogant insouciance amidst danger and attraction to intrigue, along with a colonizing gaze upon non-western regions allegedly offering vast laboratories ripe for "discovery" whose birds were harvested with gunfire, characterized attitudes by Austin, his guest Ripley, as well as McClure.

On November 24, Austin took Ripley to visit Kuroda Nagamichi, who proudly showed Ripley his collection's most prized treasures, including "two new Birds of Paradise, … two Rothschild's Starlings from Bali, … the types of the Dagelt [sic] and Saishu races of the white-backed woodpecker."[20] Notably, birds of paradise are only found in Papua New Guinea and Australia, and certain islands Austin traversed. However, out of all of Kuroda's rarities, Austin recounts that "The Pseudotadornas [from Korea] were the main show, and I got a kick out of handling them. Damnedest crest you ever saw."[21] As conquering elites, Austin and Ripley were regaled by Kuroda with his proudest possessions remarkably surviving wartime flames caused by US bombings. The morning before Ripley's November 26 Calcutta departure, Austin accompanied him to view Momiyama Tokutarô's astonishingly intact egg collection, with an afternoon visit to Yamashina's museum. Through meeting Austin's collaborators and friends, Ripley embedded himself into the American official's potentially useful intelligence networks featuring influential Japanese Peers who now ingratiated themselves with the conquerors for survival and scientific interest. Given Austin's position in the Occupation, it was unlikely these connections resulted from happenstance.

In a August 19, 1950, *New Yorker* article, staff writer Geoffrey Hellman characterized Ripley as an imposing man with an upper-class Yankee pedigree, who at "six feet three and a half and looks like a rather worldly pelican," could easily interact with men of elite social positions throughout the world.[22] Kuroda Nagamichi and Yamashina, themselves aristocrats, nearly matched Ripley's height at six feet tall and were also unusual amongst ordinary Japanese. However, for Ripley's generation experiencing the Great Depression's decade-long deprivations and Japanese suffering years of wartime rationing, his height was all the more imposing. Like General MacArthur, who at six feet towered over most Japanese and was popularly called *Gaijin shôgun* [the foreign Shôgun (barbarian-quelling generalissimo)],[23] Ripley was immensely visible and carried himself with presence.

Hellman also described how Ripley complained of greater notoriety as a spy than for his ornithological research: "he has functioned with airy competence both as a political observer and as a secret agent, but in so doing he has on occasion reversed the politics-and-espionage-through-ornithology formula."[24] To Hellman, Ripley never confirmed nor denied espionage involvement, but amicable relations with individuals he met in Japan and elsewhere—intensified through mutual interest in birds—most definitely aided in gathering useful information for his government and its scientific institutions. Certainly, Austin as an ornithologist and SCAP official did the same, and additionally placed his Japanese colleagues in positions embedded in the American military occupation apparatus to serve as useful subaltern informants employing wartime knowledge and technical expertise.

One of these individuals was Kuroda Nagahisa, Nagamichi's son and Austin's co-collaborator in a later 1953 study on Japanese birds. After his 1940 graduation from Tokyo Imperial University in zoology, he served during the war as an Army lieutenant from 1941 until 1945 in a military unit [*guntai kinmu*].[25] His precise role in Japan's

Figure 8.1 Kuroda Nagahisa, circa 1947–1950. © Oliver L. Austin Photographic Collection, permission granted by curator Annika A. Culver. https://austin.as.fsu.edu/files/original/3956b617ada2d19f125aa0e099f6ebb4.jpg.[26]

military is unclear, but his considerable ornithological expertise and zoological training was certainly deployed in the imperial capital. Initiated in 1892 as a private institute associated with the Hygiene Society of Japan, the *Institute for Infectious Diseases* was incorporated in 1914 into Tokyo Imperial University,[27] where Kuroda would study, while the Laboratory for Infectious Disease Control was formerly based at the Imperial Army Medical College in Shinjuku Ward's Toyama area.[28] The nature of Kuroda's postwar work for the Occupation posits a high likelihood of his having worked at this laboratory. In addition, declassified US sources from 1946 indicate that "BW [biological warfare] being a military activity and highly classified for security reasons, civilian scientists and facilities of civilian research institutions were not utilized for this activity."[29] Lacuna in Kuroda's wartime occupational record as an ornithologist mobilized into Japan's military suggest this.

Tokyo Imperial University–educated microbiologist Ishii Shirô (1892–1959) also used this laboratory as headquarters before moving to Manchukuo, performing research for the Imperial Japanese Army and specializing in Japanese encephalitis, also carried by birds. Later, he headed the infamous Ishii Unit in Pingfan, near Harbin, also known as Unit 731, which engaged in human experimentation to weaponize infectious diseases, amongst other projects.[30] A December 6, 1946, letter to MacArthur by Japanese informant Okada Hisashi indicates that the Infectious Disease Research Laboratory conducted bacteriological experiments on prisoners-of-war.[31] In a prevailing wartime atmosphere, individuals involved in the Unit and in Tokyo undoubtedly felt convinced of their actions' absolute necessity, and since victims were not Japanese or deemed state enemies, helped justify their crucial mission. This also enabled perceptions of human subjects as less than human—researchers used the term *maruta* [logs] to relate to them as research materials [*kenkyû zairyô*].[32] As journalist and Kyoto-based historian Hal Gold alleges, Unit 731, which relied upon the *Kempeitai* ["Constitutional Soldiers" or Japanese military police] to round up *maruta* for experimental laboratories, belonged to Manchukuo's punitive apparatus under its Japanese occupation.[33] Moreover, human experimentation lacking ethical constraints became a perfect scientific opportunity, which medical ethicists Howard Brody, Sarah Leonard, Jing-Bao Nie, and Paul Weindling call "wartime exigency" where "wartime loosens unwelcome constraint"; they argue that "[w]artime exigency does more than simply prioritize national security over human rights. It urges toughness and decisiveness in decision-making, so that a moral blindness that would be seen as a deficiency in other times is instead seen as a virtue and a necessity."[34] Researchers likely understood their mission's full implications for science and justified their actions as performed on conquered racial and political others or on those requiring subjugation. During wartime, Japanese scientists' imperial masculinity was focused on absolute state loyalty and application of scientific expertise for military victory by any means necessary.

Traces of this dark wartime past took over forty years to resurface. Nowadays, the National Institute of Infectious Diseases and the National Institute of Health and Nutrition, founded in 1947 by the Ministry of Health and Welfare, sit on land formerly occupied by the Imperial Army Medical College.[35] In 1989, a grisly discovery of bones during excavations for nearby apartment units in Shinjuku's Toyama area prompted further questions, leading to a 1992 anthropological study verifying that the remains belonged to sixty-two people of mainly East Asian origin; a later 2001 report by the Health and Welfare Ministry based on interviews with 368 "former army officers related to the medical school" determined that "some of the remains were probably bodies kept as specimens at the medical school for educational purposes," but posited that others were "bodies brought from battlefields."[36] Some of the skulls included holes and other perforations,[37] indicating interest in human brain contents and possible experimentation with encephalitis, which causes brain inflammation and intellectual disabilities. Nevertheless, a 2004 citizens group headed by Kanagawa University Professor Tsuneishi Kei'ichi believed that "Unit 731's victims' corpses were sent to the medical school for research purposes."[38]

While the precise nature of work performed at the Imperial Army Medical College remains shrouded in mystery, it certainly necessitated frequent wartime visits by Emperor Hirohito, for whom a unique air raid shelter was prepared in a bunker-like building atop Toyama Park's "Mount Hakone" near the contemporary National Institute of Infectious Diseases. According to Toyama Church's current pastor Dr. Nishi Kôsuke, the building was once the army college's meeting hall.[39] Since the Meiji period, schools often served to lodge the Emperor during national tours, so the room likely served as safe overnight accommodations if air raids prevented return to the Imperial Palace in central Tokyo. More likely, due to his microbiological studies and mentorship by Hattori Hirotarô, Hirohito's deep fascination for microscopic organisms likely prompted lengthy conversations with on-site microbiological scientists deep into the night.[40] Presently housing the Protestant Christian Toyama Yôchien [Preschool] since 1952,[41] the church was built in 1950 in response to General Headquarters' desire for a "spiritual center" for new homes in Toyama Heights; its wooden frame perches over this older, stone edifice[42] harboring a hidden basement room raised several inches over a dirt hallway featuring an ornate, square-patterned woodwork ceiling above its hardwood flooring. If Kuroda indeed worked for the College, he entered the hall for meetings and possibly admired the round bay surrounded by six art deco leaded glass windows now illuminating the current pastor's large office.[43]

Despite (or perhaps because of) his wartime activities, since Kuroda had worked with Austin from 1947 until 1950 in the NRS Wildlife Bureau, he then received Austin's favorable recommendation and thus was pre-vetted as a "loyal" collaborator within the Occupation apparatus. Arriving in Tokyo in 1950, entomologist and epidemiologist McClure, who referred to himself as an "accidental ornithologist," was tasked to work for the US Army 406th Medical Laboratory.[44] Here, he soon benefited from Kuroda's expert assistance in a mission to investigate, and possibly eradicate, infectious mosquito-borne diseases spread by avian vectors—including Japanese encephalitis. To assist McClure, Kuroda joined Yoshii Masashi, who also subsequently served as an ornithologist at the YIO, just like Kuroda.

Initially, the three men hunted birds in Tokyo to collect specimens to take blood samples later examined for parasites carrying encephalitis. At this time, permission to carry guns for the vanquished Japanese was a sign of elite and trusted status. On June 3, 1946, through an Imperial Ordinance, the Japanese government banned all firearms and swords to collect them from a possibly restive public, but still allowed licensed hunting guns.[45] In 1950, the Order Concerning Firearms and Swords allowed a broadening of activities involving such weapons to extirpate or frighten harmful wildlife.[46] Moreover, the American Military Police gave their blessing for the two Japanese ornithologists "to carry guns and collect birds everywhere and anywhere."[47]

One incident particularly illuminates the complex power dynamics between Japan's former elites and Occupation authorities. Yoshii accompanied McClure as he entered the Meiji Shrine's precincts as one of Tokyo's few undisturbed places still harboring numerous birds, whose blood samples they needed to compile a broad research pool. McClure remembered being gobsmacked by Takatsukasa's chastisement after attempting to hunt in Shrine forests with Yoshii.[48] While switching out his 12-guage shotgun intended for larger bird species, McClure laid down his smaller caliber .410;

to his dismay, the lighter shotgun disappeared despite the early morning and deserted surroundings. When Yoshii asked guards at the gate about the gun's whereabouts, they were ushered into the Shrine's headquarters, greeted by Takatsukasa, who McClure indicates "listened, and made a few comments while I noted the many bird books and artefacts about the room, and then smiling [sic] he returned the gun, wished us success in our study, but not in the Meiji Shrine."[49] Through practiced reticence and mannered restraint in a calibrated reassertion of imperial masculinity, Takatsukasa carefully reminded McClure that he was only a temporary visitor trespassing upon a sacred space with actions reminiscent of wartime. Ironically, this encounter was unnecessary, since McClure later discovered that shooting often contaminated avian blood samples, with blood more effectively collected from live specimens netted with indigenous Japanese mist-nets.

Almost fifty years after completing McClure's work, Kuroda and Yoshii revealed no illusions about the true reasons for their research. In McClure's 1999 *The Auk* obituary, a space often exposing ornithological and personal candidness, the two scientists matter-of-factly admitted that "[t]ogether, the three of us collected birds for blood samples that were smeared onto glass slides and then sent to a medical center in the United States for study and to infect test mice."[50] Clearly, they referred to Fort Detrick in Maryland. In April 1943, the US Army constructed a site on a former National Guard Airfield at Camp (later Fort) Detrick for the US Army Biological Warfare Laboratories.[51] In the early post-war era, chemical and biological weapons formed an important component of US strategic planning. Avian-borne diseases like Japanese encephalitis held great interest as potential biological weapons, since they could spread deeply into enemy territory, while retaining an element of plausible deniability. Indeed, by 1950, Fort Detrick was an active part of the US Chemical Corps' Research and Engineering Division, since "The Korean War spurred efforts to again develop BW [bioweapon] retaliatory capability based on the ominous threat of USSR involvement," but the program remained largely secret since "there was reluctance to publicize the program."[52] After 1969, when the United States under President Richard Nixon (1913–94) banned development of biological weapons for offensive purposes, Fort Detrick became known as USAMRIID (United States Army Medical Research Institute of Infectious Diseases) where military scientists engaged in national defense research.[53]

For the US Army 406th General Medical Laboratory, clearly linked to US Fort Detrick, Kuroda and Yoshii indicated that McClure "engaged in research related to arthropod-borne zoonoses [insect-borne diseases transmittable to humans] (including Japanese encephalitis)."[54] The 406th functional chart reveals that the Far East Medical Research Unit headed teams including the Epidemiological, Entomology, Bacteriology, and Virus-Rikettsial Departments, where the latter included an Ornithology Section, to which McClure, Kuroda, and Yoshii belonged.[55] Below the flow chart appear two photographs of Japanese researchers, captioned "Mr. Saburo Shibata, Medical Illustrator, prepares an illustration of a chigger," and "Miss Tomoko Shimada is shown preparing ectoparasite slide mounts"—which are headed by a curt title in bold block letters: "JAPANESE TECHNICIANS AND SCIENTISTS ARE [sic] IMPORTANT COMPLEMENT OF 406TH LABORATORY."[56] A Japanese man and woman are shown assisting the Americans' research, representing new ideas about gender equality to position the Japanese as useful, though still subordinate, allies. Kuroda certainly brought much-needed expertise to a US military operation researching transmission of illnesses spread by birds hosting parasitical insects.

Antecedents to this research, of particular interest to Fort Detrick, were parasitological experiments performed by Ishii, including on Japanese encephalitis. In the mid-to-late 1940s, three Fort Detrick military and civilian personnel arrived in Japan to investigate the then-alleged war crimes by the Ishii Unit, also known as Unit 731. They included bacteriologist Lt. Colonel Murray Sanders, Lt. Colonel Arvo T. Thompson, and Dr. Norbert H. Fell, a civilian scientist employee, who ultimately successfully extracted necessary information.[57] The English translator for Sanders' initial investigation, Kamei Kan'ichirô, a businessman and go-between with a prewar doctorate from Columbia University, told Sanders that cooperation might succeed if he focused on scientific aspects of the research, rather than unethical, criminal ones.[58] This approach eventually satisfied both sides, where the Japanese could assist the occupiers and possibly collaborate to rehabilitate their careers. Indeed, 1940s-era interrogators like Hanns-Joachim Scharff (1907–92)[59] showed high effectiveness with non-coercive, low-pressure techniques putting prisoners at ease to entice them into friendly interactions revealing useful information.[60] A former German Luftwaffe interrogator of downed Allied Anglo-American pilots during the Second World War, Scharff lectured on his methods to the US military and defense establishment at the Washington, DC-based Pentagon following a 1948 invitation to testify at Lieutenant Martin J. Monti's (1921–2001) treason trial.[61] The late 1940s represented a watershed moment in how the Occupation viewed imperial Japan's wartime scientific endeavors amidst growing Cold War exigencies and initiated colleagueship in projects where the United States could now utilize invaluable information usually inaccessible to a democratic nation purportedly following higher medical ethics.

Thus, in late April 1947, when Fell conducted his investigation, "Those behind Kamei now saw that an emerging U.S. priority was keeping biological warfare information out of Communist hands."[62] In a June 24, 1947 addendum to Fell's final report, General MacArthur kept informal consensus that "all information obtained in this investigation would be held in intelligence channels and not used for 'War Crimes' programs."[63] MacArthur thus separated the Unit's investigation from the US military, and shifted materials to intelligence authorities.[64] The State-War-Navy Coordinating Committee (SWNCC), in overseeing the Occupation's military aspects, initiated a task force that in August indicated how "The value to the U.S. of Japanese [biological warfare] data is of such importance to national security as to far outweigh the value accruing from 'war crimes' prosecution."[65] Yet, they dithered for half-a-year over the task force's findings, with the Joint Chiefs of Staff's March 13, 1948 order basically ceasing all prosecutions against the Ishii Unit, where any information was put under secret G-2 (US Military Intelligence) jurisdiction: "By that time the Tokyo war crimes trials against high Japanese military officials had been concluded, so the SWNCC review constituted a delaying tactic."[66]

Brody, Leonard, Nie, and Weindling reveal that the Americans selectively punished those involved in human experimentation. Only Japanese doctors harming US servicemen were brought to justice: "A U.S. tribunal in Yokohama in 1948 indicted nine

Japanese physician-professors and medical students for conducting vivisection upon captured American fliers."[67] Contrary to visible traces left by inhumane experiments suffered by women survivors in the Ravensbrück Nazi concentration camp, Brody, Leonard, Nie, and Weindling noted that no survivors in Japan surfaced to show their scars and elicit emotional responses of horror and disgust at the Trials.[68] Also, any victims were likely Chinese or Russian prisoners-of-war, with some Japanese leftists. US officials may have believed that, because racial, cultural, or political others had experienced these injustices, harsh punishment was not merited if their actions were not performed against US military members.[69] American prejudice and Cold War concerns possibly aided in shielding from prosecution these Japanese scientists, now viewed as honorary whites and potentially useful allies.

The US decision to maintain secrecy for individuals involved in biological weapons research, whether directly connected to the Ishii Unit or not, and to enlist them in Cold War research, parallels Operation Paperclip, where data from Nazi scientists was taken in exchange for postwar cooperation and their assimilation into American research institutes in bioweaponry and rocket science.[70] Though the Japanese incarnation of this transfer of useful past research never received an American operational moniker like Operation Paperclip, information and research generated during wartime was deemed so valuable to growing Cold War concerns about the Soviet Union that many officials felt that reports and testimonies should not fall into Russian hands.[71] From a realist geopolitical standpoint, the Japanese archipelago was indeed uncomfortably surrounded by communist countries possibly destabilizing to the Occupation's democratizing mission—notably, the Soviet Union to the north, which had even invaded Japanese islands *after* declaration of surrender; the People's Republic of China, which had initiated a communist regime in October 1949; and the divided Korean Peninsula, where the putatively democratic Republic of Korea abutted against its communist counterpart, the Democratic People's Republic of Korea, since 1948. Here, the Cold War's first proxy war between the United States and USSR raged between 1950 and 1953, only ending in an unsatisfying armistice resolving little but the DMZ's primacy.

Considering this political backdrop following Japan's Reverse Course, it was unsurprising that, in 1950, during the Korean War, handbills printed by Japanese leftists circulated in Tokyo, purporting that McClure's research was a pretext for injecting birds with harmful viruses. McClure noted that "We were supposed to be inoculating birds with viruses and freeing them to take infection to China and other lands."[72] In reality, he collected samples of bird-borne pathogens potentially transmitted by humans, and later sent them to Fort Detrick for analysis. Naturally occurring avian diseases potentially could be reproduced for weaponization. A December 15, 1950, article on "Feathers as Carriers of Biological Warfare Agents," written by Camp Detrick's Biological Department of the Chemical Corps So and C Divisions, highlights that American bioweapons experts were seriously considering ornithologists' contributions to research on alternative weapons delivery systems, defensive or otherwise.[73] Birds carry insect-borne diseases transmissible to humans, so in the early 1950s, Kuroda and Yoshii accompanied McClure to net birds and collect blood samples in "Wakanai, the northernmost point of Hokkaido, where winter birds first arrived from the north, and at Kagoshima, the southernmost point, where summer birds arrived from Asia."[74] After netting, collecting samples, and banding, the birds were released, and subsequent studies illuminated their range to theoretically reveal how far migratory birds might spread pathogens. Following the Occupation, from 1952 onwards, Kuroda's work seemingly took a more peaceful direction, when he joined the YIO where he researched "the anatomy and ecology of sea birds, and participated in maritime survey work with several international groups."[75]

Cold War Cooperation for Birds: Yamashina, Delacour, and the Rebuilding of Scientific and Intelligence Networks

Following the war's devastation, individuals and institutions first focused on survival and then slowly directed their attention toward reconstruction. Part of this process included transcending their wartime activities and seeking to rejoin an international community of researchers and scholars. Moreover, science, due to its allegedly objective quality, could play a key role in rehabilitating Japan. According to historian Laura Hein, in the late 1940s, a "concept of 'cultural science' was widely accepted in postwar Japan as a tool to unmake fascism."[76] Notably, when Tokyo University's president, Nanbara Shigeru (1889–1974), met with an American scientific delegation and held a welcome address on October 2, 1948, he proposed that "It is nevertheless a self-evident fact that the promotion of science, natural and cultural, is the *sine qua non* of Japan's rehabilitation as an enlightened and peace-loving nation."[77] Thus, through their efforts in conserving birds, Japanese ornithologists would greatly impact national and international efforts to democratize and diffuse peaceful scientific pursuits, now no longer connected to empire or the aggressive shooting and collecting of birds in expeditions.

By the mid-1950s, renovations of the political structure and society soon allowed these scientists a widening focus beyond domestic Japan. After 1952, Anglo-American officials running the Allied Occupation had generally left, allowing a more independent focus for scientific research and other activities. Moreover, the Korean War's 1953 resolution ended American usage of Japan as a staging point for the conflict, which ushered in economically valuable US military procurements jumpstarting a period of postwar prosperity termed the "Economic Miracle" (1955–91). By the late 1950s until early 1960s, a quasi-consensus emerged in Japan's political environment, with a modicum of stability achieved by the conservative Liberal Democratic Party, while the Economic Miracle's advent generated an atmosphere of economic consistency and permanent fruitful climate for scientific endeavors.

However, despite these positive developments, the early 1960s began with tension for American policy-makers engaging with Japan. On January 19, 1960, the United States and Japan renewed their Occupation-era Mutual Security Treaty [*Sôgo kyôryoku oyobi anzen hoshô jôyaku*], widely known as the Anpô Treaty, which allowed American military bases to remain in Japan. This treaty especially irked younger generations of students and trade union members; after its final approval and ratification on May 20, huge protests swelled on June 15 that prevented President Eisenhower from a three-day visit to Tokyo and the Nikkô Shrine.[78] According to historian Nick Kapur, "For a period of fifteen months, from March 1959 through June 1960, an estimated 30 million people from across the archipelago—approximately one-third of Japan's population of 92.5 million—participated in protest activities of some kind."[79] Yet, when the world's ornithologists met in Japan in late spring, a more peaceful atmosphere prevailed in pockets of calm amidst the protests.

In 1960, largely due to Yamashina's successful lobbying, the International Council for Bird Preservation (ICBP) hosted its first postwar meeting in Tokyo for a twelfth annual conference held in Japan. Yet, this coup for Japan's leading ornithologists took years, and followed Occupation-era bans preventing Japanese from traveling overseas. Yamashina's first invitation to attend an ICBP meeting arrived in 1954; however, a weak yen prevented his overseas travel. By 1958, Japan's postwar economic conditions improved rapidly, and when exchange rates finally normalized, Yamashina brought Kuroda Nagahisa, now working at the YIO, to Helsinki, where the 1958 ICBP meeting was held at the University's engineering campus, and initiated plans to link US and European migratory bird treaties with those of Asia. Yamashina was also asked to head the ICBP's Asian subsidiaries, and suggested that Tokyo host its next meeting.[80]

This historic 12th ICBP conference was held in May at the *Kokusai bunka kaikan* [International Culture Hall], also known as the "International House" established on August 27, 1952, which represented transwar connections between an influential internationalist Japanese media representative and the American philanthropist who would create the Asia Society in 1956:

> Its origins date back to an encounter between John D. Rockefeller III and [the influential journalist] Matsumoto Shigeharu at the third conference of the Institute of Pacific Relations held in Kyoto in 1929. Having experienced diverse cultures and peoples, they *recognized the importance of the cultural dimension within international affairs*, which tended to be viewed primarily in political and economic terms.[81]

After its official June 11, 1955, opening, this space combining Western modernity and Japanese tradition was billed as "a crossroads for intellectuals from the world" and became a more-than-suitable venue for Japan's ornithologists to extend their global reach in hosting the ICBP meeting. Moreover, Yamashina's laboratory was relatively close, with nearby homes of other Japanese aristocrats. Perched atop a steep hill traversed by Toriizaka Lane on industrialist Iwasaki Koyata's (1879–1945) former property, the structure was built by leading architects Maekawa Kunio (1905–1986) and Sakakura Junzô (1901–69), both prewar apprentices of Le Corbusier (1887–1965), and Yoshimura Junzô (1908–97), who had trained under Czech-American architect Antonin Raymond (1888–1976).[82] This modernist edifice was reflected in the carp-stocked ponds of a famed early Edo style walled garden designed by Ogawa Jihei VII (1860–1933), a seventh-generation landscape artist often commissioned by leading Meiji-era to early Shôwa period industrialists and politicians.[83] Its modern, yet fashionably Spartan rooms could comfortably host foreign visitors in a peaceful environment removed from the surrounding metropolis' bustle (and mass protests).

Expansive loggia outside and spacious lobbies within allowed for personal discussions to continue outside of the event's framework, held on the ground floor in the Iwasaki Koyata Hall and other banquet rooms.[84] Ornithologists came from all over, including India, Europe, and the Anglo-American world, to this location showcasing Japan's newly reconstructed capital. Prior to the 1964 Tokyo Olympics, Japan's first postwar international event emphasized the nation's emergent soft power in peaceful, collaborative scientific ventures extending globally.

At this conference under Yamashina's leadership, Japanese ornithologists actively promoted Tokyo as a central Asian locus for developing international bird conservation initiatives, whose success could be scientifically investigated in region-wide collaborative bird-banding and migration studies throughout Asia and Pan-Pacific areas,[85] while broader publics could be mobilized to care for bird populations by designating a national bird. Conference participants thus advocated naming a national bird for each of their countries, much like the *Ornithological Society* chose the green pheasant in 1947. The famed Indian ornithologist Salim Ali (1896–1987), Ripley's friend and collaborator in a late 1950s Bharatpur-based bird-banding project investigating bird-migration,[86] explained that "The purpose was to pinpoint public interest and attention to some particular species that stood in the greatest need of protection in each country, especially where it was threatened with extinction owing to public apathy or direct human persecution."[87] Ali also indicated that "The Great Indian Bustard [*Ardeotic nigriceps*] is a species that merits this distinction."[88] Yamashina and Ali's projects represented a growing trend for Asian ornithologists to work with Anglo-American conservation partners to broaden networks and tap into larger funding sources. In 1961, to fund for such massive initiatives tying together the programs of the ICBP and the International Union for Conservation of Nature (IUCN), British organizers including evolutionary biologist Julian Huxley conceived the World Wildlife Fund (WWF).[89] Notably, to recognize his Asia-wide conservation efforts, Yamashina later received the WWF's Golden Ark Award in 1978. This followed his 1977 Jean Delacour Award for lifetime achievements as an ornithologist.[90]

The celebrated French ornithologist Jean Théodore Delacour also traveled from his new New York home base, following retirement as director of the Museum of History, Science, and Art (Los Angeles County Museum), to Tokyo to attend this historic ICBP meeting, which represented the revival of transpacific ornithological connections. However, at a potential moment of reunion and celebration, he instead mourned two close friends' deaths: Hachisuka (1953) and Takatsukasa (1959). Nevertheless, his memoirs indicate that bright moments punctuated the meeting, after which Kuroda Nagamichi hosted participants for "a formal Japanese dinner-party, complete with geishas and dancers, just as in pre-war days. The restaurant had been built on a part of his former garden. House, museum and aviaries had disappeared in the bombing of the war, and he lived in a small house at one side of the grounds, the rest having had flats built on it."[91] While Delacour lamented destruction of old aristocratic properties and "lovely private gardens"[92] in this bustling newly reconstructed city, he found respite from construction and encroaching urban development within the imperial palace's secluded grounds, where he found a bird paradise that was "still an oasis in the middle of the great city."[93] This island of peace also harbored imperial laboratories:

The extensive grounds, beyond the moats and the walls, are a natural forest where birds abound, and the magnificent dwarf trees, some centuries old, make one forget the turmoil and ugliness of the surrounding world. The park is not spoilt by any large buildings, and only small pavilions show here and there. Two are devoted to the Emperor's collections: one of mammals and birds, the other of fish and invertebrates.[94]

Yamashina, as Emperor Hirohito's first cousin and Taishô Emperor Yoshihito's nephew, arranged for Delacour to meet Japan's once notorious Emperor.[95] Joining them was Ripley, now the ICBP's third president, following Delacour, who had served as its second one. Like Kuroda, Ripley kept a large collection of waterfowl on his five-acre Litchfield, Connecticut estate, praised by Delacour as "the choicest such colony in the Western hemisphere," and also directed the Avicultural Society's American branch.[96] As a Yale University professor, Ripley headed the on-campus Peabody Museum of Natural History. Cordial prewar social reciprocity still mattered, so the American visitors extended gifts: Ripley presented the Emperor with "a rare old marine worm from the bottom of L. [Long] I. [Island] Sound."[97] Hirohito was delighted.

Helman's December 17, 1960 *New Yorker* article describes the fateful meeting of Delacour and Ripley with Emperor Hirohito, presided over by Yamashina.[98] "Delacour Reobserved" contains extensive quotes by Delacour, who Hellman interviewed several months after his return to New York City, and detailed the peripatetic life of the ornithologist, whose Japan visit composed one aspect of a full year of social and professional engagements—including a trip "to Florida to advise a friend on his bird collection," who likely was Austin.[99] Helman commonly wrote articles in the magazine's "Talk of the Town" section, which catered to an elite East Coast American public interested in lengthier descriptions of broader news stories, and penned features on prominent NYC cultural institutions, including natural-history related organizations with which Austin, Ripley, and Delacour were intimately familiar, like the AMNH, New York Zoological Society, the Explorer's Club, and Audubon Society. In 1966, Hellman would publish *The Smithsonian: Octopus on the Mall*,[100] and in 1969, he issued *Bankers, Bones & Beetles*, a history of the AMNH where Delacour once worked.[101] Appealing to a cultured audience, Hellman notes that Delacour recounted how "Emperor Hirohito is himself a noted zoologist, specializing in marine invertebrates, and spends two afternoons a week, he told us, in his laboratory."[102] Delacour believed that the Emperor understood his English, but a translator filtered conversations, including replies, which caused stilted communication.[103] Hirohito also enjoyed meeting these high-positioned scientists with similar interests, and proffered a requisite tour of his labs and specimen collections.[104] This fascinating visit represents the intriguing ornithological connections maintained internationally during the Cold War.

In June 1961, a year later, Delacour went to Moscow to fraternize with prominent Soviet ornithologist Georgy Dementiev (1898–1969), a University of Moscow zoology professor he had met at numerous international conferences, but still neglected to visit.[105] Dementiev served as curator of bird collections for the Zoological Museum of Moscow University (ZMMU),[106] and headed the Ornithology Department from its founding in 1932 until 1947.[107] With Sergey A. Buturlin (1872–1938), as a top authority on the Soviet Union's bird life, Dementiev published the *Complete Key to the Birds of the USSR* from 1934 until 1941.[108] Contemporary ZMMU ornithologist Pavel Tomkovich, and ornithology secretary Mikhail Kalyakin, note how Dementiev's book still serves as a notable ornithological reference.[109] From 1951 until 1954, Dementiev and Nikolai A. Gladkov (1905–75) also edited and published six volumes of *Birds of the Soviet Union* [in Russian], with an English translation,[110] and thus, brought this knowledge into the broader Anglo-American world, and arguably, into Delacour's frame of reference.

In such meetings, did science transcend politics, or was science bolstering a nation's soft power or intelligence motives? These ornithologists arguably did both, even if not serving as official ambassadors of their countries or engaged in "intentional" intelligence gathering. Yet, while these encounters evinced revival of cordial international relations amidst colleagues in a mutual field of study, other transwar imperial continuities also resumed by reactivating knowledge gleaned during wartime.

Bird Banding and Migration Projects Intersect with American Cold War Bioweapons and Intelligence Initiatives

Amidst key postwar American and Japanese ornithological circles, similar figures re-emerged who had circulated around Austin during the Allied Occupation. In 1958, McClure's subsequent posting took him to Malaysia for five years to study tropical birds for the US Army Medical Research Unit in cooperation with the Kuala Lumpur-based Institutes of Medical Research.[111] Yet, he soon returned to Japan. When Colonel C. M. Barnes of the Washington, DC-based Walter Reed Institute turned to the YIO for information on how wild birds might transmit Japanese encephalitis viruses during migration, Kuroda, who then headed the YIO, "suggested that Barnes organize a project to capture and band birds throughout the Far East," and "recommended McClure as the most suitable person to develop such a project."[112] Thus, Kuroda helped initiate the process where McClure planned the Migratory Animals Pathological Survey (MAPS) project, active from 1963 until 1975, for which the US Army Research and Development Group and the Southeast Asian Treaty Organization (SEATO) funded an "international investigation of migratory bird patterns throughout southeast Asia, hoping to better understand the regular spread and retreat of Japanese encephalitis across the Asian landmass."[113] McClure led the MAPS project in Tokyo from 1963 until 1966, when he moved headquarters to Thailand where he worked until his 1975 retirement; in total, banding was completed for "1,165,288 individuals of 1,218 species, of which 5,601 individuals of 235 species were recovered."[114] Yoshii and his colleagues in the YIO's Bird Migration Research Center spoke favorably of this project generously funding extensive Japanese bird-banding efforts.[115] In 1967, MAPS also financed a massive Indian bird-banding project directed by Ali, following interim Smithsonian support temporarily garnered by Ripley after WHO funding initiated in 1959 had dried up.[116] Ali headed this Bombay Natural History Society (BNHS) project to study how a potentially devastating new tick-borne hemorrhagic encephalitis variant that had emerged in the South Indian state of Karnataka presented surprising similarities with Russian hemorrhagic fevers and encephalitis—specimens including ectoparasites, blood samples, and bird banding records soon circulated between the BNHS, Smithsonian, and MAPS head office, while Indian and Soviet scientists investigated the disease's virology.[117] MAPS as a collaborative project relied on overseas partners to develop bird-banding initiatives and thus became embedded in complex transnational connections between ideologically opposed countries unavoidably meshing birds and their researchers within global politics.

The year 1963 also marked inception of another six-year-long bird-banding project running until 1970, which in contrast to MAPS, exclusively relied on American scientific networks connected to the Smithsonian Institute. Called the Pacific Ocean Biological Survey Program (POBSP), it cost 2.7 million US dollars funded by Fort Detrick, and was led by the Smithsonian's Secretary, psychologist Leonard Carmichael (1898–1973), who "favored the appropriate application of science to national defense."[118] According to *Washington Post* investigative journalist Ted Gup, "In October 1961, the CIA funded a project titled 'Role of Avian Vectors in Transmission of Disease,'" and indicated that, from 1959 to 1963, Carmichael had directed the Human Ecology Fund, a CIA front organization, and key funding source for programs like MKULTRA, which engaged in chemical and biological research.[119] As environmental historian Michael Lewis substantiates, the survey clearly intersected with US Cold War bioweapons programs:

> The study was run out of Ft Detrick, Maryland, the US Army's biological warfare centre, and blood samples, ticks, birds, and data were regularly sent to both Ft Detrick and to the Deseret Test Centre in Utah, a testing site for chemical and biological warfare agents. Scientists who worked on the project were given security clearances, some of their documents and research results were classified, and some were given inoculations before being sent to the field to collect data.[120]

In 1964, none other than Ripley took over Carmichael's position as secretary until 1984; with the POSBP project under media scrutiny in the late 1960s, Ripley strove to limit public focus on the Smithsonian's unique, incredibly generous funding source for the POSBP project. Yet, in a 1969 letter to Congresswoman Patsy Mink (D-Hawaii), who expressed concern about the project's reach into the Hawaiian Islands, Ripley later denied Smithsonian involvement with military bioweapons programs.[121] According to science historian Roy MacLeod, in 1977, Ripley also "wrote to the CIA, disassociating the Smithsonian from any CIA work on behavior modification,"[122] presumably distancing it from the now-controversial MKULTRA program. Both of Ripley's denials were technically true—the funding merely paid for Smithsonian researchers to collect bird skin specimens and blood samples, and their ultimate use was not the Smithsonian's responsibility. Instead, this was left up to Fort Detrick's in-house researchers and other US military actors. Moreover, Ripley merely refuted the institution's involvement with CIA behavior modification research, but no other aspects.

The Smithsonian scientists conducted POSBP research much like work once conducted by Austin, McClure, Kuroda, and Yoshii in Tokyo—with mist-nets and shotguns on similar Pacific Islands traversed during wartime by Austin, including the Solomon Islands and Christmas Island, which encompassed several million square miles intersecting with American nuclear testing sites.[123] Not so coincidentally, the project began in 1963, when the United States, Great Britain, and Soviet Union signed the Limited Nuclear Test Ban Treaty in Moscow on July 25 to eliminate nuclear testing in the atmosphere, space, and underwater.[124] Regardless of connections to US bioweapons, nuclear concerns, or intelligence initiatives, the program's notable successes included banding of over 1,1500,000 birds by 1968.[125] These birds' migration routes could be traced, plus valuable information about Pacific Island ecology was gleaned, sometimes near nuclear testing sites, that US military authorities could use to determine environmental impacts on avifauna and flora.

For the Japanese since the Tokugawa period, collecting birds was helpful in gathering intelligence, a charge also levied against American scientists like McClure and Ripley in projects they headed or supported, including in Japan, India, and Pacific Islands. Lewis views Ripley's and Ali's participation in US military projects as composing a Cold War intelligence complex where nearly unlimited funding emerged for strategic concerns, with scientists scrambling for support of pet projects as they overlooked strategic purposes for their research findings.[126] This echoes deployment of Japanese ornithologists during wartime. However, only Kuroda Nagahisa was likely actively connected to Japanese military medical research in four years of military service, though recently-compiled declassified National Archives and Records Administration documents omit his name. However, Kuroda's qualifications and wartime expertise were clearly useful to McClure and Ripley's US military projects amidst an incipient Cold War.

Conclusion

Anecdotes involving Kuroda, Yoshii, and McClure reveal that, while many Japanese believed in American occupiers' efforts to democratize Japan, they still engaged in complex transactional power relations where US Cold War interests roped them into projects resembling their wartime support of imperial Japan. After the Soviet Union's dissolution and fifty years after the war's end, Kuroda in his early eighties could again mention collaboration with a now-deceased McClure as a US military-funded scientist with whom he worked. Moreover, in the early 1990s, Kuroda transferred from his visible YIO position to serve as curator of the newly created Abiko City Bird Museum, built next to the YIO to house Yamashina's old collections.[127] This allowed Kuroda some candidness about activities supporting American interests.

In sum, Japan's history of prewar sociability initiated with Anglo-Americans, coupled with continuity of putatively reciprocal relations, and new perceived threats by the Soviet Union and People's Republic of China, enabled Japanese ornithologists to carve space for themselves in multiple endeavors, including work intersecting with US interests in bioweapons projects and possible intelligence initiatives. Initially, many American authorities like Austin or McClure viewed Japanese scientists as "Japs," but once they came to know them well, saw them as peers in emergent colleagueship supplanting a former imperial masculinity. When these Japanese researchers were afforded opportunities to conduct US military-funded research in a manner aggrandizing their newly conquered nation's power, they revealed how they as the conquered could once again become a valued partner and ally. When the Allied Occupation finally ended, these scientists then regained their elevated social positions, which lost much luster since the war's end. Hence, these individuals could retain their status as scientists with important geopolitical goals, and mutual projects represented both individual and national restorationist aspirations.

Conclusion: Tokyo and Cambridge, UK (1960–Present), *Fledging Global Conservation Policies*

Postwar initiatives of the Tokyo and Cambridge-based ornithologists Yamashina Yoshimaro and Nakamura Tsukasa, in a purportedly more "peaceful" field of science, allowed Japan to re-emerge as a leader in the Western scientific world, including hosting the 1960 International Council on Bird Preservation (ICBP) meeting in Tokyo at a flashpoint in domestic reactions surrounding renewal of the US-Japan Mutual Security Treaty. Long-standing connections with the Anglo-American world previously fostered through specimen-exchanges and study also precipitated such scientific exchanges. Using cytology to establish precise taxonomies, transnational bird migration surveys, and global conservation served as new ornithological arenas in the internationally recognized revival of peaceful Japanese scientific endeavors, and by the 1960s, developed as a means for Japan to reassert its international impact as Cold War aims evolved into furtherance of transnational exchanges for peace.

As ornithologists, Yamashina and Nakamura represented Japan's peripatetic scientific elites exerting a global influence, and evince a trend continuing until this day. Others like Kuroda extended their impact domestically through undertakings like bird-banding activities under the Yamashina Ornithological Institute's (YIO) auspices, sponsored by the Ministry of the Environment. In 1984, Yamashina's prewar laboratory was renamed the YIO and moved to Tokyo's outskirts, with the ornithologist receiving in the late 1970s the Jean Delacour Award and World Wildlife Fund's (WWF) prestigious membership in the Order of the Golden Ark. For Nakamura, decades of achievement led to his own 2005 receipt of the International Peace Prize by the US-Japan Conference on Cultural and Educational Interchange (CULCON). Kuroda, Yamashina, and Nakamura's stories represent how scientific research, including ornithology, was still propagated with political import by Japan's elite cosmopolitan circles enjoying privileges to migrate freely to areas of innovation, higher learning, and dynamic scientific exchange. Such activities highlight the persistence of transwar imperial continuities, where both Japanese and Anglo-American scientists enjoyed embedded networks of knowledge collaboration and competition enabling meaningful contributions upon the world stage into the present.

Bird-Banding and Conservation Efforts for a Peaceful Postwar Nation, 1960–Present

Amidst crisp November air suffusing grassy meadows bordering Ibaraki and Chiba prefectures near Tokyo, both regions hard-hit by a September typhoon's rains and floods, I observed professional Japanese ornithologists banding songbirds in their annual fall migration from the Siberian Far East, flying through Northeast China, down the Korean Peninsula, and across the East Sea into Japan's relatively mild-weathered Kantô Area to stay the winter. Lured by cheerful chirping "decoys" emanating from CD recordings, birds flew into pockets of mist-nets once so detrimental to Japanese wild bird populations from the Meiji Era into the mid-twentieth century, now banned for commercial use since 1947. When delicate birds fell into pockets with a thwack, researchers carefully untangled and gently folded them into cotton drawstring bags hanging from their waists. With a suitable number collected, unpacking of each bird occasioned much expectation regarding representation of new species. We raptly observed Eurasian Tree Sparrows (*Passer montanus*), Oriental Greenfinches (*Chloris sinica*), Black-Faced Buntings (*Emberiza spodocephala*), White Wagtails (*Motacilla alba*), and Lapwings (*Vanellinae*), all migrating southwards from Russia's Far East and Hokkaidô or through China and the Koreas. Each passerine bird was then tenderly held between thumb, index-, and forefinger, with its skull softly palpitated to determine age, and gonads checked to determine sex, while tail and head feathers, along with eyes and beak, were examined to judge health. With a pencil, researchers compiled lists of data in a ledger, including the birds' sex and species; then, they chose a proper-sized leg band, where a number indicated its retrieval location and other information later entered into databases. But, most exciting of all was its release—a carefully held bird, heart pumping furiously, quickly unfurled its wings, and then, in a thrilling whir, flew far into the sky.

These observations date from November 21, 2015, when I viewed professional Japanese ornithologists at work. I joined YIO bird-banding efforts near Moriya in Ibaraki Prefecture under the guidance of Odaya Yoshiya, an energetic young scientist acceding to Kuroda's same position as Curator of Birds at the Abiko Bird Museum, an organization adjoining the YIO. Odaya's mentor Hiroi Tadakazu, a visiting YIO researcher and retired Ministry of Agriculture and Forestry scientist, accompanied us, along with Kawahara Takayuki, a Forest and Forestry Products Research Institute botanist; and Mochizuki Michito, a PREC Institute environmental consultant. In Ibaraki's marshy grasslands, I helped Odaya set up his mist-nets, and viewed his process of untangling, bagging, cataloguing, banding, and releasing birds. Most bird banders, as well as hunters, are over sixty, aging like Japan's general population, so Odaya, in his late twenties, viewed it as his mission to train a younger generation, which today included me and Mochizuki in his early twenties. Since mist-nets are illegal except for research, the YIO hung flags indicating today's *Chôrui hyôshiki chôsa* [Investigation of Bird Species Through Banding].

During our work, we met two elderly hunters shooting ducks with aging pointer dogs hobbling alongside. In Japan, receiving a hunting license is a complex and expensive undertaking that includes a long process of applying for permission to acquire a hunting rifle and proper training to use it, along with police registration.[1] According to *The Japan Times*, "As of 2010, there were about 190,000 people with hunting licenses (including those for using nets and traps), of which 122,000 were 60 years old or above. This is down from 518,000 in 1975."[2] Most licensed hunters are well-to-do seniors—very different from the past, when hunting crossed generations and social classes.

Odaya disclosed that his group planned to band at least 300 birds from November 21 to 23 and performed the same operation monthly. Under YIO auspices, Japan's Ministry of the Environment administers data processed by volunteers to observe species diversity and determine environmental impacts on bird populations. That year, birds arrived sooner on their migration, with fewer numbers: Odaya blamed this upon North Korean food scarcity and Northeast Chinese air pollution. Here, environmental concerns again intersected with international rivalries echoing Cold War anxieties heightening suspicions against neighboring communist countries.

Our efforts arose from a long history of Japan's bird-banding, which flourished until the Pacific War's derailment. Bird-banding began in 1924, when researchers banded 6,007 individuals from 42 species under the Ministry of Agriculture and Commerce's Wildlife Research Division, and continued until 1943 where 15,382 recoveries were made until 1944.[3] In 1960, during the Tokyo ICBP meeting, Japan's leading ornithologists decided to restart bird-banding; it resumed in 1961 under Ministry of Agriculture and Forestry auspices until 1962, when the YIO was commissioned to analyze findings for three years of preliminary research and provide technical guidance, after which the project ended until 1967, when MAPS, understood by Japanese as a project with financial assistance "of the US Army," provided valuable funding to continue.[4] After MAPS ended, in 1977, the Ministry of the Environment took over supervision of the YIO's bird-banding research, now dubbed the "Commissioned Project of Managing Bird Observation Stations by Environment Agency."[5] Hence, we flew the Ministry's red flags indicating support for our bird-banding, which continues into the present.

In the 1960s, in connection with employing scientific methods to assess wild bird populations' ecological conditions, Kuroda Nagahisa began seriously promoting bird-banding, and subsequently served as the Japanese Bird-Banding Society's president.[6] While heading research at the YIO under Yamashina's mentorship and directing the ICBP's Japanese section, he tirelessly advocated for bird protection while leading as president numerous ornithological societies, including Japan's *Ornithological Society* (1970 to 1976 and 1980 to 1989), *Wild Bird Society*, and *Japanese Association for the Preservation of Birds*.[7] In 1966, Kuroda published findings from the 1965 bird census conducted via banding and other methods in the relatively secluded Imperial and Akasaka Palaces in central Tokyo,[8] with a bird census repeated in 1966.[9] To understand prewar conditions for urban songbirds, he also analyzed banding data for the Japanese tree sparrow (*Passer montanus*) from banding's 1924 inception until wartime in 1943.[10]

These studies informed Kuroda's influential 1967 Japanese-language monograph, *The Study of Birds: Ecology*; it featured an extensive English and Japanese bibliography, which revealed his desire to reach an English-speaking world.[11] Tellingly, Kuroda sent a copy to Austin, who served as *The Auk*'s editor from 1969 until 1977, overlapping his tenure as Curator of Ornithology, with an office in now Dickenson Hall, where he worked at the Gainesville-based Florida Museum of Natural History on the University of Florida campus.[12] He included a handwritten note to "Dr. O. L. Austin with the compliments of Nagahisa Kuroda" and added a typewritten English-translated table of contents for the American ornithologist. After Austin's receipt, with a square Japanese-English name-stamp[13] acquired in Tokyo, he affixed his seal on the book's front endpaper, and promptly absorbed his former colleague's book into his extensive personal library.[14] Such enduring transnational exchanges characterized Japanese ornithology's postwar landscape, now increasingly focused on peaceful conservation efforts initially prompted by American military financing for dual-use purposes. Yet, by the 1970s, the Japanese abandoned connections suffused by earlier militarisms.

Reactivating Global Ornithological Networks for a New Generation

On October 5, 1948, in the elegant town of Fontainebleau near Paris where French elites own weekend homes, British evolutionary biologist Julian Huxley founded the International Union for Conservation of Nature (IUCN) while directing the United Nations Educational, Scientific, and Cultural Organisation [UNESCO].[15] In 1964, the IUCN created the Red List of Threatened Species, while in 1966, its partner organization, the ICBP initiated the Red Data Book for birds, now termed the IUCN Red List, to promote the Convention on the Conservation of Migratory Species.[16] Contemporary Japan closely supports this tracking project, where currently 50, or 12%, of its birds are threatened, with 4 species extinct, 6 critically endangered, 13 endangered, 31 vulnerable, 33 near-threatened, and 360 of least concern.[17] Of the IUCN Red List's critically endangered birds, most notable is the Crested Shelduck (*Tadorna cristata*), described by Kuroda Nagamichi in 1917, and lacking confirmed records since 1964, despite numerous sightings recorded in northeast China; it possibly exists in Russia's Far East and North Korea.[18]

Japan's contemporary bird conservation efforts enjoy a long transwar history of ornithological endeavors. The quaint English town of Cambridge, boasting the University of Cambridge's world-renowned colleges, first hosted the ICBP's international headquarters, where Hachisuka and influential twentieth-century Japanese scientists studied or interacted with British peers. The ICBP, as a global bird conservation organization now known as Bird Life International (BLI) since 1993, includes national organizations as partners, like Japan's *Wild Bird Society*, established by Nakanishi Gotô in 1934, which tracks Japanese bird populations and ecologies. According to the BLI's Data Zone, Japan currently harbors 447 bird species, with 221 land birds, 98 sea birds, 179 water birds, and 361 migratory birds, with 21 breeding endemically.[19] These facts illuminate Japanese ecological conditions in bird diversity; however, they captured little of the passion and resolve of the leading scientist studying and protecting them.

Hence, I was keenly interested in meeting and interviewing this modest ornithologist who commanded great international stature. On a hot August 2015 afternoon, I meandered through Shinjuku station's heavy crowds to meet Professor

Nakamura Tsukasa, the last individual working with scientists discussed in this book, and Hachisuka's most loyal disciple. After initial introductions, we stopped by a nearby cafe, where we engaged in small talk prior to our interview. A short, distinguished pale elderly man in a suit, with dark sunglasses and white sunhat, Nakamura shared that he avoided the sun after a recent skin cancer diagnosis. He likely mentioned this because older *yakuza* bosses also favored such sunshades, sometimes passing through this area in Shinjuku station. Nakamura proudly noted that his skin condition was due to a 2005 Galapagos Island trip where he observed Darwin's finches, with the equatorial sun burning his skin and contributing to his illness. Such was my introduction to this illustrious, yet humble, ornithologist whose global peregrinations mirrored his peripatetic mentors. Nakamura also kindly invited me to visit his home in Yamanashi prefecture.

As a former *shosei* mentored first by Austin, and then by a brilliant, mercurial Hachisuka, Nakamura's ornithological work emerged from the postwar period's bright promises. His research indicated how Japanese now employed nuclear technology for peaceful purposes: three months at the Japan Atomic Energy Research Institute taught him to employ radiological isotopes to understand lipid metabolism in migratory birds before conducting research at University of California at Berkeley. Nakamura's work on bird migration, using radioisotopes to pinpoint its exact beginning, led him to study nuclear physics in the late 1940s when the atom's power was now harnessed for peace. In the mid-1950s, he acquired a position as biology professor at Yamanashi University, where he taught for half-a-century.[20]

From 1978 to 1982, Nakamura followed Hachisuka's footsteps and took a sabbatical-leave to serve as the ICBP's director, headquartered in Cambridge. Beginning in 1978, he also was a committee member of the International Ornithological Congress (IOC), a meeting series held every four years by the International Ornithologists' Union (IOU), a group of two hundred distinguished international ornithologists. In 2014, the IOC held its first meeting in Tokyo, honoring Japanese ornithologists' global contributions, and highlighting international organizational efforts of scientific luminaries like Nakamura who paved the way.[21] Awarded the American government's prestigious International Peace Prize by the US-Japan Conference on Cultural and Educational Exchange in 2005,[22] Nakamura's fifty-year career as a Yamanashi University ornithologist and biology professor spans the ornithological world's key international developments and illuminates experiences of Japanese scientists whose research and personal lives embodied transwar imperial continuities.

Before his 2018 passing away, Nakamura sought to commemorate Hachisuka's remarkable life and achievements as his much-indebted mentor [*onshi*] and honored the aristocrat's patronage of his father Yoshio. In 2003, the Bio-Geographical Society of Japan, co-founded by Hachisuka, held a symposium to celebrate the ornithologist's would-be hundredth birthday, and published proceedings in its bulletin.[23] Nakamura provided the Society with information and materials for this event. Recently, Hachisuka was recognized for his bold ornithological contributions as a Japanese innovator extensively spanning Anglo-American ornithological networks, a narrative encouraged by Nakamura's support of the 2016 publication of Hachisuka's biography by former broadcast media journalist Murakami Kimio.[24] Whatever scandals and ill temperaments plagued Hachisuka during his life, his reputation as a pioneering ornithologist flourishes in Japan more than fifty years posthumously. Regardless of his past reputation, Hachisuka was a pioneering trailblazer; his story now resonates strongly with contemporary issues as a scientist of color struggling for a voice in the international arena. Because of Hachisuka and others, including a world-renowned Yamashina, Japanese ornithologists like Nakamura attained global stature as respected scientists and innovative researchers.

Fledging Future Generations: YIO Aristocratic Lineages and Ornithology's Popularization

Arising from collections housed at Yamashina's Shibuya laboratory on his estate and converted into an informal research center in the 1930s, the contemporary YIO enjoys an almost century-long history. In 1984, amidst Japan's astonishing 1980s-era economic bubble, the laboratory's Tokyo site was sold, with its land used for apartments. That same year, the YIO, along with collections including 18,000 books, 59,000 bird specimens, and 7,400 mammal or animal specimens, was moved to Abiko in Chiba prefecture, an hour from Tokyo by train amidst bucolic rice fields and suburban homes.[25]

When I arrived in November 2015 to meet Dr. Tsurumi Miyako, head curator of the YIO's natural history collections, numerous researchers and interns concentrated intently on their work. In the large open staff room minus cubicles characterizing American offices, I noticed the distinguished gray-haired and mustachioed figure of (now Crown) Prince Akishino Fumihito (b. 1965), currently first in line to Japan's imperial throne, and the YIO's president since 1986, quietly working at his desk. Like his predecessors from the late nineteenth century onwards, Dr. Akishino's intellectual pedigree featured connections with the UK. As a graduate student at Oxford University's St. John's College studying ichthyology and fish taxonomy from 1988 until 1990, he served as a research assistant at the university's Museum of Natural History and London-based British Museum, whose bird specimen collections he encountered.

Following his British sojourn, Prince Akishino enrolled in doctoral studies in ornithology at *Sôkendai*, or the Graduate University for Advanced Studies, an institution once housed at the Tokyo Institute of Technology in Yokohama, now located in Kanagawa prefecture's beach town of Hayama. This quiet surfing town also hosts the Imperial Family's beachside summer residence and the Museum of Modern Art. In 1996, after writing a dissertation thesis on the Southeast Asian jungle origins of domestic fowl, he received his ornithological doctorate.[26]

Prince Akishino's decades-long interest in chickens likely was generated by his mentor Yamashina, who bred them on his ruined estate after the war, and pioneered in Japan cytology experiments to establish these domesticated birds' lineages. In the mid-1990s, as a graduate student, Akishino published in the prestigious US-based *Proceedings of the National Academy of Sciences* papers proposing the red jungle fowl [*Gallus gallus gallus*] as modern domestic chickens' direct ancestor.[27] Like Takatsukasa Nobusuke in the 1930s and 1940s, he began co-editing volumes targeting general Japanese audiences, including *The Illustrated Encyclopedia of European Poultry* 1994),[28]

Chickens and Human Beings—Viewed from the Perspective of Ethnobiology (2000),[29] and *The Encyclopedia of Ornithology* (2008).[30] This desire to reach broader audiences beyond scientific colleagues mirrors earlier contributions by Austin, who in 1961 published the best-selling *Birds of the World*, with illustrations by prize-winning bird artist Arthur Singer (1917–90).[31] Recently, Akishino's data-driven research follows global scientific trends featuring biostatistics, microsatellite DNA analysis, and computer modeling to understand genetic variations and lineages.

Other imperial family members share interests in zoology, biology, and the natural sciences, including Crown Prince Akishino's older brother, current Emperor Naruhito (b. 1960), who wrote his doctoral thesis *A Study of Navigation and Traffic on the Upper Thames in the 18th Century* at Oxford University.[32] His father, abdicated Heisei (1989–2019) Emperor Akihito (b. 1933), still pursues enthusiasm for ichthyology, focusing on the goby [*gobiidae* family], of whom he named five species, while two other goby species, *Exyrias akihito* and *Platygobiopsis akihito*, were named after him in the 1990s.[33] His grandfather, Shôwa Emperor Hirohito, was educated as a marine biologist and became a leading hydrozoan (jellyfish) expert.[34] Prince Akishino's sister, Princess Kuroda Sayoko (b. 1969), first majoring in Japanese language and literature at Gakushuin University, served as a YIO researcher specializing in kingfishers from 1998 until her 2004 marriage.

Renowned University of Tokyo cell biologist Hideo Mohri recently published books on the imperial family's contributions to biological research, where three generations and descendants were actively involved in biological studies or other scientific activities. *The Imperial Family and the Study of Biology* (2015)[35] also appeared in English as *Imperial Biologists* (2019).[36] Since the late nineteenth century, no other country besides Japan boasts ruling classes with such active patronage and personal engagement in scientific research involving biology and various forms of zoology.

Currently, Princess Takamado, widow of Prince Takamado Norihito (1954–2002), and Emperor Akihito's first cousin, serves as the BLI's honorary president since March 2004, when she succeeded Queen Noor of Jordan (b. 1951). Officially, she defines the organization's role and her own position as leader of a conservation body receptive to government, corporate, and local interests based on scientific data.[37] In 2019, to honor her 15th anniversary of leadership, the organization created the BirdLife International Japan Fund for Science, "with the aim of supporting the scientific research and analysis for threatened bird species globally and for the foreseeable future."[38] Since then, BLI has been active in sixteen countries, and developed the Japan Professional Football League's bird mascot characters to raise awareness for bird conservation amongst children and fans.[39] Because of enterprising individuals like Princess Takamado, who further popularized bird conservation and birding, the organization has clearly garnered broad-based allies.

In particular, Princess Takamado has promoted ornithology's accessibility to broader general audiences—especially women. As previous chapters have revealed, in prewar and immediate postwar Japan, aristocratic men dominated ornithology, with homosocial forms of sociability and specimen exchanges, including performances of past imperial masculinity and later colleagueship, largely excluding women. According to BLI's website, "Since April 2011 [Takamado] has been writing 'Opening Serials' in

Fujin gahô [Ladies' Pictorial] called 'Through the Lens', where she documents her experiences as a birdwatcher and photographer."[40] Her heart-warming observations of birds and empathetic gaze toward the natural world endeared her to readers of this popular women's magazine, and likely inspired many young women to venture forth into Japan's fields, parks, and woods with binoculars and birding notebook in hand.

Popular interest in bird watching and birding has certainly risen amongst the Japanese population, with the *WBSJ* boasting an impressive 2019 membership of 34,499 people in eighty-seven urban or regional groups throughout the nation with 16,510 sponsors, including individuals, groups, and corporations.[41] Bird watching has also become quite commercialized in Japan, with larger bookshops purveying official WBSJ products in their ornithology sections, including bird-banding gloves, along with books. The well-stocked ornithology sections of Tokyo-based Japanese megabookstores like Junkudô in Ikebukuro and Maruzen in Shinjuku attest to this growing hobby for both women and men, where practitioners can find bird-banding gloves in multiple colors, handlists with birding checklists, and laminated folders imprinted with common bird varieties, amongst other items. In Tokyo's Gotanda area in Meguro Ward, the WBSJ boasts an entire store of its own, filled with ornithological accouterments and much-coveted rubber birding boots in diverse colors, including trendy camouflage.[42] One can also apply for a Master Card, Visa, or JCB credit card where .5 percent of every purchase is donated by the credit card company to the WBSJ.[43] Certainly, outfitting amateur ornithologists with books and gear is now a lucrative business.

Big Japanese corporations also began to fund bird conservation in the 1980s and 1990s. These forms of philanthropic activities, predating and intersecting with emerging ideas of global Anglo-American corporate responsibility and sustainability, were adopted by companies to appeal to consumers' desired ethics and became integral to corporate branding. In 2019, five different Shintô shrines were also BLI corporate members, with one Buddhist temple's support[44]—revealing longstanding connections between religious organizations and amateur bird appreciation in Japan. Other corporate BLI supporters include Japan Tobacco, Kaigai [Overseas] Fine Wine Asia, the US subsidiary Dow Chemical Japan, Ltd., and the French jewelry subsidiary, Chopard Japan, Ltd.[45]

Notably, the large multi-national Suntory corporation, which gained renowned in producing Japanese-made Western alcoholic beverages like whiskey, has worked closely with BLI on various projects. To further popular enjoyment of bird-watching, Suntory's company-funded bird identification website features a hundred common Japanese birds, with calls and detailed descriptions complete with physical markers, videos, and sounds, allowing amateurs and specialists an audio-visual exploration of wild birds using only a hand-held smartphone.[46] Since 1973, Suntory also manages a bird sanctuary on its Hakushû distillery grounds, indicating that "Wild birds are said to be an indicator of natural environment."[47] In 1990, the corporation established the Suntory Fund for Bird Preservation charitable trust, with a website depicting birds as barometers of environmental change, and has "granted [a] total of ¥538.46 million to 425 organizations up to 2020."[48]

During my 2015 YIO visit, Dr. Tsurumi emphasized how her organization was grateful for corporate support from sources including Suntory for bird conservation and other YIO activities. She indicated how the Mitsui Company also helped fund post-disaster restoration of salt-water soaked bird specimens from Fukushima.[49] While ornithologists in Japan once generated popular interest in birds through *Wild Bird Society* outings, tourist-oriented bird-watching manuals, and royal endorsement, Japanese birds now find staunch protectors in corporations, who view their conservation as crucial to corporate missions ensuring a natural patrimony for future generations.

Conclusion

Since ornithology began as a formal scientific field in Japan, it underwent a transformative journey throughout a long historical trajectory of over one hundred and fifty years. British, American, and Japanese imperialisms intersected in the proverbial ornithological field and highlighted the enduring importance of class status and imperial masculinities continuing beyond wartime. A mutual obsession with birds translated into an exclusive language for its practitioners which transcended cultural or linguistic differences amongst a highly Westernized cosmopolitan elite interacting with British and American counterparts. As revealed in the enduring nature of transwar imperial continuities in Japanese relationships with Anglo-Americans in ornithological endeavors, key figures circulated amidst larger narratives in uncanny, seemingly coincidental apparitions patterning Freud's return of the repressed. These individuals included American military authorities, British aristocrats, and multiple Anglo-American colonial officials and scientists enmeshed in imperialist endeavors, much like the Japanese who encountered them in London, Cambridge, Paris, New York, Los Angeles, Tokyo, Manchukuo, or the Philippines. Because of practitioners' high-class status embedded within elite political power structures, Japanese ornithological obsessions notably paralleled broader trends in international relations.

An internationalist cosmopolitanism reflected in ornithological activities overlaid inherent strains of avian imperialism evolving from formal ornithological inquiry initiated in the late nineteenth century. During wartime, ornithologists' particular talents were incorporated into the Japanese state's militaristic apparatus of total war with all available means mobilized for strategic goals. Yet, in the postwar, Japanese ornithological authorities were subsumed into a greater Cold War political environment despite putative commitments to peace; as an American ally, Japan and its broader global initiatives became subordinate to US interests in funding structures often tied to military aims. Hence, wartime research patterns were unearthed amidst a climate favoring anti-communism and Cold War competition, where seemingly peaceful practices like bird-banding and transregional collection of blood samples from avian species hid potential applications for bioweapons initiatives and intelligence activities monitoring Soviet counterparts. Nevertheless, by the 1960s, earlier foundations of beneficial Anglo-American scientific relationships and formal or informal specimen exchanges between Japanese and foreign counterparts contributed to productive re-activations of previous informal alliances and friendships soon useful in steering efforts toward ecology and award-winning bird conservation activities.

More recently, with corporate sponsorship, online websites facilitating bird identification, and groups increasingly welcoming amateur birdwatchers like the WBSJ, ornithology in Japan has finally undergone the postwar democratization so desired by Anglo-American observers like Austin. Yet, contemporary Japanese ornithology still enjoys the imperial family's strong support and relies upon funding from wealthy elites, often through corporate donations. These individuals migrate throughout the Anglo-American world within influential circles spanning the Pacific and beyond Asia. Because of such long-term elite connections, ornithology in Japan also harbors strong connections to broader political trends caching important continuities with earlier imperialisms.

Thus, as the transwar lives and careers of its most prominent scientists reveal, Japanese ornithology is not just for the birds, but very much about politics.

Notes

Acknowledgments

1. "The Oliver L. Austin Photographic Collection," https://austin.as.fsu.edu/, accessed June 19, 2020.
2. Palmitos Park Maspalomas, *Conoce el Parque* [Get to Know the Park], https://www.palmitospark.es/zoo/, accessed August 18, 2020.

Introduction

1. Juan Siliezar, "Going the Distance for Himself and a Larger Purpose: Prof Cycles across U.S. to Raise Awareness for BLM, Black Birders Week," *Harvard Gazette* (July 31, 2020), https://news.harvard.edu/gazette/story/2020/07/harvard-ornithology-professor-bicycles-across-the-us/, August 10, 2020.
2. Scott V. Edwards, https://twitter.com/scottvedwards1?lang=en, August 10, 2020.
3. Siliezar, "Going the Distance," July 31, 2020.
4. These include Subic Bay, a decommissioned US military base in the Philippines. Mel White, "Getting Familiar with the Philippines" (April 15, 2011), *All about Birds—The Cornell Lab of* Ornithology, https://www.allaboutbirds.org/getting-familiar-with-the-philippines/, accessed February 22, 2018.
5. Chapter 2, "The University and the Undercommons," in Stefano Harney and Fred Moten, ed., *The Undercommons: Fugitive Planning & Black Study* (New York: Minor Compositions, 2013), 22–43.
6. According to Praseeda Gopinath, "imperial masculinities are configured differently in each empire and in each phase of empire. They are dialectically constructed; that is, they are an imperial social formation, produced in the interstices and overlaps of national and imperial cultures." Praseeda Gopinath, "Imperial Masculinities," in *Literary and Critical Theory* (July 26, 2017), in Oxford Bibliographies.
7. Ann Laura Stoler, *Along the Archival Grain: Epistemic Anxieties and Colonial Common Sense* (Princeton, NJ: Princeton University Press, 2009), 50.
8. Vicki L. Ruiz, Professor, History and Chican@/Latinx Studies, University of California-Irvine, question and answer session, Florida State University, History Department, February 18, 2016.
9. Marc Augé, *Oblivion* (Minneapolis: University of Minnesota Press, 2004), 20. Translated by Marjolijn de Jager.
10. Augé, *Oblivion*, 21.
11. Augé, *Oblivion*, 21.
12. I thank retired physicist Dr. Yoshi Nakada for photos of the original camera.
13. Beginning in 2013, 951 slides for the Oliver L. Austin Photographic Collection were digitized, organized, and captioned under my curation, assisted by FSU student interns: https://austin.as.fsu.edu/.

14 Robert Rosenstone, *Mirror in the Shrine: American Encounters with Meiji Japan* (Cambridge, MA: Harvard University Press, 1988).
15 Robert Rosenstone, "History in Images/History in Words: Reflections on the Possibility of Really Putting History onto Film," *The American Historical Review*, Vol. 93, No. 5 (December 1988), 1181.
16 Hayden White, "Historiography and Historiophoty," *The American Historical Review*, Vol. 93, No. 5 (December 1988), 1193–4.
17 Raymond Williams, "Chapter Nine: Structures of Feeling," in Raymond Williams, *Marxism and Literature* (New York: Oxford University Press, 1977), 128–35, 134.
18 Williams, "Structures of Feeling," 132.
19 Greg Grandin, "Can the Subaltern Be Seen? Photography and the Affects of Nationalism." *Hispanic American Historical Review*, Vol. 84, No. 1 (February 2004): 83–111.
20 Asians still construct race within varying social contexts, like the LA-based Filipino community at once Asian, "white," and Latino. Anthony Christian Ocampo, *The Latinos of Asia: How Filipino Americans Break the Rules of Race* (Stanford: Stanford University Press, 2016).
21 Thomas K. Nakayama, and Robert L. Krizek, "Whiteness: A Strategic Rhetoric," in Carl Burgchardt, ed., *Readings in Rhetorical Criticism* (3rd edition) (State College, PA: Strata Publishing, 2005), 291–309; and Thomas K. Nakayama and Judith N. Martin, eds., *Whiteness: The Communication of Social Identity* (Thousand Oaks, CA: Sage Publications, 1998).
22 In 2012, social psychologists Hajo Adam and Adam D. Galinsky developed "'enclothed cognition' to designate clothing's systematic influence upon wearers' psychological processes and behavioral tendencies." Hajo Adam and Adam D. Galinsky, "Enclothed Cognition," *Journal of Experimental Social Psychology*, Vol. 48 (2012), 918–19.
23 Austin was one of two civilian Occupation employees receiving personal commendations from the General: https://austin.as.fsu.edu/about.
24 John M. MacKenzie coined "collecting imperialism" in *Museums and Empire: Natural History, Human Cultures, and Colonial Identities* (Manchester: Manchester University Press, 2009), 60.
25 Doreen Massey, *Space, Place, and Gender* (Minneapolis: University of Minnesota Press, 1994).
26 Doreen Massey, "A Global Sense of Place," *Marxism Today*, 38 (June 24, 1991), 29; "Power-geometry and a Progressive Sense of Place," in J. Bird, B. Curtis, T. Putnam, G. Robertson, and L. Tickner, eds., *Mapping the Futures: Local Cultures, Global Change*, London: Routledge, 1993), 59–69; and "Imagining Globalization: Power-Geometries of Time-Space," in Avtar Brah, Mary J. Hickman and Mairtin Mac An Ghaill, eds., *Global Futures: Migration, Environment, and Globalization* (New York: Palgrave Macmillan, 1999), 27–44.
27 Massey, Space, Place, and Gender, 19.
28 Takie Sugiyama Lebra, *Above the Clouds: The Status Culture of the Modern Japanese Nobility* (Berkeley: University of California Press, 1995).
29 Oliver L. Austin, Hachisuka Masauji, Takashima Haruo, and Kuroda Nagahisa, "Japanese Ornithology and Mammalogy during World War II (An Annotated Bibliography)," Natural Resources Section (NRS) Report Number 102 (January 30, 1948). U.S. Fish and Wildlife Service.

30 Austin notes its creation "with the full approval and backing of the military." Austin et al., "Japanese Ornithology and Mammalogy during World War II," 3.
31 Odaya Yoshiya, the Abiko Bird Museum's curator and YIO volunteer bird bander, recounted this story.
32 Austin and Kuroda, *The Birds of Japan: Their Status and Distribution* (1953), and Kuroda Nagahisa, *The Ornithologists and Avifauna in Early History of Japanese Field Ornithology: Chiefly Until 1950—the Draft Prepared for "The Birds of Japan: Their Status and Distribution" by Oliver L. Austin and Kuroda Nagahisa, 1953* (Tokyo: Self-Published Memorial Publication of 88 Copies, 2004).
33 Nakamura uses *kansen-shi,* "persistent (traversing) war history," to refer to imperial continuities from the prewar into postwar periods. Nakamura Masanori, *Sengoshi* (Tokyo: Iwanami shoten, 2005).
34 Andrew Gordon, ed., *Postwar Japan as History* (Berkeley: University of California Press, 1993); *The Evolution of Labor Relations in Japan: Heavy Industry, 1853–1955* (Cambridge, MA: Harvard University Press, 1988); and *A Modern History of Japan: From Tokugawa Times to the Present* (New York: Oxford University Press, 2002). In "Consumption, Leisure, and the Middle Class in Transwar Japan," Gordon claims transwar continuity beyond politics. Andrew Gordon, *Social Science Japan Journal*, Vol. 10, No. 1 (April 2007), 4.
35 Quoted in Gopinath, "Imperial Masculinities."
36 Leo T. S. Ching, *Becoming "Japanese": Colonial Taiwan and the Politics of Identity Formation* (Berkeley: University of California Press, 2001), 4–6.
37 Gopinath, "Imperial Masculinities."
38 J. A. Mangan, *"Manufactured" Masculinity: Making Imperial Manliness, Morality and Militarism* (London: Routledge, 2012), 9.
39 Joan Wallach Scott, "Gender: A Useful Category of Historical Analysis," *The American Historical Review*. Vol. 91, No. 5 (December 1986), 1053–75.
40 Judith Butler, "Performative Acts and Gender Constitution: An Essay in Phenomenology and Feminist Theory," *Theatre Journal*, Vol. 40, No. 4 (1988), 519–31.
41 Joseph A. Vandello, Jennifer K. Bosson, Dov Cohen, Rochelle M. Burnaford, and Jonathan R. Weaver, "Precarious Manhood," *Journal of Personality and Social Psychology*, American Psychological Association, Vol. 95, No. 6 (2008), 1325–39.
42 Adam and Galinsky, "Enclothed Cognition," 918–19.
43 Vandello et al., "Precarious Manhood," 1325–39.
44 Vandello et al., "Precarious Manhood," 1326.
45 Vandello et al., "Precarious Manhood," 1327.
46 Vandello et al., "Precarious Manhood," 1325.
47 Vandello et al., "Precarious Manhood," 1337.
48 Amanda Mull, "Health: Psychology Has a New Approach to Building Healthier Men—A controversial set of guidelines aims to help men grapple with 'traditional masculinity,'" *The Atlantic* (January 10, 2019), https://www.theatlantic.com/health/archive/2019/01/traditional-masculinity-american-psychological-association/580006/, accessed June 19, 2020.
49 Barbara Maloney and Kathleen Uno, *Gendering Modern Japanese History* (Cambridge, MA: Harvard University Press, 2005), 8.
50 Rajyashree Pandey, "Gender in Pre-Modern Japan," in Jennifer Coates, Lucy Fraser, and Mark Pendleton, eds., *The Routledge Companion to Gender and Japanese Culture* (New York: Routledge, 2019), 22–30.

51 Donald Roden, "Thoughts on the Early Meiji Gentleman," 61–98; Moloney, "The Quest for Women's Rights in Turn-of-the-Century Japan," 463–92; and "Womanhood, War, and Empire: Transmutations of 'Good Wife, Wise Mother' before 1931," 493–519; in Moloney and Uno, eds., *Gendering Modern Japanese History*.
52 Maloney and Uno, "Introduction to Gendering Modern Japanese History," in *Gendering Modern Japanese History*, 8.
53 Bird Life International website, "Our History," https://www.birdlife.org/worldwide/partnership/our-history, accessed June 26, 2020.
54 Anthropologist Takie Sugiyama Lebra examines the Japanese aristocracy's lifestyles and everyday practices, but fails to situate these into broader historical contexts. See Lebra, *Above the Clouds*.
55 Barbara Fuchs, *Mimesis and Empire: The New World, Islam, and European Identities* (Cambridge, UK: Cambridge University Press, 2001), 5.
56 Anthropologist Aiwa Ong asserts that individuals use "honorary whiteness" to attain privileges in colonial systems presupposing white superiority. Aiwa Ong, *Neoliberalism as Exception: Mutations in Citizenship and Sovereignty* (Durham, NC: Duke University Press, 2006), 169.
57 John Dower, *Embracing Defeat: Japan in the Wake of World War II* (New York: W. W. Norton, 1999).
58 Naoko Shibusawa, *America's Geisha Ally: Reimagining the Japanese Enemy* (Cambridge, MA: Harvard University Press, 2010).
59 Morris Low, *Science and the Building of a New Japan* (London: Palgrave Macmillan, 2005).
60 Hiromi Mizuno, *Science for the Empire: Scientific Nationalism in Modern Japan* (Stanford: Stanford University Press, 2008).
61 Hiromi Mizuno, Aaron S. Moore, and John DiMoia, *Engineering Asia: Technology, Colonial Development, and the Cold War Order* (London: Bloomsbury, 2018).
62 Miriam L. Kingsberg Kadia, *Into the Field: Human Scientists of Transwar Japan* (Palo Alto, CA: Stanford University Press, 2020).
63 Andrea Wulf, *The Invention of Nature: Alexander Humbolt's New World* (New York: Vintage Books, 2016).
64 Andrea Wulf, *Founding Gardeners: The Revolutionary Generation, Nature, and the Shaping of the American Nation* (New York: Vintage Books, 2012).
65 Massey, "A Global Sense of Place," 24–9; "Power-geometry," 59–69; and "Imagining Globalization," 27–44.
66 Massey, *Space, Place, and Gender*, 19.
67 Ann Laura Stoler, *Along the Archival Grain: Epistemic Anxieties and Colonial Common Sense* (Princeton, NJ: Princeton University Press, 2010), 32–3.
68 Neil L. Whitehead, "Post-Human Anthropology," *Identities: Global Studies in Culture and Power*, 16 (2009), 4.
69 Stoler, *Along the Archival Grain*, 34.

Chapter 1

1 Defined later in the chapter.
2 Sachiko Koyama, "History of Bird-Keeping and the Teaching of Tricks Using *Cyanistes Varius* (Varied Tit) in Japan," *Archives of Natural History*, Vol. 42,

Notes

3   No. 2 (Edinburgh University Press, 2015), 211–25, file:///Users/aculver/Downloads/ANH.2015.0306.pdf, accessed on March 31, 2020.
3   Koyama, "History of Bird-Keeping," 215.
4   Koyama, "History of Bird-Keeping," 215.
5   Nicholas Fiévé and Paul Whaley, eds., *Japanese Capitals in Historical Perspective: Place, Power, and Memory in Kyoto, Edo and Tokyo* (Abingdon, UK: Routledge Curzon, 2013) 285.
6   Oliver L. Austin, "Mist Netting for Birds in Japan" (Supreme Commander for the Allied Powers, Natural Resources Section, Report 88) August 1947, 9–10.
7   Photographs in Horiuchi Sanmi, *Nippon dentô shuryô-hô: shashin kiroku* (Tokyo: Shuppan kagaku kenkyû-jo, 1984), 155.
8   Austin, "Mist Netting," 10–11.
9   Austin, "Mist Netting," 13–14.
10 Koyama Sachiko, *Yamagara no gei: bunkashi to kôdôgaku no shiten kara* (Tokyo: Hôsei Daigaku shuppankyoku, 1999).
11 Koyama, "History of Bird-Keeping," 212, 222.
12 Koyama, "History of Bird-Keeping," 221.
13 Animal-based entertainments existed in premodern Japan and continue into the present. On November 15, 2015, at a Shintô shrine near the Asakusa Kannon Buddhist temple, during the *Shichi-go-san* Festival, I observed a *saru-geki* or "monkey-play" with my son and daughter.
14 Noriko Otsuka, "Falconry: Tradition and Acculturation," *International Journal of Sport and Health Science*, Vol. 4 (2006), 199.
15 Otsuka, "Falconry," 200.
16 Morgan Pitelka, *Spectacular Accumulation: Material Culture, Tokugawa Ieyasu, and Samurai Sociability* (Honolulu: University of Hawaii Press, 2015), 95–6.
17 Otsuka, "Falconry," 198.
18 Pitelka, *Spectacular Accumulation*, 97–8.
19 Hanami Kaoru, *Tennô no takajô* (Tokyo: Sôshi-sha, 2002).
20 Oliver L. Austin, "Waterfowl of Japan," *Natural Resources Section Report No. 118* (Tokyo: General Headquarters, Supreme Commander for the Allied Powers, 1949), 81. Austin includes a schematic diagram of the duck decoy pond.
21 Pamphlet, "Hama-rikyu Gardens: Special Place of Scenic Beauty and Special Historic Site." Tokyo: Hamarikyu Garden Office, 2015. A 1939 map of the Hamarikyû imperial gardens appears in Hanami, *Tennô no takajô*, 32.
22 Otsuka, "Falconry: Tradition and Acculturation," 202. Hugh Alexander Macpherson, *A History of Fowling: Being an Account of the Many Curious Devices by Which Wild Birds Are or Have Been Captured in Difference Parts of the World* (London: D. Douglas, 1897), 266–9.
23 Images of these hunting practices appear in Horiuchi, *Nippon dentô shuryô-hô*, 70.
24 Austin, "Waterfowl of Japan," 82.
25 Austin, "Waterfowl of Japan," 82. Tony Austin, Interview 12, November 14, 2014, interviewer Annika A. Culver, Florida State University History Department, 56.
26 The Imperial Household Agency website, *Kamoba* (Imperial Wild Duck Preserves), https://www.kunaicho.go.jp/e-about/shisetsu/kamoba.html, accessed July 2, 2020.
27 Horiuchi Sanmi, *Nippon chôrui shuryô-hô: shashin kiroku* (Tokyo: Sansho-dô, 1939), 1. This first edition included an English translation, likely by Uchida.

28 Natural History Books indicates that this book sold for $850. In July 2020, a copy appeared from Austin's personal library: https://www.naturalhistorybooks.com/products/bird-hunting-methods-japan-photographic-record, accessed July 4, 2020.
29 Horiuchi, *Nippon chôrui shuryô-hô*, 1.
30 Horiuchi Sanmi, *Nippon chôrui shuryô-hô: shashin kiroku* (Tokyo: Sansho-dô, 1942), https://yomitaya.co.jp/?p=85706, accessed July 4, 2020.
31 Horiuchi, *Nippon dentô shuryô-hô*.
32 Horiuchi, *Nippon chôrui shuryô-hô* (1939), 1.
33 Greg Gillespie, *Hunting for Empire: Narratives of Sport in Rupert's Land, 1840–70* (Vancouver: University of British Columbia Press, 2008).
34 This term is inspired by "canine imperialism." Aaron Skabelund, *Empire of Dogs: Canines, Japan, and the Making of the Modern Imperial World* (Ithaca, NY: Cornell University Press, 2011), 2.
35 Carol Grant Gould, *The Remarkable Life of William Beebe: Explorer and Naturalist* (Washington, DC: Island Press, 2004), 135.
36 Setoguchi Akihisa, Shuryô to dôbutsugaku no kindai–Tennôsei to "shizen" no poriteikkusu, *Seibutsugaku-shi kenkyû*, Vol. 84 (October 1, 2010), 76.
37 Setoguchi Akihisa, Shuryô to kôzoku–zasshi "Ryôyû" ni miru dôbutsu wo meguru seiji. kagaku. gendaa, *Dôbutsu-kan kenkyû*, Vol. 13 (December 2008), 39.
38 Nihon shuryô kyôkai, *Ryôyû*. Vol. 1, No. 1 (February 1, 1900), frontismatter.
39 Iijima Isao, *Ryôyû no hakkan wo shukusu*, *Ryôyû* (February 1, 1900), 6.
40 Image of Iijima, *Ryôyû* (February 1, 1900), frontismatter.
41 Samuel D. I. Emerson, "The Makigari," Fuji no makigari, in *Ryôyû rinji hakkan* (April 25, 1900), 88.
42 Emerson, "The Makigari," 89.
43 Emerson, "The Makigari," 90.
44 Gordon Smith, "[To] T. ICHINOI, Esq., The Tokio [sic] Gun Club," *Ryôyû*, Vol. 2, No. 2 (February 1, 1901), 24.
45 Advertisement, Kinga shinnen for the Kuwahara Gun Shop, *Ryôyû*, Vol. 2, No. 1 (January 1, 1901), n.p.
46 National Diet Library online archives, *Nihon hôrei sakuin*, Jûhô torishimari kisoku: Daijôkan fukoku dai 28 gô (January 29, 1872), https://hourei.ndl.go.jp/simple/detail?lawId=0000000013&current=-1, accessed October 28, 2021, and Library of Congress, Law Library website, Sayuri Umeda, "Firearms—Control Legislation and Policy: Japan," https://www.loc.gov/law//help/firearms-control/japan.php, accessed July 15, 2020.
47 Setoguchi, *Shuryô to dôbutsugaku no kindai*, 76.
48 Setoguchi, *Shuryô to dôbutsugaku no kindai*, 77.
49 Smith, "[To] T. ICHINOI, Esq.," 24.
50 Smith, "[To] T. ICHINOI, Esq.," 25.
51 Smith, "[To] T. ICHINOI, Esq.," 24–5.
52 David B Kopel, "Japanese Gun Control," *Asia Pacific Law Review*, Vol. 2, No. 2 (1993), 32.
53 David L. Howell, "The Social Life of Firearms in Tokugawa Japan," *Japanese Studies*, Vol. 29, No. 1 (2009), 67.
54 Howell, "The Social Life of Firearms," 67.
55 Howell, "The Social Life of Firearms," 66.
56 Ôtsuka Takeyuki, *Chôjû gyôsei no ayumi* (Tokyo: Forestry Agency, 1969), 537–47.
57 Quoted in Austin, "Waterfowl of Japan," 77.
58 Kopel, "Japanese Gun Control," 35.

59 Noted in graph (Figure 33, p. 95) and text in Austin, "Waterfowl of Japan," 94–5.
60 Contemporary rifles for sale, https://www.hollandandholland.com/all-guns/holland-holland-deluxe-bolt-action-magazine-rifle-cm1107/
61 Terry Wieland, "What Happened to the 16 Gauge Shotgun?" *Gun Digest* (August 19, 2013), at https://gundigest.com/more/classic-guns/the-16-gauge-shotgun-what-happened.
62 From 1976 to 1998, the Japanese company Miroku, established in 1893 and now in Kochi prefecture, was commissioned to produce the Browning A-5. Browning website, "Where Are Browning Firearms Manufactured?" https://www.browning.com/support/frequently-asked-questions/firearms-manufactured.html, accessed July 9, 2020. See also Browning website, Support, Date Your Firearm, "Auto-5 Semi-Automatic Shotgun," https://www.browning.com/support/date-your-firearm/auto-5-semi-automatic-shotgun.html, accessed July 15, 2020.
63 Guns and Ammo Staff, "Browning A-5 Shotgun History," *Guns and Ammo* (November 29, 2017), https://www.gunsandammo.com/editorial/history-of-the-browning-a5-shotgun-gun-stories/248006, accessed July 10, 2020.
64 Author's field notes, Horseshoe Plantation, Tallahassee, Florida, November 5, 2016.
65 Skabelund, *Empire of Dogs*, 24.
66 *Ryôyû*, Vol. 1, No. 10 (November 1, 1900), frontismatter.
67 *Ryôyû*, Vol. 1, No. 11 (December 1, 1900), frontismatter.
68 Ikuta Shigeru, *Shuryôkai no kashitsu bôshi to ryôjû sôsahô teigi ichimei shuryô jôshiki sôjû doku* (Tokyo: Ikuta Shigeru shôten, 1939).
69 NRA Online Hunter Education Course, https://nra.yourlearningportal.com/Course/HuntersEdActivityInfoPage, accessed July 10, 2020.
70 MIT Visualizing Cultures website, "Selling Shiseido—II: Cosmetics Advertising, & Design in Early 20th-Century Japan," Visual Narratives, "Leisure & the Smart Set: Shiseido Graph (*monthly*), 1933–1937," https://visualizingcultures.mit.edu/shiseido_02/sh_visnav05.html, [sho2_GR_1936_03_3211].
71 Ikuta, *Shuryôkai no kashitsu bôshi*, 1.
72 Ikuta, *Shuryôkai no kashitsu bôshi*, frontismatter near table of contents.
73 Ikuta, *Shuryôkai no kashitsu bôshi*, 13.
74 Author's field notes, George Fuller, retired manager, Horseshoe Plantation, Tallahassee, Florida, November 2016.
75 Akio Baba, "The History of Sociology in Japan," *The Sociological Review*, Vol. 10, No. 1 (May 1, 1962), 7.
76 Timon Screech, *The Lens within the Heart: The Western Scientific Gaze and Popular Imagery in Later Edo Japan* (Honolulu: University of Hawaii Press, 2002).
77 Timon Screech, *Obtaining Images: Art, Production and Display in Edo Japan* (Honolulu: University of Hawaii Press, 2012), 316–42.
78 Frederico Marcon, *The Knowledge of Nature and the Nature of Knowledge in Early Modern Japan* (Chicago: University of Chicago Press, 2015), 6.
79 Marcon, *The Knowledge of Nature*, 7.
80 Robert T. Singer and Masatomo Kawai, *The Life of Animals in Japanese Art* (Princeton, NJ: Princeton University Press, 2019).
81 "Scholars engaged in the study of birds, insects, fish, and beasts, and illustrations played a fundamental role in documenting research, conveying information, and aiding taxonomical precision." Curator's introduction in Frederico Marcon, "The Roles and Representations of Animals in Japanese Art and Culture, Part 6—Art, Science, and the Representation of Nature," NGA (National Gallery of Art) Notable

Lectures podcast, https://www.nga.gov/audio-video/audio/japanese-animals-symposium-6.html, accessed May 3, 2021.
82 Marcon, *The Knowledge of Nature*, 5.
83 Marcon, *The Knowledge of Nature*, 6.
84 Marcon, *The Knowledge of Nature*, xi.
85 Marcon, *The Knowledge of Nature*, 5.
86 Marcon, *The Knowledge of Nature*, 28, 33.
87 Positing a binary between religion and science is essentially a modern, constructed concept formulated in the West.
88 Yajima M., "Franz Hilgendorf (1839–1904): Introducer of Evolutionary Theory to Japan Around 1873," in Wyse Jackson, ed., *Four Centuries of Geological Travel: The Search for Knowledge on Foot, Bicycle, Sledge, and Camel, Geological Society (Special Publications)*, 287 (London: Geological Society of London, 2007), 389–93.
89 Isono Naohide, "Contribution of Edward S. Morse to Developing Japan," in Edward R. Beauchamp and Akira Iriye, eds., *Foreign Employees in Nineteenth Century Japan* (New York: Routledge, 1990), 196.
90 Frederico Marcon, "Honzôgaku after Seibutsugaku: Traditional Pharmacology as Antiquarianism after the Institutionalization of Modern Biology in Early Meiji Japan," in Benjamin Elman, ed., *Antiquarianism, Language, and Medical Philology: From Early Modern to Modern Sino-Japanese Medical Discourses* (Leiden: Brill, 2015), 152.
91 L. O. Howard, "Biographical Memoir of Edward Sylvester Morse, 1838–1925," National Academy of the Sciences of the United States of America Biographical Memoirs, Vol. XVII, First Memoir (1935), 5–6.
92 Howard, "Biographical Memoir," 1, 9.
93 Isono, "Contribution of Edward S. Morse to Developing Japan," 193–4.
94 Julia Adeney Thomas, *Reconfiguring Modernity: Concepts of Nature in Japanese Political Ideology* (Berkeley: University of California Press, 2002), 86–7.
95 Morse, *Japan Day by Day, 1877, 1878–1879, 1882–1883* (Boston: Houghton Mifflin, 1917), 138–9.
96 Morse, *Japan Day by Day*, 326.
97 Isono in Beauchamp and Iriye, eds., *Foreign Employees in Nineteenth Century Japan*, 196.
98 Edward S. Morse, Iijima Isao, and Sasaki Chûjirô, *Shell Mounds of Omori: Volume 1, Part 1 of Memoirs of the Science Department* (Tokyo: Tôkyô Daigaku, 1879).
99 Morse, *Japan Day by Day*, 3.
100 Iijima Isao and Sasaki Chûjirô, *Okadaira Shell Mound at Hitachi, Being an Appendix to Memoir, Vol. 1, Part 1 of the Science Department* (Tokyo: University of Tokio [sic], 1882).
101 Many ancient peoples scraped flesh from bones in preparation for ossuaries or other entombments, while animal carcasses and mollusk shells appeared as afterlife provisions for the deceased, so scholars' suppositions of cannibalism betray an Orientalist gaze.
102 Edward S. Morse, *Japan Day by Day, 1877, 1878–1879, 1882–1883* (Atlanta: Cherokee, 1990) and *Japan Day by Day, 1877, 1878–1879, 1882–1883* (Boston: Houghton Mifflin, 1917).
103 Kerim Yasar, *Electrified Voices: How the Telephone, Phonograph, and Radio Shaped Modern Japan, 1868–1945* (New York: Columbia University Press, 2018), 62.
104 Edward S. Morse, *Japanese Homes and Their Surroundings* (New York: Harper, 1885).

105 Edward S. Morse, *Museum of Fine Arts, Catalogue of the Japanese Collection of Pottery* (Cambridge, MA: Riverside Press, 1901).
106 Isono in Beauchamp and Iriye, eds., *Foreign Employees in Nineteenth Century Japan*, 200. University of Tokyo Library System website, "About the Library: Special Collections," https://www.lib.u-tokyo.ac.jp/en/library/contents/about/all_collection, accessed July 15, 2020.
107 Morioka Hiroyuki, Edward C. Dickenson, Hiraoka Takashi, Desmond Allen, Yamasaki Takeshi, *Types of Japanese Birds. National Science Museum Monographs no. 28*. (Tokyo: National Science Museum, 2005), 22, 126.
108 "Isao Ijima: The Father of Parasitology in Japan (With Potrait [sic] Plate)," *The Journal of Parasitology*, Vol. 10, No. 3 (1924), 165–7.
109 Setoguchi Akihisa, "Control of Insect Vectors in the Japanese Empire: Transformation of the Colonial/Metropolitan Environment, 1920–1945," *East Asian Science, Technology, and Society: An International Journal*, Vol. 1 (2007), 170, https://link.springer.com/article/10.1007/s12280-007-9024-3, accessed on March 28, 2020
110 Setoguchi, "Control of Insect Vectors," 170.
111 Amateur Entomologists' Society (AES), "Biting and Sucking Lice (Order Phthiraptera)," https://www.amentsoc.org/insects/fact-files/orders/phthiraptera.html, accessed July 10, 2020.
112 Dr. Tsurumi Miyako, director of the YIO's Division of Natural History, trained in parasitology at Tokyo University of Agriculture. Besides articles on bird censuses and ecology, Tsurumi wrote about ticks in Albatrosses, chewing lice [*Mallophaga*] in Japanese pigeons and doves, and DNA of Japanese ticks: http://www.yamashina.or.jp/hp/gaiyo/staff/tsurumi_miyako_gyoseki.html, accessed July 10, 2020.
113 Uchida Seinosuke, "On a second collection of Mallophaga from Formosan birds," *Annotationes Zoologicae Japonenses*, Vol. 9, No. 5 (1920), 635–52, 1920; "Mallophaga from birds of the Ponape I. (Carolines) and the Palau Is. (Micronesia)," *Annotationes Zoologicae Japonenses*, Vol. 9, No. 4 (1918), 481–93; "Bird-infesting Mallophaga of Japan (III). (Genus Lipeurus)," *Annotationes Zoologicae Japonenses*, Vol. 9, No. 3 (1917), 201–215; "Mallophaga from birds of Formosa," *Journal of the College of Agriculture, Tokyo*, Vol. 3 (1917), 171–88; "Bird-infesting Mallophaga of Japan (II). (Genera Goniodes and Goniocotes)," *Annotationes Zoologicae Japonenses*, Vol. 9, no. 2, pp. 81–8, 1916; "Bird-infesting Mallophaga of Japan. (Genus Physostomum)," *Annotationes Zoologicae Japonenses*, Vol. 9, No. 1 (1915), 67–72.
114 Marcon, in Elman, *Antiquarianism, Language, and Medical Philology*, 152.
115 Hachisuka Masauji, "List of Birds Described by the Japanese Authors," *Tori*, Vol. 11, No. 53–54 (1942), 276.
116 Tôzawa Kôichi, *Rekidai kaichô no kotoba*, in *Nihon Chôgakkai—100 nen no rekishi, Nihon Chôgakkai 100 shûnen kinen tokubetsu gô, Nihon Chôgakkai-shi*, Vol. 61 (November 2012), 8–9.
117 *Nihon chôgakkai* website, *Mokuteki to katsudô*, http://ornithology.jp/aboutOSJ.html. See also Ornithology Exchange website, Journals Database, Ornithological Journals, *Tori*, https://ornithologyexchange.org/resources/journals/database/ornithological-journals/tori-r282/, accessed June 29, 2020.
118 Harriet Ritvo, *The Animal Estate: The English and Other Creatures in the Victorian Age* (Cambridge, MA: Harvard University Press, 1987), 3.
119 Lynn L. Merrill, *The Romance of Victorian Natural History* (New York: Oxford University Press, 1989), 7.

120 Peter D. A. Boyd, "Pteridomania—The Victorian Passion for Ferns," *Antique Collecting* Vol. 28, No. 6 (1993a), 9–10.
121 Barbara T. Gates, "Introduction: Why Victorian Natural History?" *Victorian Literature and Culture* Vol. 35, No. 2 (2007), 541.
122 Steven Cowan, *The Growth of Public Literacy in Eighteenth-Century England* (Doctoral Dissertation, Institute of Education, University of London, 2012), 3, 17–18.
123 "Advertisement," William MacGillivray, *Manual of British Ornithology: Being a Short Description of the Birds of Great Britain and Ireland, including the essential characters of the species, genera, families, and orders: Part I: The Land Birds* (London: Scott, Webster, and Geary, 1840), 7–8.
124 MacGillivray, *Manual of British Ornithology*, Frontispiece.
125 L. B. Holthuis and Tsune Sakai, *Ph. F. von Siebold and* Fauna Japonica: *A History of Early Japanese Zoology* (Tokyo: Academic Press of Japan, 1970).
126 MacGillivray, *Manual of British Ornithology: Part I*, 7.
127 Irby Lovett and, *Cornell Laboratory of Ornithology Handbook of Bird Biology: Third Edition* (Hoboken, NJ: Wiley Blackwell, 2016), 1.
128 Natural History Museum website, "History of the Natural History Museum at Tring," https://www.nhm.ac.uk/visit/tring/about-the-natural-history-museum-at-tring.html, accessed June 23, 2020.
129 Smithsonian Institution Archives, Smithsonian History, "General History," https://siarchives.si.edu/history/general-history, accessed on June 21, 2020.
130 The British Museum website, "The British Museum Story," https://www.britishmuseum.org/about-us/british-museum-story/history, accessed on June 21, 2020.
131 Steven Lubar, *Inside the Lost Museum: Curating, Past and Present* (Cambridge, MA: Harvard University Press, 2017), 14–15.
132 Skabelund, *Empire of Dogs*, 2.
133 Skabelund, *Empire of Dogs*, 13.
134 Skabelund, *Empire of Dogs*, 14–16.
135 A. F. R. Wollaston, *The Life of Alfred Newton: Late Professor of Comparative Anatomy, 1866–1907, with a Preface by Sir Archibald Geikie* (New York: Dutton, 1921), 61.
136 Wollaston, *The Life of Alfred Newton*, 66.
137 Wollaston, *The Life of Alfred Newton*, vii–viii.
138 Wollaston, *The Life of Alfred Newton*, 69–70.
139 The British Ornithologists' Club website, "A History of the BOC," https://boc-online.org/about/history, accessed June 28, 2020.
140 USGS website, "Biological Survey Unit: Patuxent Wildlife Research Center," https://www.usgs.gov/centers/pwrc/science/biological-survey-unit?qt-science_center_objects=0#qt-science_center_objects, accessed June 28, 2020.
141 American Ornithological Society website, "History," https://americanornithology.org/about/history/, accessed June 28, 2020.
142 Ridgeway's 1929 death delayed completion of the eleven volumes until 1950. Robert Ridgeway, *The Birds of North and Middle America. A Descriptive Catalogue of the Higher Groups, Genera, Species, and Subspecies of Birds Known to Occur in North America, from the Arctic Islands to the Isthmus of Panama, the West Indies and Other Islands of the Caribbean Sea, and the Galapagos Archipelago.* No. 50, Part I. (Washington, DC: U.S. National Museum, 1901); *The Birds of North and Middle America.* No. 50, Part II. (Washington, DC: U.S. National Museum, 1902); *The Birds of North and Middle America.* No. 50, Part III. (Washington, DC: U.S.

National Museum, 1904); *The Birds of North and Middle America*. No. 50, Part IV. (Washington, DC: U.S. National Museum, 1907); *The Birds of North and Middle America*. No. 50, Part V. (Washington, DC: U.S. National Museum, 1911); *The Birds of North and Middle America*. No. 50, Part VI. (Washington, DC: U.S. National Museum, 1914); *The Birds of North and Middle America*. No. 50, Part VII. (Washington, DC: U.S. National Museum, 1916); and *The Birds of North and Middle America*. No. 50, Part VIII. (Washington, DC: U.S. National Museum, 1919).

143 Mark Barrow, *A Passion for Birds: American Ornithology After Audubon* (Princeton, NJ: Princeton University Press, 1998), 69.
144 American Ornithological Society website, "History."
145 Barrow, *A Passion for Birds*, 70.
146 Barrow, *A Passion for Birds*, 70–1.
147 Barrow, *A Passion for Birds*, 68.
148 Frontispiece, Hachisuka Masauji, *The Birds of the Philippine Islands: With Notes on the Mammal Fauna, Vol. I Parts I & II Galliformes to Pelecaniformes* (London: H. F. & G. Witherby, 1931–1932).
149 Hiromi Mizuno, *Science for the Empire: Scientific Nationalism in Modern Japan* (Palo Alto, CA: Stanford University Press, 2008), 12.
150 Mizuno, *Science for the Empire*, 12.
151 Jean Delacour, "Hachisuka Masauji," at http://www.avibushistoriae.com/Hachisuka%20Masauji.htm (accessed January 18, 2018).
152 Barney G. Glaser, "The Local-Cosmopolitan Scientist," *American Journal of Sociology* Vol. 69, No. 3 (November 1963), 255. Glaser notes that he conducted his study at "a large government medical research organization devoted to basic research," 250.
153 Glaser, "The Local-Cosmopolitan Scientist," 250–1.
154 The terms "local" and "cosmopolitan" were first used by Robert K. Merton in his study of community leaders. See Merton, *Social Theory and Social Structure* (Glencoe, IL: Free Press, 1957), 387–420. Cited in Glaser, "The Local-Cosmopolitan Scientist," 249.
155 See chapter three.
156 In 1932, Yamashina Yamashina, Inukai Tetsuo, and Natori Bukô published "A List of Bird Skins Presented by Captain Blakiston in the University Museum of Natural History of Sapporo with a Brief Account of His Life in Hokkaidô" (Sapporo: Sapporo University Museum, December 1932).
157 Hachisuka, *A Comparative Handlist of Birds from Japan and the British Isles*.
158 Hachisuka, *Birds of Egypt*.
159 Hachisuka Masauji, *Le Sahara* (Paris: Société d'éditions géographiques maritimes et coloniales, 1932).
160 Murakami, *Zetsumetsu dori dôdô wo oi motometa otoko*, 224–36; and Hachisuka, *The Dodo and Kindred Birds: The Extinct Birds of the Mascarene Islands*.
161 Thomas Wright Blakiston and Henry James Stovin Pryer, *Catalogue of the Birds of Japan* (Yokohama: Lane, Crawford & Co. and London: Trübner and Co., 1880) and *Birds of Japan* (Tokyo: Asiatic Society of Japan, 1882); Thomas Wright Blakiston, Henry James Stovin Pryer, Leonhard Stejneger, and Alexander Wetmore, *Birds of Japan (Revised to 1882)* (Yokohama: Asia Society of Japan, 1882); Thomas Wright Blakiston, *Amended List of the Birds of Japan, According to Geographical Distribution: With Notes Concerning Additions and Corrections Since January 1882* (London: Taylor and Francis, 1884) and *Water Birds of Japan* (Washington, DC: United States National Museum, 1887).

162 Blakiston and Pryer, *Catalogue of the Birds of Japan*.
163 Blakiston, and Pryer, "Birds of Japan." *Transactions of the Sapporo Natural History Society* Vol. XII, Pt. 4 (December 1932), 256, 260.
164 Hachisuka Masauji, *The Birds of the Philippine Islands: With Notes on the Mammal Fauna, Volume 1 [Parts I & II, Galliformes to Pelecaniformes]* (London: H. F. & G. Witherby, 1931), v.
165 Hachisuka Masauji, *Contributions to the Birds of the Philippines* (Tokyo: Ornithological Society of Japan, 1929–30); *Notes sur les Oiseux des Philippines* [Notes on the Birds of the Philippines] (Paris: publisher unspecified, 1931–3); *The Birds of the Philippine [sic] Islands: With Notes on the Mammal Fauna* (London: H. F. and G. Whitherby, 1931–5).
166 Takatsukasa Nobuhisa, Hachisuka Masauji, Kuroda Nagamichi, Uchida Seinôsuke, Yamashina Yoshimaro, *Birds of Jehol* (Tokyo: Office of the Scientific Expedition to Manchoukuo, Waseda University, 1935).
167 Hachisuka Masauji, *Contributions to the Birds of Hainan* (Tokyo: Ornithological Society, 1939).
168 Hachisuka Masauji, *Birds of South and West China* (Tokyo: Ornithological Society, 1940).
169 Hachisuka Masauji, *Birds of Indochina* (Tokyo: Ornithological Society, 1941).
170 Hachisuka Masauji, *Birds of Thailand* (Tokyo: Ornithological Society, 1941).
171 Hachisuka Masauji, Kuroda Nagamichi, and Takatsukasa Nobusuke, *A List of the Birds of Micronesia under Japanese Mandatory Rule* (Tokyo: Ornithological Society, 1942).
172 Yamashina Yoshimaro, *Yamashina Yoshimaro hakase no ayumi* (Tokyo: Sôbun shinsha, 1984).
173 Yamashina Yoshimaro, *How to Breed Fancy Birds* (Tokyo: Takezono-kai, 1926).
174 Yamashina Yoshimaro, Inukai Tetsuo, and Natori Bukô, *A List of Bird Skins Presented by Captain Blakiston in the University Museum of Natural History of Sapporo with a Brief Account of His Life in Hokkaidô* (Sapporo: Sapporo University Museum, December 1932).
175 Michael S. Quinn, "Fleming, James Henry," in *Dictionary of Canadian Biography*, Vol. 16, University of Toronto/Université Laval, 2003, http://www.biographi.ca/en/bio/fleming_james_henry_16E.html, accessed May 22, 2020.
176 Kevin P. Johnson and Michael D. Sorenson, "Phylogeny and Biogeography of Dabbling Ducks (Genus: Anas): A Comparison of Molecular and Morphological Evidence," *The Auk* Vol. 116, No. 3 (July 1999), 792–805, https://sora.unm.edu/sites/default/files/journals/auk/v116n03/p0792-p0805.pdf, accessed June 12, 2020.

Chapter 2

1 Ong, *Neoliberalism as Exception*, 169.
2 Fuchs, *Mimesis and Empire*, 5.
3 Fuchs, *Mimesis and Empire*, 5.
4 Lebra, *Above the Clouds*, 173–4.
5 Jordan Sand, *House and Home in Modern Japan: Architecture, Domestic Space, and Bourgeois Culture* (Cambridge, MA: Harvard University Press, 2005), 16.
6 Sand, *House and Home*, 17–18.
7 Mitsubishi Corporation website, "About Us, MC Library, Our Roots, Vol. 15, Josiah Condor," https://www.mitsubishicorp.com/jp/en/mclibrary/roots/vol15/, accessed August 16, 2020.

8   Keio University website, "History: Keio University, 1912 Keio University Library Completed," https://www.keio.ac.jp/en/about/history/, accessed August 16, 2020.
9   Mitsubishi Corporation website, "About Us, MC Library, Our Roots, Vol. 15, Josiah Condor," https://www.mitsubishicorp.com/jp/en/mclibrary/roots/vol15/, accessed August 16, 2020.
10 Mitsubishi Corporation website, "About Us, MC Library, Our Roots, Vol.14 Yanosuke Resolves to Build Japan's First Modern Business District," https://www.mitsubishicorp.com/jp/en/mclibrary/roots/vol14/, accessed August 16, 2020.
11 Seikado Bunko Art Museum, "About the Museum," http://www.seikado.or.jp/en/about/, accessed August 16, 2020.
12 Teien Art Museum website, "Former Residence of Prince Asaka/Brief History," https://www.teien-art-museum.ne.jp/en/museum/index03.html, accessed August 16, 2020.
13 Takanami Machiko, *Kôun no kenchiku Asaka gutei–kenbutsu no chikara*, in *Kagu dogu shitsunai-shi gakkai, Tokushû Asaka gutei to Aaru.Deko, Kagu dôgu shitsunai-shi*, Vol. 6 (May 2014), 9–10.
14 Teien Art Museum website, "Architects and Designers I," https://www.teien-art-museum.ne.jp/en/museum/index05.html, accessed August 16, 2020.
15 *Tokushû Asaka gutei to Aaru.Deko*.
16 Daqing Yang, "Convergence or Divergence? Recent Historical Writings on the Rape of Nanjing," *American Historical Review*, Vol. 104, No. 3 (June, 1999), 855.
17 Teien Art Museum website, "Former Residence of Prince Asaka/Brief History," at http://www.teien-art-museum.ne.jp/en/museum/index03.html, accessed August 16, 2020.
18 Tony Austin, oral history interview transcript, interviewed by author, FSU History Department (November 12–13, 2013), 3.
19 http://japan.embassy.gov.au/tkyo/aboutus.html
20 Australian Information Office guidebook, *Teien–Oosutoraria taishikan* (Tokyo: Shunhôsha, possibly 1988), 1.
21 Australian Information Office, *Teien–Oosutoraria taishikan*, 1.
22 Author's visit to Mita Ward archives, November 19, 2015.
23 "Fusing East and West," in Sand, *House and Home*, 304–6.
24 Sand, *House and Home*, 332.
25 Miriam Silverberg, "Constructing the Japanese Ethnography of Modernity," *The Journal of Asian Studies*, Vol. 51, No. 1 (February, 1992), 38.
26 Nakamura Tsukasa, *Hachisuka Masauji seidan hyakunen ni yosete: Hachisuka Masauji hakase ni manabu*, Bulletin of the Biogeographical Society of Japan, Vol. 58, No. 2 (December 2003), 101–2.
27 Murakami, *Zetsumetsu dori dôdô wo oimotometa otoko*, 238.
28 Nakamura, *Hachisuka Masauji seidan hyakunen ni yosete*, 101.
29 Murakami, *Zetsumetsu dori dôdô wo oimotometa otoko*, 239.
30 Rexford Newcomb and Marc Appleton, *Mediterranean Domestic Architecture for the United States* (Twentieth Century Landmarks in Design, Vol. 9) (New York: Acanthus Press, 1999).
31 Tang Yingxian, "Shanghai's Little-Known Spanish Style," *Shanghai Daily* online (May 10, 2011), B4, https://archive.shine.cn/feature/Shanghais-littleknown-Spanish-style/shdaily.shtml, accessed September 9, 2020.
32 Ogasawara-Tei website, "Contents of Ogasawara-Tei: History," https://www.ogasawaratei.com/en/ogasawara/history/, accessed September 9, 2020.

33 World Biographical Encyclopedia, "Nagayoshi Ogasawara: Politician," https://prabook.com/web/nagayoshi.ogasawara/3745266, accessed September 9, 2020.
34 Toyo Eiwa Jogakuin, "History," https://www.toyoeiwa.ac.jp/english/history/index.html, accessed August 15, 2020.
35 Nakanishi Gotô, *Chôeisha*, in *Zenshû Nippon yachô ki 6* (Tokyo: Kodansha, 1986); and Murakami, *Zetsumetsu dori dôdô wo oimotometa otoko*, 240.
36 Murakami, *Zetsumetsu dori dôdô wo oimotometa otoko*, 242–3.
37 Patrick W. Gailbraith, *Otaku Spaces* (Seattle: Chin Music Press, 2012).
38 Yuji Gushiken and Tatiane Hirata, "Processes of Cultural and Media Consumption: The Image of 'Otaku' from Japan to the World," in *Intercom: Revista Brasiliera de Ciências da Communicaçao*, Vol. 37, No. 2 (July/December 2014), found at http://www.scielo.br/scielo.php?pid=S1809-, accessed on January 18, 2018.
39 Alex Martin, "Defining the Heisei Era: Part 7—Obsession: Examining the Rise of Otaku Culture," *The Japan Times*, November 24, 2018, https://features.japantimes.co.jp/heisei-moments-part-7-obsession/, accessed February 26, 2020.
40 Walter Benjamin (trans. Howard Eiland and Kevin McLaughlin), *The Arcades Project* (Cambridge, MA: Harvard University Press, 2002), 204–5.
41 Lubar, *Inside the Lost Museum*, 14–15.
42 Samuel J. M. M. Alberti, "Placing Nature: Natural History Collections and Their Owners in Nineteenth-Century England," *The British Journal for the History of Science*, Vol. 35, No. 3 (September 2002), 293.
43 Pitelka, *Spectacular Accumulation*, 5.
44 Pitelka, *Spectacular Accumulation*, 11, 13.
45 It sold for USD $11,701,000 to a Chinese collector on September 15, 2016, after being deregistered as "Important Art Object" in Japan on September 4, 2015. Christie's website, "The Classic Age of Chinese Ceramics: The Linyushanren Collection, Part II, New York, September 15, 2016," Lot 707: "The Kuroda Family *Yuteki Tenmoku*: a Highly Important and Very Rare 'Oil Spot' *Jian* Tea Bowl, Southern Song Dynasty (1127–1279)," https://www.christies.com/lotfinder/Lot/the-kuroda-family-yuteki-tenmokua-highly-important-6019217-details.aspx, accessed July 14, 2020.
46 Oliver L. Austin and Kuroda Nagahisa, "Birds of Japan: Their Status and Distribution," *Bulletin of the Museum of Comparative Zoology*, Vol. 109, No. 4 (1953), 350.
47 Noriko Aso, *Public Properties: Museums in Imperial Japan* (Durham, NC: Duke University Press, 2013).
48 Ian Miller, *The Nature of the Beasts: Empire and Exhibition at the Tokyo Imperial Zoo* (Cambridge, MA: Harvard University Press, 2013), 81–5.
49 Alberti, "Placing Nature," 296.
50 Smithsonian Institution, Report on the Progress and Condition of the United States National Museum for the Year Ended June 30, 1935 (Washington, DC: US Government Printing Office, 1936).
51 The Report of the National Museum, 69.
52 Smithsonian Institution Archives, Mr. Theodore T. Belote, 1935, caption for image taken by Ruel P. Tolman, September 1935, Smithsonian Institution Archives, Record Unit 7433, Box 3, Folder: Scrapbook A-N, https://siarchives.si.edu/collections/siris_sic_13996, accessed August 18, 2020.
53 The Report of the National Museum, 88.
54 Bird Life International, IUCN Red List of Threatened Species (2012), https://www.iucnredlist.org/species/22681531/125513230, accessed April 15, 2020.

55 Brigette C. Kammsler, "The Gripsholm Exchange and Repatriation Voyages," The Burke Library Blog, Columbia University (September 17, 2012), https://blogs.cul.columbia.edu/burke/2012/09/17/the-gripsholm-exchange-and-repatriation-voyages-2/, accessed September 9, 2020.
56 Oliver L. Austin, letter to Jimmy Peters, November 25, 1946, Tokyo.
57 Yamashina Yoshimaro, *Watashi no rirekisho, dai jûhakkai sengo no seikatsu*, at http://www.yamashina.or.jp/hp/yomimono/rirekisho/rirekisho18.html, accessed February 28, 2018.
58 Alexander Wetmore, "In Memoriam: James L. Peters," *The Auk*, Vol. 74, No. 2 (April 1957), 167–73.
59 Likely Lloyd R. Wolfe, who Hachisuka befriended in the Philippines, but it might be Lieutenant General Kenneth Bonner Wolfe (1896–1971), who directed bombing raids on Japan from western China, beginning on June 15, 1944. After the war, he served as chief of staff, then commanding general of Okinawa's Fifth Air Force. From his Nagoya-based headquarters, he directed transitioning a wartime fighting force to occupational air arm. From May 1926 to August 1928, he served as plans and operations officer at Clark Field in the Philippines. See "Lieutenant General Kenneth Bonner Wolfe," at http://www.af.mil/About-Us/Biographies/Display/Article/105222/lieutenant-general-kenneth-bonner-wolfe/.
60 My translation. Yamashina Yoshimaro, *Watashi no rirekisho, dai jûhakkai sengo no seikatsu*, found at http://www.yamashina.or.jp/hp/yomimono/rirekisho.
61 Geoffrey T. Hellman, "Profiles: Curator Getting Around," *The New Yorker* (August 26, 1950), https://www.newyorker.com/magazine/1950/08/26/curator-getting-around, accessed September 6, 2020.
62 Marcel Mauss, *The Gift: The Form and Reason for Exchange in Archaic Societies* (London: Routledge, 1990), 6–7.
63 Mauss, *The Gift*, 6–7.
64 Mauss, *The Gift*, 9.
65 Jean Delacour, *The Living Air: The Memoirs of an Ornithologist* (London: Country Life, 1966), 42.
66 BirdLife International website, "Our History: BirdLife—The World's Oldest International Conservation Organization," https://www.birdlife.org/worldwide/partnership/our-history, August 18, 2020.
67 *Asahi Shimbun*, evening edition (April 20, 1924), 2.
68 "Notes and News," *The Auk*, Vol. XLI (July 1924), 514.
69 The Avicultural Society website, http://www.avisoc.co.uk/, accessed on July 4, 2020.
70 "Recent Literature," *The Auk*, Volume XXIX (July 1922), 446–7.
71 Delacour, *The Living Air*, 49.
72 "I had met in Europe Prince Nobosuke [sic] Taka-Tsukasa, who had come over to work on birds. We soon became friends, having common interests. One evening when he came to dine at our Paris house he brought a young cousin Masauji Hachisuka, who had just arrived to perfect his knowledge of English and French and also to undertake zoological studies at Cambridge University." Delacour, *The Living Air*, 135.
73 Lindholm of Delacour, "M. Delacour's New Birds," *The Avicultural Magazine*, Series IV, Vol. IV (July 1926), 194–5. Found on The Avicultural Society website, Josef Lindholm III, "Jean Delacour and the Avicultural Magazine, Part II: 1920–1944," first published in *The Avicultural Magazine*, Vol. 100, No. 2 (1994), http://www.avisoc.co.uk/table-of-contents/jean-delacour-and-the-avicultural-magazine-part-ii/, accessed July 4, 2020.

74　Lindholm of Delacour, "M. Delacour's New Birds," 194–5.
75　Delacour, *The Living Air*, 41.
76　Lindholm of Delacour, "M. Delacour's New Birds," 194–5.
77　Lindholm of Delacour, "M. Delacour's New Birds," 194–5.
78　Jean Delacour, "Japanese Aviculture," *The Avicultural Magazine*, Vol. IV, Series IV (August 1926), 213–18. The Avicultural Society website, Josef Lindholm III, "Jean Delacour and the Avicultural Magazine, Part II: 1920–1944," first published in *The Avicultural Magazine,* Vol. 100, No. 2 (1994), http://www.avisoc.co.uk/table-of-contents/jean-delacour-and-the-avicultural-magazine-part-ii/, accessed July 4, 2020.
79　Lindholm of Delacour, "Japanese Aviculture (August 1926)," 213–18.
80　Lindholm of Delacour, "Japanese Aviculture (August 1926)," 213–18.
81　Lindholm of Delacour, "Japanese Aviculture (August 1926)," 213–18.
82　Jean Delacour, "Japanese Aviculture," Vol. IV, Series IV (September 1926), 247–53. The Avicultural Society website, Josef Lindholm III, "Jean Delacour and the Avicultural Magazine, Part II: 1920–1944," first published in *The Avicultural Magazine,* Vol. 100, No. 2 (1994), http://www.avisoc.co.uk/table-of-contents/jean-delacour-and-the-avicultural-magazine-part-ii/, accessed July 4, 2020.
83　Lindholm of Delacour, "Japanese Aviculture (September 1926)," 247–53.
84　Takashi Fujii, of the Japanese Society for Preservation of Birds, believes rose-ringed parakeets are escaped descendants of 1960s-1970s pet trades. Alice Gordenker, "Feral Parakeets," *Japan Times* (March 19, 2009), https://www.japantimes.co.jp/news/2009/03/19/reference/feral-parakeets/, accessed July 16, 2020.
85　Lindholm of Delacour, "Japanese Aviculture (September 1926)," 247–53.
86　Gen Anderson, "Surviving Against All Odds," *Galliformes TAG* Newsletter (January 2016), 5, http://aviansag.org/Newsletters/Galliformes_TAG/Galliformes_2016.pdf, accessed July 5, 2020.
87　Koyama, *Japanese Students*, 81.
88　Koyama, *Japanese Students*, 82.
89　Koyama, *Japanese Students*, 91.
90　Koyama, *Japanese Students*, 83.
91　Cornell Lab of Ornithology, All About Birds, Eurasian Widgeon Range Map, https://www.allaboutbirds.org/guide/Eurasian_Wigeon/maps-range, accessed July 5, 2020.
92　Lindholm of Delacour, "Japanese Aviculture (September 1926)," 247–53.
93　Delacour, *The Living Air*, 136.
94　Lindholm of Delacour, "Japanese Aviculture (September 1926)," 247–53.
95　Bird Life International Website, Datazone, Species Factsheet: Mikado Pheasant *Syrmaticus mikado*, http://datazone.birdlife.org/species/factsheet/mikado-pheasant-syrmaticus-mikado, accessed July 5, 2020.
96　Lindholm of Delacour, "Japanese Aviculture (September 1926)," 247–53.
97　Christos Sokos and Periklis K. Birtsas, "The Last Indigenous Pheasant Black-Necked Pheasant Population of Europe," *G@llinformed [Newsletter of the Galliformes Specialist Group]*, Vol. 8 (April 2014), 13–22; https://www.researchgate.net/publication/283788773_The_last_indigenous_Black-necked_Pheasant_population_of_Europe, accessed September 9, 2020.
98　Noted in "Recent Literature," *The Auk*, Vol. XXIX (July 1922), 449. Takatsukasa Nobusuke and Kuroda Nagamichi, "On a New Genus Proposed for Mikado Pheasant of Formosa," *Tori*, Vol. 3, No. 12–13 (1922), 37.
99　Kuroda Nagamichi, *The Pheasants of Japan Including Korea and Formosa* (Tokyo: Self-Published, 1926).

100 Witmer Stone, "Recent Literature: Kuroda's Monograph of the Pheasants of Japan," *The Auk*, Vol. 43, No. 3 (July 1, 1926), 394.
101 William Beebe, *A Monograph of the Pheasants: Four Volumes* (London: George Witherby and Company, 1918–22).
102 Stone, "Recent Literature," 394.
103 "NEW EXPERT AT ZOO LOST BIRDS IN WAR; Captain Jean Delacour, Whose Aviaries Nazis Destroyed, Gets Post in Bronx TO TEST NATURAL HABITAT Ornithologist Plans to Make Over Undeveloped Parts of Park Under New Plan," *The New York Times*, Obituaries Section (January 27, 1941), 17.
104 Jean Delacour, *The Pheasants of the World* (London: Country Life, 1951).
105 Carol G. Gould, *The Remarkable Life of William Beebe: Explorer and Naturalist* (Washington, DC: Island Press, 2004), 131–3.
106 Beebe, *A Monograph of the Pheasants*, Frontispiece, Plate 1.
107 Gould, *The Remarkable Life of William Beebe*, 170.
108 Gould, *The Remarkable Life of William Beebe*, 171.
109 Gould, *The Remarkable Life of William Beebe*, 172.
110 Gould, *The Remarkable Life of William Beebe*, 171.
111 Delacour also published Jean T. Delacour, *Pheasants: Their Care and Breeding* (Neptune, NJ: T. F. H. Publishing, 1978).
112 Lindholm of Delacour, "Japanese Aviculture (September 1926)," 247–53.
113 Lindholm of Delacour, "Japanese Aviculture (September 1926)," 247–53.
114 Kuroda Nagamichi, "On a Third Specimen of Rare Pseudotadorna cristata Kuroda: In Japanese and English," reprinted from *Tori*, Vol. iv, No. 18 (October 1924).
115 Lindholm of Delacour, "Japanese Aviculture (September 1926)," 247–53.
116 Lindholm of Delacour, "Japanese Aviculture (September 1926)," 247–53.
117 Advertisement, *Asahi Shimbun*, morning edition (August 17, 1923), 1; for Takatsukasa Nobusuke, *Kaidori* (Tokyo: Shôkabô, 1923).
118 *Asahi Shimbun*, morning edition (July 28, 1924), 1; *Asahi Shimbun*, morning edition (April 25, 1925), 3; *Asahi Shimbun*, morning edition (May 28, 1925), 1; *Asahi Shimbun*, morning edition (May 23, 1926), 6; for Takatsukasa Nobusuke, *Kaidori* (Tokyo: Shôkabô, 1924).
119 Takatsukasa Nobusuke, *Chakushoku zuhen kaidori shusei* (Tokyo: Yokendô, 1930).
120 Prince Takatsukasa Nobusuke, *Illustrated, Colored, Bird Collection, with Care and Feeding* (Tokyo: Yokendô Bookstore, 1930), 320 pp.
121 Hachisuka Masauji, *The Birds of the Philippine Islands: With Notes on the Mammal Fauna: Volume I, Parts I & II, Galliformes to Pelecaniformes* (London: H. F. & G. Witherby, 1931–2), V.
122 Takatsukasa Nobusuke, *Japanese Birds* (Tokyo: Maruzen, 1941).
123 Kingsberg, *Into the Field*, 7.
124 Kuroda Nagahisa, "Preface," in *The Ornithologists and Avifauna in [sic] Early History of Japanese Field Ornithology (Chiefly Until 1950)* (Tokyo: Maruzen, 2004), V.
125 Oliver L. Austin, Jr., letter to Jimmy (James L.) Peters, September 19, 1946, 1.
126 Suginami Ward Regional Museum staff, *Nakanishi Gotô seitan 120-nen: Yachô no chichi, Nakanishi Gotô wo meguru hitobito–tenji zuryoku* (Tokyo: Suginami Ward Regional Museum, 2015), 15.
127 Kuroda, "A History of Field Ornithology in Japan," in *The Ornithologists and Avifauna*, 37.
128 Hidefumi Imura and Miranda Alice Schreurs, eds., *Environmental Policy in Japan* (Cheltenham, UK: Edward Elgar Press, 2005), 322.

129 My translation. Takatsukasa Nobusuke, *Chô no hôgo*, *Asahi Shimbun* (March 11, 1927), 9.
130 Suginami Ward Regional Museum staff, *Nakanishi Gotô seitan 120-nen*, 10.
131 Hiratsuka Raichô (translated by Teruko Craig), *In the Beginning, Woman Was the Sun: Autobiography of a Japanese Feminist* (New York: Columbia University Press, 2006), 166.
132 Yamashina Institute for Ornithology website, "About Us: The Founder, Dr. Yamashina Yoshimaro," http://www.yamashina.or.jp/hp/english/about_us/founder.html, accessed September 9, 2020.
133 Suginami Ward Regional Museum staff, *Nakanishi Gotô seitan 120-nen*, 54.
134 Recounted by her niece, Dr. Hasegawa Michiko, professor emerita, Saitama University, personal interview, March 23, 2018, Association for Asian Studies Annual Meeting, Washington, DC.
135 Pëtr Kropotkin, *Mutual Aid: A Factor of Evolution* (1902), 14–15; https://theanarchistlibrary.org/library/petr-kropotkin-mutual-aid-a-factor-of-evolution, accessed September 9, 2020.
136 Kuroda, *The Ornithologists and Avifauna*, 37.
137 Author's interview with Hasegawa.
138 Author's interview with Hasegawa.
139 Suginami Ward Regional Museum staff, *Nakanishi Gotô seitan 120-nen*, 42–2.
140 Murakami, *Zetsumetsu-chô dôdô wo oimitometa otoko*, 242.
141 Kuroda, *The Ornithologists and Avifauna*, 37.
142 Yamashina Yoshimaro, *Yamashina Yoshimaro hakase no ayumi* (Tokyo: Sôbunshinsha, 1984), 15.
143 Yamashina, *Yamashina Yoshimaro hakase no ayumi*, 17.
144 *Reichs Archiv–seikai teiô jiten*, Ie risuto, Yamashina ie, https://reichsarchiv.jp/家系リスト/山階宮家#sankai02b, accessed September 3, 2020.
145 Yamashina Yoshimaro, *Dai 25-kai tsuma no shi, Yamashina Yoshimaro: Watakushi no Rirekisho*, http://www.yamashina.or.jp/hp/yomimono/rirekisho/rirekisho25.html, accessed August 30, 2020.
146 Tony Austin, interview with author, November 12 to 14, 2013, Florida State University History Department, interview transcript, 40–1.
147 Austin, interview transcript, 40–1.
148 Austin, interview transcript, 40–1.
149 Referencing connotations of "queer," Cook writes that "[t]hese varied yet loosely connected associations of oddity, badness, malformation or foreignness resonate through the proliferating and then narrowing meanings of queer in the years that followed. In the last part of the nineteenth century and the first half of the twentieth these touched eccentricity, Bohemianism, and exoticism. Thereafter they signaled homosexual difference more distinctly." Matt Cook, *Queer Domesticities: Homosexuality and Home Life in Twentieth Century London* (New York: Palgrave Macmillan, 2014), 7.
150 Murakami, *Zetsumetsu-chô dôdô wo oi mitometa otoko*, 238.
151 Yamashina, *Yamashina hakase no ayumi*, 15.

Chapter 3

1 Early twentieth-century marine biologists, especially those in the Pacific basin, enthusiastically sought international connections and research exchanges. Antony

Adler, *Neptune's Laboratory: Fantasy, Fear, and Science at Sea* (Cambridge, MA: Harvard University Press, 2019), 82.

2   Masa U. Hachisuka, F. Z. S., *A Comparative Hand List of the Birds of Japan and the British Isles* (Cambridge, UK: Cambridge University Press, 1925), n.p.

3   In mid-October 2014, Clare Welford-Elkin of University of Cambridge Library's Rare Books Department showed me Hachisuka's student record card and materials on former Japanese students, while Noboru Koyama of the Japanese and Korean Department illuminated lives of the University's prewar Japanese aristocrats, and noted how patterns of informal study were extremely common. I thank both librarians for extensive assistance.

4   Individuals use "honorary whiteness" to achieve privileges in a colonial system presupposing white superiority. Ong, *Neoliberalism as Exception*, 169.

5   Edward W. Said, *Orientalism* (New York: Pantheon Books,1978) and *Out of Place: A Memoir* (New York: Vintage Books, 2000). Sir Hugh Cortezzi, *Japan in Late Victorian London: The Japanese Native Village in Knightsbridge and "The Mikado", 1885.* (London: Sainsbury Institute, 2009) and Koyama Noboru, *Rondon Nihonjin-mura o tsukutta otoko: nazo no Kōgyōshi Tanakā Buhikurosan, 1839–94* (Tokyo: Fujiwara shoten, 2015). At Humphrey's Hall in Knightsbridge, a fashionable London shopping area, Tannaker Buhicrosan built a Japanese village, featuring one hundred Japanese inhabitants hired for a tea-house, Buddhist temple, houses, and artisans' shops in an exhibition from January 10, 1885 to June 1887. W. S. Gilbert (1836–1911) visited the village and employed Japanese workers to train his *Mikado* cast in Japanese social customs and gestures.

6   Hilary Perraton, *A History of Foreign Students in Britain* (New York: Springer, 2014), 162.

7   Perraton, *A History of Foreign Students in Britain*, 162.

8   Adam and Galinsky, "Enclothed Cognition," 918–19.

9   Henry Poole & Co., "Heritage: Hall of Fame," https://henrypoole.com/individual/crown-prince-hirohito-of-japan/ and https://henrypoole.com/individual/hih-prince-takamatsu-japan/, accessed May 31, 2020.

10 University of Cambridge Office of Communications, "Filming the Life of a Clare College Alumnus," (October 9, 2008) https://www.cambridgenetwork.co.uk/news/filming-the-life-of-a-clare-college-alumnus, accessed May 31, 2020; and Edan Corkill, "Redefining Defiance for a Modern Japan: Yusuke Iseya plays Japan's Most Singular Gentleman, Jiro Shirasu, Whose Sturdy Principles Helped Rebuild a Postwar Nation," *Japan Times* (March 6, 2009), https://www.japantimes.co.jp/culture/2009/03/06/general/redefining-defiance-for-a-modern-japan/#.XtRvQRNKiIY, accessed May 31, 2020.

11 Koyama Noboru (trans. Ian Ruxton), *Japanese Students at Cambridge University in the Meiji Era, 1868–1912: Pioneers for the Modernization of Japan* (Morrisville, NC: Lulu Press, 2004), 44.

12 Norman Boyd Kinnear, "Obituary: The Marquess Hachisuka," *Ibis* 96 (1954), 150.

13 Murakami, *Zetsumetsu dori dôdô wo oimotometa otoko*, 324.

14 Pernille Rudlin, *The History of Mitsubishi Corporation in London: 1915 to Present Day* (London: Routledge, 2014), 45.

15 Murakami, *Zetsumetsu dori dôdô wo oimotometa otoko*, 324.

16 Michael S. Quinn, "Fleming, James Henry," *Dictionary of Canadian Biography*, Vol. XVI (1931–40) (Toronto: University of Toronto/Université Laval, 2017), http://www.biographi.ca/en/bio/fleming_james_henry_16E.html, accessed May 22, 2020.

17 E. J. H. Corner, "His Majesty Emperor Hirohito of Japan, K. G.: April 29, 1901–January 7, 1989, Elected F.R.S. 1971," *Biographical Memoirs of Fellows of the Royal Society*, Vol. 36 (December 1990), 244, 246–7, or https://royalsocietypublishing.org/doi/pdf/10.1098/rsbm.1990.0032, accessed May 11, 2021.

18 Antony Adler, "Emperor Hirohito: The Marine Biologist Who Ruled Japan," *History of Oceanography* (August 15, 2015), https://oceansciencehistory.com/2015/08/15/emperor-hirohito-the-marine-biologist-who-ruled-japan/, accessed May 9, 2021.

19 Yves Samyn, "Return to Sender: Hydrozoa Collected by Emperor Hirohito of Japan in the 1930s and Studied in Brussels," *Archives of Natural History*, Vol. 41, No. 1 (2014): 1–9.

20 Steven Large, *Emperor Hirohito and Showa Japan: A Political Biography* (London: Routledge, 1992, 2013), 16.

21 Corner, "His Majesty Emperor Hirohito of Japan," 247.

22 Corner, "His Majesty Emperor Hirohito of Japan," 257.

23 Murakami, *Zetsumetsu dori dôdô wo oimotometa otoko*, 42.

24 Corner, "His Majesty Emperor Hirohito of Japan," 245. For condensed details of Hirohito's life, see also Robert Trumbull, "A Leader Who Took Japan to War, to Surrender, and Finally to Peace: [Obituary]," *The New York Times* Section 1 (January 7, 1989), 6.

25 Murakami, *Zetsumetsu dori dôdô wo oimotometa otoko*, 43.

26 Murakami, *Zetsumetsu dori dôdô wo oimotometa otoko*, 42.

27 Murakami, *Zetsumetsu dori dôdô wo oimotometa otoko*, 324.

28 Great Britain's oldest and most prestigious universities hold the three-term academic year system reflecting Anglican underpinnings. Michaelmas Term (named for the September 29 Feast of Saint Michael) traditionally begins late September or early October and runs to early December; Lent Term runs from mid-January to mid-March; and Easter Term runs from late April to mid-June. See "About the University: Term Dates," https://www.cam.ac.uk/about-the-university/term-dates-and-calendars, accessed on May 27, 2020.

29 Selwyn achieved full collegiate status on March 14, 1958. See The Master, Fellow and Scholars, "Selwyn College, 1882–1973: A Short History," https://www.sel.cam.ac.uk/about/selwyn-history, accessed June 1, 2020.

30 Richard Bowring, "The Selwyn Swastika," in Peter Fox et al., eds., *Calendar: Selwyn College Cambridge Vol. 125 (2017-2018)* (Cambridge, UK: Selwyn College, 2018), 63.

31 Interview with author, University of Cambridge Library, October 16, 2014.

32 Americans call this "private school."

33 The Leys School, http://www.theleys.net/history, accessed May 30, 2020.

34 Koyama, *Japanese Students at Cambridge*, 50.

35 Kinnear, "Obituary: The Marquess Hachisuka," 150.

36 Koyama, *Japanese Students at Cambridge University in the Meiji Era*, 193.

37 "Dr. F. H. H. Guillemard," *Nature* 133 (February 3, 1934), 167.

38 Hachisuka Masauji, *The Birds of the Philippine Islands: With Notes on the Mammal Fauna, Volume II [Parts III & IV, Acciptriformes to Passeriformes (Timaliidae)]* (London: H. F. & G. Witherby, 1934–1935), VI.

39 Kinnear, "Obituary: The Marquess Hachisuka," 150.

40 The inside cover page of the 1925 text *A Comparative Hand List of the Birds of Japan and the British Isles*, is Hachisuka's first publication to list F. Z. S. after his name. Lord Rothschild was not elected to the Society until 1911, about three years after quitting family bank employment to work full time at his museum.

41 Masa U. Hachisuka M. B. O. U., "XXXIX: Notes on Some Birds from Egypt," *Ibis* Vol. 66, Issue 4 (October 1924), 771.
42 Flower was well-connected; from 1927 until 1929, he was Vice President of the (Royal) Zoological Society and Chairman of the BOC from 1930 until 1933. J. L. Chaworth-Musters, "Obituary. Major S. S. Flower, O.B.E.," *Ibis*, Vol. 88, No. 2 (1946), 244.
43 Chaworth-Musters, "Obituary. Major S. S. Flower," 244.
44 Hachisuka Masauji, *Birds of Egypt, with Introductory Note by D. Kuroda* (Tokyo: Ornithological Society, 1926).
45 Witmer Stone, "Hachisuka on Egyptian Birds," *The Auk*, Vol. 44, No. 1 (January 1, 1927), 132.
46 Masa U. Hachisuka, *A Handbook of the Birds of Iceland* (London: Taylor and Francis, 1927), 3.
47 Hachisuka, *A Handbook of the Birds of Iceland*, iii.
48 Hachisuka, *A Handbook of the Birds of Iceland*, 4.
49 Hachisuka, *A Handbook of the Birds of Iceland*, Plate VII.
50 Samantha Galasso, "When the Last of the Great Auks Died, It Was by the Crush of a Fisherman's Boot: Birds Once Plentiful and Abundant, Are the Subject of a New Exhibition at the Natural History Museum," *Smithsonian Magazine* (July 10, 2014), https://www.smithsonianmag.com/smithsonian-institution/with-crush-fisherman-boot-the-last-great-auks-died-180951982/, accessed June 5, 2020.
51 W. E. H. B., *Nature*, Vol. 121, No. 3052 (April 28, 1928), 669.
52 Murakami, *Zetsumetsu dori dôdô wo oimotometa otoko*, 324.
53 Ernst Mayr, "In Memoriam: Jean (Theordore) Delacour," *The Auk*, Vol. 103 (July 1986), 604.
54 Niall Ferguson, *The House of Rothschild: Volume 2: The World's Banker: 1849–1999* (New York: Penguin, 2000).
55 These were later padlocked after attacking too many guests.
56 Kirk Wallace Johnson, *The Feather Thief: Beauty, Obsession, and the Natural History Heist of the Century* (New York: Penguin, 2018), 38–9.
57 Louis Wain, "The Hon. Walter Rothschild's Pets: A Visit to Tring Museum," *The Windsor Magazine* (December 1895) in Ward, Lock, and Bowden, Limited, *The Windsor Magazine: An Illustrated Monthly for Men and Women, Vol. II, July to December* (London: William Clowes and Sons, 1895), 661.
58 Wain, "The Hon. Walter Rothschild's Pets," 663.
59 Hachisuka Masauji, *The Dodo and Kindred Birds: The Extinct Birds of the Mascarene Islands* (London: H. F. & G. Witherby, 1953).
60 Murakami, *Zetsumetsu dori dôdô wo oimotometa otoko*, 47–8. See also Hachisuka Masauji, *Minami no tanken* (Tokyo: Heibonsha, 2006).
61 Ernst Hartert, F. C. R. Ticehurst H. F. Witherby, *A Hand-List of British Birds: With an Account of the Distribution of Each Species in the British Isles and Abroad* (London: Witherby & Co, 1912.)
62 Hachisuka, *A Comparative Hand List of the Birds of Japan and the British Isles*, n.p.
63 Hachisuka, *A Comparative Hand List of the Birds of Japan and the British Isles*, n.p.
64 Hachisuka, *A Comparative Hand List of the Birds of Japan and the British Isles*, n.p.
65 J. A. Venn, *Alumni Cantabrigienses: A Biographical List of All Known Students, Graduates, and Holders of Office at the University of Cambridge, from the Earliest Times to 1900* (Cambridge, UK: Cambridge University Press, 1951), 71.
66 Murakami, *Zetsumetsu dori dôdô wo oimotometa otoko*, 92.

67 Murakami, *Zetsumetsu dori dôdô wo oimotometa otoko*, 92. See also Hachisuka Toshiko, *Daimyô kazoku* (Tokyo: Mikasa shobô, 1957).
68 Murakami, *Zetsumetsu dori dôdô wo oimotometa otoko*, 92.
69 Murakami, *Zetsumetsu dori dôdô wo oimotometa otoko*, 221, 236-7, 329.
70 Kristin Johnson, *Ordering Life: Karl Jordan and the Naturalist Tradition* (Baltimore: Johns Hopkins University Press, 2012), 128.
71 Miriam Rothschild, *Dear Lord Rothschild: Birds, Butterflies, and History* (Glenside, UK: Balaban Publishers, 1983), 92, 139, 302. Miriam Rothschild, *Walter Rothschild: The Man, the Museum, and the Menagerie* (London: Natural History Museum, 2008).
72 Johnson, *The Feather Thief*, 40; and Tim Burkhead, Jo Wimpenny, and Bob Montgomerie, *Ten Thousand Birds: Ornithology Since Darwin* (Princeton, NJ: Princeton University Press, 2014), 77.
73 Walter J. Bock, "Ernst Mayr at 100: A Life Inside and Outside of Ornithology," *The Auk*, Vol. 121, No. 3 (July 1, 2004), 642.
74 Bock, "Ernst Mayr at 100," 642.
75 Burkhead, Wimpenny, and Montgomerie, *Ten Thousand Birds*, 77-8.
76 Hachisuka, *The Birds of the Philippine Islands*, v.
77 Hachisuka, *The Birds of the Philippine Islands*, v.
78 H. S. Barlow, "John Waterstradt, 1869–1944," *Journal of the Malaysian Branch of the Royal Asiatic Society*, Vol. 42, No. 2 (216) (December, 1969), 127.
79 Herbert A. Giles and Masa U. Hachisuka, *Record of Strange Nations: From the Chinese of 1392 A.D.–Publishers' Announcement* (London: Percy Lund, Humphries and Co. Ltd., early 1927). (forward by Alfred C. Haddon), front cover.
80 Giles and Masa U. Hachisuka, *Record of Strange Nations*, Order Form.
81 CPI Inflation Calculator, UK Inflation Calculator, https://www.in2013dollars.com/uk/inflation/1927?amount=3.15, accessed on April 14, 2020. This amount is equivalent to $250.24 US dollars in currency rates.
82 Giles and Hachisuka, *Record of Strange Nations*, 5.
83 Giles and Hachisuka, *Record of Strange Nations*, 5.
84 Charles Aylmer, "The Memoirs of H. A. Giles," *East Asian History* Numbers 13/14 (June/December 1997), 6.
85 Aylmer, "The Memoirs of H. A. Giles," 4.
86 CPI Inflation Calculator, UK Inflation Calculator, https://www.in2013dollars.com/uk/inflation/1927?amount=3.15, accessed on April 14, 2020.
87 Koyama Noboru, *Japanese Students at Cambridge*, 89.
88 Koyama Noboru, *Japanese Students at Cambridge*, 89.
89 Koyama Noboru, *Japanese Students at Cambridge*, 91.
90 Aylmer, "The Memoirs of H. A. Giles," 56.
91 Aylmer, "The Memoirs of H. A. Giles," 56.
92 My italics. Aylmer, "The Memoirs of H. A. Giles," 77-8.
93 Aylmer, "The Memoirs of H. A. Giles," 77-8.
94 My italics. Aylmer, "The Memoirs of H. A. Giles," 8.
95 Memories of Backhouse's forgeries still irked Giles, who despised falsifications of qualifications or scholarly work.
96 Giles and Hachisuka, *Record of Strange Nations*, front cover.
97 Norman Parley to Herbert G. Giles, "*Record of Strange Nations*" (letter), on behalf of Percy Lund Humphries & Co Ltd., London, January 14, 1929, 1.
98 Parley to Giles, 1.
99 Murakami, *Zetsumetsu dori dôdô wo oimotometa otoko*, 324-5.

100 Koyama, interview with the author, October 16, 2014.
101 Bowring, "The Selwyn Swastika," 33.
102 Bowring, "The Selwyn Swastika," 33.
103 Asada Sadao, "Cherry Blossoms and the Yellow Peril: American Images of Japan in the 1920s," in *Culture Shock and Japanese-American Relations: Historical Essays* (Columbia: University of Missouri Press, 2007), 27–52.

Chapter 4

1 John M. MacKenzie, *Museums and Empire: Natural History, Human Cultures, and Colonial Identities* (Manchester: Manchester University Press, 2009), 60.
2 Sherrie Cross defines "Meiji Social Darwinism," as a quasi-scientific philosophy imported by American zoologist Edward S. Morse, characterized by an "idiosyncratic combination of Spencerism, hereditarianism, and Darwinian selectionism applied to human affairs," 335. In Sherrie Cross, "Prestige and Comfort: The Development of Social Darwinism in Early Meiji Japan, and the Role of Edward Sylvester Morse," *Annals of Science*, Vol. 53, No. 4 (January 1, 1996), 323–44.
3 Matthew Karp, *This Vast Southern Empire: Slaveholders at the Helm of American Foreign Policy* (Cambridge, MA: Harvard University Press, 2016).
4 Franz Von Liszt, *Das Völkerrecht: Systematisch Dargestellt* (Berlin: Verlag von O. Haering, 1898), https://archive.org/details/dasvlkerrechtsy00liszgoog, accessed June 17, 2020.
5 Nadin Hée, *Imperiales Wissen und koloniale Gewalt: Japans Herrschaft in Taiwan 1895–1945* (Frankfurt: Campus Verlag, 2012).
6 Mark Rice, "His Name Was Don Francisco Muro: Reconstructing an Image of American Imperialism," *American Quarterly*, Vol. 62, No. 1 (March 2010), 50.
7 Reo Matsuzaki, *Statebuilding by Imposition: Resistance and Control in Colonial Taiwan and the Philippines* (Ithaca, NY: Cornell University Press, 2019).
8 Gregory Gillespie, *Hunting for Empire: Narratives of Sport in Rupert's Land, 1840–1870* (Vancouver: University of British Columbia Press, 2008), 14.
9 John MacKenzie, *The Empire of Nature: Hunting, Conservation, and British Imperialism* (Manchester, UK: Manchester University Press, 1997), 7.
10 Designated a Forest Reserve in 1910, the colonial Bureau of Forestry administered it for forestry land management study and education. In 1933, Governor General Theodore Roosevelt Junior (III) (1887–1944) named it Makiling National Park game preserve, as the first such national park in the Philippines. See Makiling Center for Mountain Ecosystems (MCME), "Laws Governing the Makiling Forest Reserve," https://web.archive.org/web/20130618214849/http://mountmakiling.org/index.php?page=gov_law, accessed via the Internet Archive Wayback Machine, August 2, 2020.
11 S. M. Durant et al., "Fiddling in Biodiversity Hotspots while Deserts Burn? Collapse of the Sahara's Megafauna," *Diversity and Distributions*, Vol. 20, No. 1 (January 2014), 114–22.
12 Robert Winstanley-Chesters, *Environment, Politics, and Ideology in North Korea: Landscape as Political Project* (Lanham, MD: Lexington Books, 2014), 45–6.
13 David Fedman, *Seeds of Control: Japan's Empire of Forestry in Korea* (Seattle: University of Washington Press, 2020), 7.

14 Conrad Totman, *Japan's Imperial Forest Goryôrin, 1889–1946* (Folkestone, UK: Global Oriental, 2007), 1.
15 The first exploration of the Marianas began under Royal Navy Admiral Baron George Anson (1697–1762), who captured the Spanish Galleon Nuestra Senora de Covadonga in 1743. Frank Knight, *Captain Anson and the Treasure of Spain* (London: Macmillan, 1959).
16 Fuchs, *Mimesis and Empire*, 10. Africans appropriated and inverted Spanish fears of black slave rebellions. María Elena Martínez, "The Black Blood of New Spain: *Limpieza de Sangre*, Racial Violence, and Gendered Power in Early Colonial Mexico," *The William and Mary Quarterly*, Vol. 61, No. 3 (2004).
17 Christian Perez, "A Short History of Philippine Bird Books–Part 6 American Period," *eBON Official* Newsletter *of the Wild Bird Club of the Philippines* (March 2, 2015), https://ebonph.wordpress.com/2015/03/02/a-short-history-of-philippine-bird-books-part-6-american-period/, accessed February 16, 2018.
18 Dean C. Worcester, *The Philippine Islands and Their People: A Record of Personal Observation and Experience, with a Short Summary of the More Important Facts in the History of the Archipelago* (New York: The Macmillan Company, 1898), viii–ix.
19 Dean C. Worcester papers: 1887–1925, Biography–Worcester Chronology, Bentley Historical Library, University of Michigan, https://quod.lib.umich.edu/b/bhlead/umich-bhl-86354?byte=143295236;focusrgn=bioghist;subview=standard;view=reslist, accessed February 16, 2018.
20 Dean C. Worcester Photographic Collection, University of Michigan, Museum of Anthropology website, http://webapps.lsa.umich.edu/umma/exhibits/Worcester%202012/index.html, accessed August 2, 2020.
21 Worcester, *The Philippine Islands and Their People*.
22 Dean C. Worcester Photographic Collection, University of Michigan, Museum of Anthropology website, http://webapps.lsa.umich.edu/umma/exhibits/Worcester%202012/biography.html, accessed February 16, 2018.
23 Mark Rice, *Dean Worcester's Fantasy Islands: Photography, Film, and the Colonial Philippines* (Ann Arbor: University of Michigan Press, 2014), 1.
24 Richard Crittenden McGregor and Dean Conant Worcester, *A Hand-list of the Birds of the Philippine Islands* (Manila: Bureau of Printing, 1906).
25 Richard C. McGregor, *A Manual of Philippine Birds* (Manila: Bureau of Science, 1909).
26 Joseph Grinnell, "In Memoriam: Richard C. McGregor Ornithologist of the Philippines," *The Auk* Vol. 55, No. 2 (April 1938), 169.
27 Grinnell, "In Memoriam," 169. In McGregor's obituary, Grinnell thanks Hachisuka, Wolfe, and others for sharing reminiscences: "Then I had essential helps [sic] from McGregor's junior colleague in the Philippines, Professor Canuto G. Manuel, and from The Marquess Hachisuka whose own work in those islands gives him authority to speak in technical matters. Additionally, I received very much in the way of reminiscence from Major Leon L. Gardner, Captain Lloyd R. Wolfe and Doctor Wilfred H. Osgood," 170.
28 McGregor and Worcester, *A Hand-list of the Birds of the Philippine Islands*, 4. See also Frank S. Bourns and Dean Worcester, Proceedings of the United States National Museum, Vol. XX (1898), 549–66.
29 McGregor and Worcester, *A Hand-list of the Birds of the Philippine Islands*, Introduction, 3.
30 Grinnell, "In Memoriam," 167.

31 See chapter three.
32 Kuroda Nagamichi, *Birds of the Island of Java*, Volume One: Passeres (Tokyo: Self-Published, 1933), frontispage. See Abe Books website, "Birds Island Java [sic] by Kuroda Nagamichi," at https://www.abebooks.com/servlet/SearchResults?an=kuroda%20nagamichi&tn=birds%20island%20java&cm_sp=mbc-_-ats-_-all, accessed July 13, 2020; and *Birds of the Island of Java, Volume One: Non-Passeres* (Tokyo: Self-Published, 1936), https://www.amazon.com/Birds-Island-Java-volumes-complete/dp/B00HL3NT8O, accessed July 25, 2020.
33 Pauline Rose Clance and Suzanee Ament Imes, "The Imposter Phenomenon in High Achieving Women: Dynamics and Therapeutic Intervention," *Psychotherapy: Theory, Research and Practice*, Vol. 15, No. 3 (Fall 1978), 241.
34 World Digital Library, Library of Congress, "Situation in Manchuria: Report of the Lytton Commission of Inquiry," https://www.wdl.org/en/item/11601/, accessed July 25, 2020.
35 This seems an error, since McGregor published *A Hand-list of the Birds of the Philippine Islands* with Worcester in 1906. In 1926, McGregor only published short articles on shore birds and raptors in *Philippine Education Magazine*. Though he enjoyed life in the Philippine until his 1938 death, McGregor lamented that his bureaucratic position prevented extensive writing and fieldwork.
36 "Ornithology of the Philippines," Review of *The Birds of the Philippine Islands: With Notes on the Mammal Fauna*. By the Hon. Masauji Hachisuka. Part 2. Pp. 169–439 + plates 25–39. (London: H. F. and G. Witherby, 1932), in *Nature*, Vol. 134, No. 3386 (September 22, 1934), 438–9.
37 "Ornithology of the Philippines," 439.
38 "Small Buttonquail (masaaki),"Avibase: The World Bird Database, https://avibase.bsc-eoc.org/species.jsp?lang=EN&avibaseid=872267F4BCE8A65C&sec=map, accessed May 10, 2021.
39 Delacour, *The Living Air*, 135.
40 Ernst Mayr, "In Memoriam: Jean (Theodore) Delacour," *The Auk*, Vol. 103 (July 1986), 605.
41 Delacour, *The Living Air*, 136.
42 Delacour, *The Living Air*, 136.
43 Delacour, *The Living Air*, 136.
44 Delacour, *The Living Air*, 135.
45 Hachisuka Toshiko, *Daimyô kazoku* (Tokyo: Mikasa shobô, 1957).
46 Nakamura Tsukasa, interview with author, November 17, 2015, Shinjuku, Tokyo, Japan.
47 Nakamura, interview.
48 Murakami, *Zetsumetsu dori dôdô wo oimotometa otoko*, 183.
49 Murakami, *Zetsumetsu dori dôdô wo oimotometa otoko*, 326.
50 Delacour, *The Living Air*, 135. Delacour notes Hachisuka's father's death on "1938" and not 1932.
51 Oi Miyoko, *Hachisuka Masa'aki to Fueko–Shimoda Utako kenkyû (1)*, *Jissen kokubungaku*, No. 91 (August 2013), 222.
52 Mentioned in letter by Oliver L. Austin, Junior, to his father, Oliver Austin, Senior, October 23, 1946, Tokyo, Japan.
53 Kinnear, "Obituary," 150.

54 Emily K. Abel, *Tuberculosis and the Politics of Exclusion: A History of Public Health and Migration to Los Angeles* (New Brunswick, NJ: Rutgers University Press, 2007), 86–107.
55 Hadley Meares, "The Sunshine Cure," *Curbed Los Angeles* (updated June 17, 2019), https://la.curbed.com/2015/9/30/9916132/southern-california-rehab-capital-sanitariums, accessed August 3, 2020.
56 Abel, *"Tuberculosis and the Politics of Exclusion,"* 1.
57 Murakami, *Zetsumetsu dori dôdô wo oimotometa otoko*, 328.
58 Jean Delacour, "Obituaries: Masauji, 18th Marquess Hachisuka," *The Auk* (October 1953), 521.
59 Delacour, "Obituaries: Masauji, 18th Marquess Hachisuka," 521.
60 Greg Flugfelder, *Cartographies of Desire: Male-Male Sexuality in Japanese Discourse, 1600–1950* (Berkeley: University of California Press, 1999), 142–92. After 1882, sexual acts between men in Japan were not criminalized.
61 Hachisuka was likely twenty-one, since Takatsukasa mentions his 1924 to 1925 travels in *The Avicultural Magazine*.
62 Delacour, *The Living Air*, 135.
63 Delacour, *The Living Air*, 15.
64 Delacour, *The Living Air*, 29.
65 Mishima Yukio's (1925–70) *Kamen no kokuhaku* (1949) is a thinly veiled memoir of boyhood love for a teenager named Omi. Meredith Wheatherby translated *Confessions of a Mask* (New York: New Directions, 1958).
66 Donald Roden, *Schooldays in Imperial Japan: A Study in the Culture of a Student Elite* (Berkeley: University of California Press, 1980).
67 Flugfelder, *Cartographies of Desire*, 40–1.
68 Flugfelder, *Cartographies of Desire*, 149.
69 Matt Cook, ed., *A Gay History of Britain: Love and Sex between Men Since the Middle-Ages* (Oxford, UK: Greenwood World Publishing, 2009), 176; and Charles Upchurch, *Beyond the Law: The Politics of Ending the Death Penalty for Sodomy in Britain* (Philadelphia: Temple University Press, 2021).
70 In 1991, following posthumous discovery of Prince Takamatsu's diaries, his wife published them, revealing his opposition to the war and personal struggles. Chihaya Masatake et al., *Takamatsu no miya nikki* (Tokyo: Chûô kôronsha, 1995).
71 J. Keith Vincent, *Two-Timing Modernity: Homosocial Narrative in Modern Japanese Fiction* (Cambridge, MA: Harvard University Asia Center Press, 2012), 87.
72 Vincent, *Two-Timing Modernity*, 87.
73 Grinnell, "In Memoriam," 168.
74 Grinnell, "In Memoriam," 167.
75 Robert Aldrich, *Colonialism and Homosexuality* (London: Routledge, 2002), 1.
76 Aldrich, *Colonialism and Homosexuality*, 2.
77 Murakami, *Zetsumetsu dori dôdô wo oimotometa otoko*, 324.
78 Hokkaido University Museum website, "WATASE Shozaburo," https://www.museum.hokudai.ac.jp/english/watase-shozaburo/, accessed July 31, 2020.
79 Janet Browne, "A Science of Empire: British Biogeography before Darwin," *Revue d'Histoire de Sciences*, Vol. 45, No. 5 (1992), 453–4.
80 Hachisuka, *The Birds of the Philippine Islands: Volume I*, V.
81 These include references to animals and foods evincing a British perspective.
82 Hachisuka Masauji, *Separata of Papers, Philippine Islands (Chiefly Birds) 1890–1934: Reprinted from Various Journals* (London: Natural History Museum, 1934); and Henry Fleming, letter to Hachisuka Masauji, 1932.

83 L. L. Snyder, "In Memoriam: James Henry Fleming," *The Auk*, Vol. 58, No. 1 (January 1941), 9.
84 In May 2020, four volumes cost $1,999.00 on Amazon! Hachisuka Masauji, *The Birds of the Philippine Islands, with Notes on the Mammal Fauna*, Parts I-IV in 4 volumes, complete, https://www.amazon.com/Philippine-Islands-Mammal-volumes-complete/dp/B004FOULCY, accessed May 24, 2020.
85 Hachisuka Masauji, "A Short Account of the Author's Journey to the Philippines," *The Birds of the Philippine Islands: With Notes on the Mammal Fauna: Volume I, Parts I & II, Galliformes to Pelecaniformes* (London: H. F. & G. Witherby, 1931–2), 53–95.
86 Hachisuka, "Geography and Climate," *The Birds of the Philippine Islands: Volume I*, 5–14.
87 Hachisuka, "Ornithological History," *The Birds of the Philippine Islands*, 15–52.
88 Hachisuka, *The Birds of the Philippine Islands: Volume I*, 53.
89 Hachisuka, *The Birds of the Philippine Islands: Volume I*, 54.
90 This animal must be a pangolin, since armadillo species only inhabit southern portions of North and Central America—possibly an uninformed edit by Hachisuka's typist Ms. Lawson. See Hachisuka, *The Birds of the Philippine Islands: Volume I*, V.
91 Hachisuka (via Ms. Lawson) probably referred to how mutton is prepared in England.
92 Hachisuka, *Birds of the Philippine Islands: Volume I*, 54.
93 Hachisuka, *Birds of the Philippine Islands Vol. 1*, 55.
94 Hachisuka, *Birds of the Philippine Islands Vol. 1*, 55.
95 He appears as "L. T." Wolfe in the text, which also is a typo. Hachisuka, *Birds of the Philippine Islands Vol. 1*, 55.
96 This appears as "Hooper" Ornithological Club, clearly a typo. The COC of California, which published the scientific journal *The Condor*, was the West coast counterpart to the AOU, which published *The Auk* and boasted an East coast membership. Hachisuka, *Birds of the Philippine Islands Vol. 1*, 55.
97 Hachisuka, *Birds of the Philippine Islands Vol. 1*, 55.
98 Grinnell, "In Memoriam," 168.
99 Grinnell, "In Memoriam," 168.
100 Quoted in Grinnell, "In Memoriam," 165.
101 Hachisuka, *Birds of the Philippine Islands Vol. 1*, 56.
102 Hachisuka, *Birds of the Philippine Islands Vol. 1*, 56.
103 University of Toronto Libraries, RPO Representative Poetry Online, "The White Man's Burden (*The United States and the Philippine Islands*) Kipling, Rudyard (1865–1936)," from the original text Rudyard Kipling, *Rudyard Kipling's Verse: Definitive Edition* (London: Hodder and Stoughton, 1940), 323–4, https://rpo.library.utoronto.ca/poems/white-mans-burden, accessed August 1, 2020.
104 Ann Kaplan, *The Anarchy of Empire in the Making of US Culture* (Harvard University Press, 2005), 93.
105 Jonah Walters, "An Empire of Patrolmen: An Interview with Stuart Schrader," *Jacobin* (October 18, 2019), https://www.jacobinmag.com/2019/10/badges-without-borders-stuart-schrader-imperialism-policing-cold-war, accessed July 26, 2020.
106 Stuart Schrader, *Badges without Borders: How Global Counterinsurgency Transformed American Policing* (Berkeley: University of California Press, 2019), 66–7.

107 Macario C. Tiu, *Davao 1890–1910: Conquest and Resistance in the Garden of the Gods* (Davao, Philippines: UP Center for Integrative and Development Studies, 2002).
108 Oliver Charbonneau, "'A New West in Mindanao': Settler Fantasies on the U.S. Imperial Fringe," *The Journal of the Gilded Age and Progressive Era*, Vol. 18, No. 3 (2019), 304–5.
109 Dee Brown, *Bury My Heart at Wounded Knee* (New York: Holt, Rinehart, & Winston, 1970).
110 Patricia Irene Dacudao, *ABACA: The Socio-Economic and Cultural Transformation of Frontier Davao, 1898–1941* (Master's Thesis, Department of History, Ateneo de Manila University, 2017), 71.
111 Charbonneau, "A New West in Mindanao," 305.
112 Macario D. Tiu, *Davao: Reconstructing History from Text and Memory* (Davao, Philippines: Ateneo de Davao University Research and Publication Office for the Mindanao Coalition of Development NGOs, 2005).
113 Hachisuka, *Birds of the Philippine Islands Vol. 1*, 84.
114 Charbonneau, "A New West in Mindanao," 307.
115 Charbonneau, "A New West in Mindanao," 307.
116 Lydia N. Jose and Patricia Irene Dacudao, "Visible Japanese and Invisible Filipino: Narratives of the Development of Davao, 1900s to 1930s," *Philippine Studies: Historical and Ethnographic Viewpoints*, Vol. 63, No. 1 (2015), 101–29.
117 Charbonneau, "A New West in Mindanao," 310, 312.
118 Hachisuka, *Birds of the Philippine Islands Vol. 1*, 84.
119 Charbonneau, "A New West in Mindanao," 312.
120 Hachisuka, *Birds of the Philippine Islands Vol. 1*, 83.
121 Hachisuka, *Birds of the Philippine Islands Vol. 1*, 83.
122 Hachisuka, *Birds of the Philippine Islands Vol. 1*, 83.
123 Charbonneau, "A New West in Mindanao," 313.
124 Charbonneau, "A New West in Mindanao," 314.
125 Lydia N. Yu-Jose, *Japan Views the Philippines, 1900–1944* (Davao, Philippines: Ateneo de Manila University Press, 1999).
126 Charbonneau, "A New West in Mindanao," 315.
127 Dacudao, "ABACA," 77.
128 Dacudao, "ABACA," 77.
129 Charbonneau, "A New West in Mindanao," 315.
130 Anderson Villa, "Visiting Filipino Scholar Rediscovers Hoover's Archival Sources on Japanese Presence in Prewar Davao," *Stanford University Hoover Library and Archives News* (December 17, 2018), https://www.hoover.org/news/visiting-filipino-scholar-rediscovers-hoovers-archival-sources-japanese-presence-prewar-davao, accessed August 1, 2020.
131 Hachisuka, *Birds of the Philippine Islands*, insert.
132 Tiu, *Davao: Reconstructing History from Text and Memory*, 237.
133 W. R. Ogilvie Grant, "On the Birds collected by Mr. Walter Goodfellow on the Volcano of Apo and in its Vicinity, in South-east Mindanao, Philippine Islands," *Ibis*, Vol. 48, No. 3 (July 1906), 465–505.
134 Hachisuka, *Birds of the Philippine Islands Vol. 1*, 64.
135 Hachisuka, *Birds of the Philippine Islands Vol. 1*, 64.
136 Hachisuka, *Birds of the Philippine Islands Vol. 1*, 92.
137 Hachisuka, *Birds of the Philippine Islands Vol. 1*, 78.
138 Hachisuka, *Birds of the Philippine Islands Vol. 1*, 92.

Notes

139 Hachisuka, *Birds of the Philippine Islands Vol. 1*, 92.
140 Bird Life International (2018), *Pithecophaga jefferyi* (amended 2017 assessment), The IUCN Red list of Threatened Species website, https://www.iucnredlist.org/species/22696012/129595746, accessed August 2, 2020. Incidentally, this website is partially funded by the Japanese Toyota Motors Foundation.
141 Hachisuka, *Birds of the Philippine Islands Vol. 1*, Plate 9.
142 Hachisuka, *Birds of the Philippine Islands Vol. 1*, Plate 10.
143 Hachisuka, *Birds of the Philippine Islands Vol. 1*, 74.
144 Hachisuka, *Birds of the Philippine Islands Vol. 1*, 74–5.
145 Hachisuka, *Birds of the Philippine Islands Vol. I*, Plate 13.
146 Hachisuka, *Birds of the Philippine Islands Vol. 1*, 77.
147 Hachisuka, *Birds of the Philippine Islands Vol. 1*, 77.
148 Hachisuka, *Birds of the Philippine Islands Vol. 1*, 77.
149 Kenneth J. Ruoff, *The People's Emperor: Democracy and the Japanese Monarchy, 1945–1995* (Cambridge, MA: Harvard University Asia Center, 2001), 21–3.
150 Hachisuka, *Birds of the Philippine Islands Vol. 1*, 78.
151 Hachisuka, *Birds of the Philippine Islands Vol. 1*, 78.
152 Hachisuka, *Birds of the Philippine Islands Vol. 1*, 78.
153 Samuel Gaabucayan, "The Medicine Men of Agusan in Mindanao, Philippines," *Asian Folklore Studies* 30, 1 (1971), 39–54; and Francisco R. Demetrio, "Shamans, Witches and Philippine Society," *Philippine Studies* 36, 3 (Third Quarter 1988), 372–80.
154 Hachisuka, *Birds of the Philippine Islands Vol. 1*, 81.
155 Hachisuka, *Birds of the Philippine Islands Vol. 1*, 82.
156 Laura E. Matthew, *Memories of Conquest: Becoming Mexicano in Colonial Guatemala* (Chapel Hill: University of North Carolina Press, 2014).
157 Hachisuka, *Birds of the Philippine Islands Vol. 1*, 82.
158 Laura E. Matthew and Michel R. Oudijk, editors, *Indian Conquistadors: Indigenous Allies in the Conquest of Mesoamerica* (Tulsa: University of Oklahoma Press, 2012.)
159 Hachisuka, *Birds of the Philippine Islands Vol. 1*, 82.
160 Hachisuka, *Birds of the Philippine Islands Vol. 1*, 84.
161 Robinson A. Herrera, historian of colonial Latin America, comments on the image, Florida State University, May 23, 2020.
162 Rice, *Dean Worcester's Fantasy Islands*.
163 Hachisuka, *Birds of the Philippine Islands Vol. 1*, 86.
164 Hachisuka, *Birds of the Philippine Islands Vol. 1*, 58.
165 Hachisuka, *Birds of the Philippine Islands Vol. 1*, 59.
166 Hachisuka, *Birds of the Philippine Islands Vol. 1*, 58.
167 Hachisuka, *Birds of the Philippine Islands Vol. 1*, 59.
168 Jeffrey Wheatley, "US Colonial Governance of Superstition and Fanaticism in the Philippines," *Method and Theory in the Study of Religion*, Vol. 30, No. 1 (2018), 21–36.
169 Hachisuka's explorations were contemporaneous to Japan's "Taishô Democracy" (1918–32), marked by increased popular political participation and 1928 promulgation of universal manhood suffrage.
170 Hachisuka, *Birds of the Philippine Islands Vol. 1*, 86.
171 Hachisuka, *Birds of the Philippine Islands Vol. 1*, 86.
172 Hachisuka, *Birds of the Philippine Islands Vol. 1*, 84.
173 "Dan Inosanto and the Filipino Memorial Bolo Sword," https://www.youtube.com/watch?v=MsD9z4kVVVE and "Don Inosanto—Filipino Martial Arts Demo at the Smithsonian," https://www.youtube.com/watch?v=tKKZuS8c7rM.

252   *Notes*

174 Hachisuka, *Birds of the Philippine Islands Vol. 1*, 86.
175 Hachisuka, *Birds of the Philippine Islands Vol. 1*, 85.
176 Hachisuka, *Birds of the Philippine Islands Vol. 1*, 84.
177 Hachisuka, *Birds of the Philippine Islands Vol. 1*, 87.
178 Hachisuka, *Birds of the Philippine Islands Vol. 1*, 89.
179 Hachisuka, *Birds of the Philippine Islands Vol. 1*, 90.
180 Hachisuka, *Birds of the Philippine Islands Vol. 1*, 93.
181 Hachisuka, *Birds of the Philippine Islands Vol. 1*, 94.
182 Hachisuka, *Birds of the Philippine Islands Vol. 1*, 94.
183 Hachisuka, *Birds of the Philippine Islands Vol. 1*, 88.
184 Hachisuka, *Birds of the Philippine Islands Vol. 1*, 90.
185 Hachisuka, *Birds of the Philippine Islands Vol. 1*, 89.
186 Hachisuka, *Birds of the Philippine Islands Vol. 1*, 89.
187 Hachisuka, *Birds of the Philippine Islands Vol. 1*, 89.
188 Mel White, "Getting Familiar with the Philippines," (April 15, 2011), *All About Birds–The Cornell Lab of* Ornithology, https://www.allaboutbirds.org/getting-familiar-with-the-philippines/

Chapter 5

1 For historian Prasenjit Duara, "The Japanese domination of Manchukuo represented a new form of imperialism ….Japan, like the later Soviet Union and the United States, sought to bring its *client-states* into a structure of governance that not only permitted dominance but integrated them into a regional and ultimately, global, game plan" (my italics). Prasenjit Duara, "The New Imperialism and the Post-Colonial Developmental State: Manchukuo in Comparative Perspective," *The Asia-Pacific Journal: Japan Focus*, Vol. 4, No. 1 (January 4, 2006), 410.
2 With considerable funding from *Asahi Shimbun*, this included the Tokyo Foreign Office's Cultural Work Bureau, and the Japanese Society for the Promotion of Science's precursor, the Foundation for the Promotion of Scientific and Industrial Research of Japan. Tokunaga Shigeyasu, *Dai'ichi ji ManMô gakujutsu chôsa kenkyû-dan hôkoku* (Tokyo: Office of the Scientific Expedition to Manchoukuo, Faculty of Science and Engineering, Waseda University, October 1934), 76.
3 Tokunaga Shigemoto, *Tokunaga Shigeyasu shôden*, Chigaku zasshi, 94–3 (1985), 55.
4 Morris Low, *Japan on Display: Photography and the Emperor* (London: Routledge, 2006), 64–5; and "The Japanese Colonial Eye: Science, Exploration, and Empire," in Christopher Pinney et al., *Photography's Other Histories* (Durham, NC: Duke University Press, 2003),105–15.
5 Yamashina Yoshimaro, "On the Specimens of Korean Birds, collected by Mr. Hyojiro [sic] Orii," *Tori*, Vol. 7, Nos. 33–4 (1932), 213–52, at https://www.jstage.jst.go.jp/article/jjo1915/7/33-34/7_33-34_213/_pdf/-char/en, accessed on October 29, 2019.
6 Janice Mimura, *Planning for Empire: Reform Bureaucrats and the Japanese Wartime State* (Ithaca, NY: Cornell University Press, 2011), 12–14, and 70–106.
7 Sadako N. Ogata, *Defiance in Manchuria: The Making of Japanese Foreign Policy, 1931–1932* (Berkeley: University of California Press, 1964).
8 Prasenjit Duara, *Sovereignty and Authenticity: Manchukuo and the East Asian Modern* (Lanham, MD: Rowman and Littlefield, 2003), 250.

9   Annika A. Culver, *Glorify the Empire: Japanese Avant-Garde Propaganda in Manchukuo* (Vancouver: University of British Columbia Press, 2013), 1–10.
10 "Carrier Pigeons Army," *Manchuria Daily News* (October 24, 1933), 2. See also Ben Breen, "Behind the Rocket Cat: Animals in Warfare from Hannibal to World War One," *The Appendix* (March 6, 2014), http://theappendix.net/posts/2014/03/behind-the-rocket-cat-animals-in-warfare-from-ancient-india-to-world-war-one, accessed October 22, 2020.
11 Louise Young, *Japan's Total Empire: Manchuria and the Culture of Wartime Imperialism* (Berkeley: University of California Press, 1999), 55–114.
12 Funabashi Yoichi, "Foreword," in The Asahi Shimbun Company (trans. Barak Kushner), *Media, Propaganda, and Politics in 20th Century Japan* (London: Bloomsbury Press, 2015), xi.
13 The Asahi Shimbun, "History," https://www.asahi.com/corporate/english/11053815, accessed September 6, 2020.
14 Natsumé Sôseki, *Travels in Manchuria and Korea* (Folkestone, Kent, UK: Global Books, 2000). Translated by Inger S. Brody and Sammy I. Tsunematsu.
15 Donald Richie, "Miserable Every Step of the Way," *The Japan Times* (November 28, 2000), https://www.japantimes.co.jp/culture/2000/11/28/books/miserable-every-step-of-the-way/, accessed September 30, 2020.
16 "Strength of the Asahi Shimbun" (December 2018), https://adv.asahi.com/adv/english/pdf/strength_of_the_asahi_shimbun/Strength_of_The_Asahi_Shimbun_201812.pdf, accessed September 15, 2020.
17 "World Currency Exchange Rates and Currency Exchange Rate History," https://www.exchange-rates.org/Rate/JPY/USD/12-31-2018, accessed September 15, 2020.
18 Library of Congress online archive, Newspaper, *Manchuria Daily News*, https://www.loc.gov/item/sn89049306/, accessed September 5, 2020.
19 Young, *Total Empire*, 81.
20 Young, *Total Empire*, 82.
21 Honda Shûgo, *Tenkô bungaku-ron* (Tokyo: Mirai-sha, 1957). Quoting Honda, Donald Keene estimates that 95 percent of Japanese left-wing intellectuals changed their political orientations to favor the imperial state by the mid-1930s. *Dawn to the West: Japanese Literature of the Modern Era—Fiction* (New York: Holt, Rinehart, and Winston, 1984), 847.
22 K. Mizuno, "On a Few Birds Peculiar to Manchuria: About a Few Birds Found in Manchuria That Are Rare in Japan Proper," *Manchuria Daily News* (September 18, 1932), 3–4. Part Two, with the same title in the same newspaper, appears in September 21, 1932, on page 4. I thank Norman smith for this reference.
23 Matsuda Michio, *Yo no naka, igai to semai—Manshû chôrui genshoku dai-zukan no Mizuno Kaoru-san, Syrinx burogu-hen: hobo yachô rokuon no hanashi* (September 8, 2012), http://syrinxmm.cocolog-nifty.com/syrinx/2012/09/post-c212.html, accessed September 30, 2020.
24 K. Mizuno, "On a Few Birds Peculiar to Manchuria," *Manchuria Daily News* (September 18, 1932), 3–4.
25 Mizuno, "On a Few Birds Peculiar to Manchuria" (September 19, 1932), 3.
26 Only the Liaodong Peninsula's leased area near Dairen and Port Arthur was Japan's colony since the US-effectuated 1905 Portsmouth Treaty; the rest of Manchukuo was considered "independent" though subjected to strong Kantô Army and Japanese governmental influence.
27 Mizuno, "On a Few Birds Peculiar to Manchuria," 4.

28 Mizuno, "On a Few Birds Peculiar to Manchuria," 4.
29 Mizuno, "On a Few Birds Peculiar to Manchuria," 4.
30 Mizuno, "On a Few Birds Peculiar to Manchuria," 4.
31 Kuroda Nagamichi, "A Small Collection of Birds from South Manchuria," *Tori*, Vol. 6, No. 30 (1930), 135.
32 Kuroda Nagamichi, "The Second Lot of Bird-Skins from South Manchuria," *Tori*, Vol. 7, No. 31 (1931), 42–6; "The Third Lot of Bird-Skins from Manchuria," *Tori*, Vol. 7, No. 32 (1931), 184–7; "A Fourth Lot of Bird-Skins from Manchuria," *Tori*, Vol. 7, No. 33–34 (1932) 348–51; "A Fifth Lot of Bird-Skins from Manchuria," *Tori*, Vol. 7, No. 35 (1932), 421–4; "A Sixth Lot of Bird-Skins from Manchuria," *Tori*, Vol. 8, No. 37 (1933), 168–74; "A Seventh Lot of Bird-skins from Manchuria," *Tori*, Vol. 8, No. 38 (1934), 277–80.
33 Witmer Stone, "Recent Literature: *Hand-List of Japanese Birds*," *The Auk*, Vol. 50, Issue 1 (January 1933), 136.
34 Stone, "Recent Literature," 136.
35 Stone, "Recent Literature," 136.
36 Stone, "Recent Literature," 136.
37 Soffer Ornithology Collection Notes: Takatsukasa, https://www.amherst.edu/library/archives/holdings/soffer/t#takatsukase2, accessed September 4, 2020.
38 AbeBooks, "About This Item," description of *The Birds of Nippon* (Tokyo: Maruzen, 1967), https://www.abebooks.com/Birds-Nippon-Taka-Tsukasa-Prince-Nobusuke-Maruzen/7232801539/bd, accessed September 25, 2020.
39 Witmer Stone, "Taka-tsukasa's 'The Birds of Nippon,'" *The Auk*, Vol. 50, No. 2 (April 1933), 244.
40 Takatsukasa Nobusuke and Kano Tadao, *Kôtôsho*, *Tori*, Vol. 8, No. 38 (April 1934), 192–218.
41 Takatsukasa Nobusuke and Kano Tadao, *Kashōtō*, *Tori*, Vol. 8, No. 38 (April 1934), 211–17.
42 National Museum of Ethnology in Taiwan, "The Indigenous Cultures of Taiwan: KANO Tadao Biographical Sketch," https://www.minpaku.ac.jp/english/museum/exhibition/main/china/taiwan_03, accessed July 20, 2020.
43 Witmer Stone, "Recent Literature," *The Auk*, Vol. 5, No. 1 (January 1934), 108.
44 These are common ornithological bird species categorizations. See Scott V. Edwards and John Harshman, "Passeriformes: Perching Birds, Passerine Birds," Tree of Life Web Project, Harvard University (February 6, 2013), http://tolweb.org/Passeriformes, accessed October 22, 2020.
45 Kuroda Nagamichi, *Birds of the Island of Java, Volume One: Passeres* (Tokyo: Self-Published, 1933), frontispage. Abe Books website, "Birds Island Java [sic] by Kuroda Nagamichi," at https://www.abebooks.com/servlet/SearchResults?an=kuroda%20nagamichi&tn=birds%20island%20java&cm_sp=mbc-_-ats-_-all, July 13, 2020.
46 These included lifetime member, *Ornithological Society*, Japan (1907), fellow member, BOU (1918), honorary fellow, AOU (1921); member, Bombay Natural History Society (1921); *korrespondanten mitglied*, *Ornithologische Gesellschaft*, Bayern (1924); corresponding member, Peking Society for Natural History (1926); charter member, *Biogeographical Society*, Japan (1928); member, BOC (1928); honorary member, *Royal Hungarian Institute of Ornithology* (1929); honorary member, Boston Society of Natural History (1929); honorary member, *Club Nederland, Vogelkundigen* (1930); and honorary member, *Societé d'Ornithologie et Mammalogie*, France (1931). Kuroda, *Birds of the Island of Java, Volume One*, frontispiece.

47 Kuroda Nagamichi, *Birds of the Island of Java, Volume One: Non-Passeres* (Tokyo: Self-Published, 1936), https://www.amazon.com/Birds-Island-Java-volumes-complete/dp/B00HL3NT8O, accessed July 25, 2020.
48 Geoffrey T. Hellman, "Profiles: Curator Getting Around," *The New Yorker* (August 26, 1950), https://www.newyorker.com/magazine/1950/08/26/curator-getting-around, accessed September 6, 2020.
49 Stone, "Recent Literature," 108.
50 Stone, "Recent Literature," 108.
51 Stone, "Recent Literature," 108.
52 Stone, "Recent Literature," 108.
53 Mizuno Kaoru, *Manshû chôrui bunpu mokuroku* (Tokyo: Kuroda Nagamichi, October 1934).
54 Matsuda Michio, O takara-gazô—"Manshû chôrui genshoku dai-zukan" no Mizuno Kaoru-san, *Hobo yachô rokuon no hanashi* (October 8, 2012), http://syrinxmm.cocolog-nifty.com/syrinx/2012/10/post-00b4.html, accessed September 28, 2020.
55 Mizuno Kaoru, *Manshû chôrui genshoku dai-zukan* (Tokyo: Purosesu-sha, 1940).
56 Matsuda Michio, *Birds Songs of Japan: CD nakigoe gaido—Nihon no yachô* (Tokyo: Wild Bird Society of Japan, 2016).
57 Matsuda Michio, Kyô ha jôkigen—"Manshû chôrui genshoku dai-zukan" wo nyûshu, *Hobo yachô rokuon no hanashi* (March 13, 2012), http://syrinxmm.cocolog-nifty.com/syrinx/2012/03/post-37e6.html, accessed September 30, 2020.
58 On the *Nihon no kôshoya* site, *Manshû chôrui genshoku dai-zukan*, sold for 143,000 yen, https://www.kosho.or.jp/products/detail.php?product_id=334353686, accessed September 30, 2020.
59 Matsuda, *Kyô ha jôkigen* (March 13, 2012).
60 "ManMô shigen no kagaku zandankai—kaku hômen ken'i no shusseki wo koi—kinô honsha ni hiraku," *Asahi Shimbun* (March 8, 1933), p. 3, col. 4.
61 "400 mannen mae no ebi, jûrai no *gakusetsu* wo mattaku kutsugaeshita mezurashii kaseki no hakken," *Asahi Shimbun* (October 10, 1928), p. 7, column 5.
62 "ManMô shigen no kagaku zandankai," *Asahi Shimbun* (October 10, 1928), page 7, column 5.
63 Tokunaga, *Dai'ichi ji ManMô gakujutsu chôsa kenkyû-dan hôkoku*, 50.
64 "ManMô shigen kagaku zadankai—9 bumon yori miru NichiMan shigen no kankei tanen kenkyu no keika happyô," *Asahi Shimbun* (March 26, 1933), p. 4, column 1.
65 Low, "The Japanese Colonial Eye," 114–15.
66 "A Japanese Scientific Expedition to Jehol in 1933," in "The Monthly Record" of *The Geographical Journal*, Vol. 86, No. 1 (July 1935), 82.
67 Low, "The Japanese Colonial Eye," 113.
68 Tokunaga, *Dai'ichi ji ManMô gakujutsu chôsa kenkyû-dan hôkoku*, 45.
69 Tokunaga, *Dai'ichi ji ManMô gakujutsu chôsa kenkyû-dan hôkoku*, 45–6.
70 "Description," *The China Journal*, https://catalog.hathitrust.org/Record/006094626, accessed September 12, 2020.
71 "Scientific Party to Study W. Jehol and E. Mongolia," *Manchuria Daily News* (June 8, 1933), 2.
72 In 1926, Japan's Foreign Ministry created Rengo News Agency to introduce "news as propaganda," while "the ministry's network consisted of Rengo, the *Japan Times*, the *Far East*, and the *Herald of Asia*." Yoshie Takamitsu, "Improving US–Japanese Relations through the News Media: Roy W. Howard, Dentsu, and the Osaka Mainichi," *The Japanese Journal of American Studies*, Vol. 29 (June 2018), 114.

73 Associated Press, "Dr. Roy Chapman Andrews Dies; Explorer and Naturalist Was 76; He Discovered Dinosaur Eggs in Asia in 1920's—Headed Natural History Museum Roy Chapman Andrews, Explorer and Naturalist, Dies," *New York Times* (March 12, 1960), 1, https://www.nytimes.com/1960/03/12/archives/dr-roy-chapman-andrews-dies-explorer-and-naturalist-was-76-he.html, accessed October 9, 2020.

74 "Travel and Exploration Notes: Projected Exploration of Inner Mongolia and Jehol," *The China Journal of Science and Arts*, Vol. 19, No. 1 (July 1933), 22.

75 "Travel and Exploration Notes," *The China Journal*, 22.

76 "Travel and Exploration Notes," *The China Journal*, 22.

77 I thank Norman Smith for indicating this article during email communication on August 20, 2020. "Scientific Party to Study W. Jehol and E. Mongolia," *Manchuria Daily News* (June 8, 1933), 2.

78 Kate McDonald, *Placing Empire: Travel and the Social Imagination in Imperial Japan* (Oakland, CA: University of California Press, 2017), 7.

79 The hotel looks similar, though an electronic billboard was placed atop its roof. When I overnighted in February 2005, décor likely resembled prewar Japanese guests' lodgings: burgundy velvet plush armchairs with a sitting room couch festooned with lace antimacassars and a king-sized bed flanked by night tables. Author's fieldwork, Dalian, People's Republic of China, February 2005.

80 "Scientific Party," *Manchuria Daily News*, 2.

81 "Tankentai'in ni hôken: seimei hôken kyôkai ga sanjoshi, ichi-nin ichi-man en zutsu fudan," *Asahi Shimbun* (July 1, 1933), p. 3, column 4.

82 "ManMô gakujutsu kenkyû chôsa-dan: kagaku no senshi no shuto wo iwau—sakuya, honsha de sôbetsuen," *Asahi Shimbun* (July 1, 1933), p. 3, column 2.

83 Tokunaga, *Dai'ichi ji ManMô gakujutsu chôsa kenkyû-dan hôkoku*, 50.

84 The article mentions 16-mm guns, likely a typo or misunderstanding of caliber. "Bihin wa torakku 67-dai, 16 miri mo keikō–shuppatsu wa raigetsu 23-nichi," *Asahi Shimbun* (June 25, 1933), Tokyo morning edition, p. 2, column 3.

85 Tokunaga, *Dai'ichi ji ManMô gakujutsu chôsa kenkyû-dan hôkoku*, 51.

86 "Iki-agaru Manmō gakujutsu chōsa-dan–akueki yobō-zai o fukuyō–shin danfuku ni dai man'etsu—sakuya dai 1-kai no kao-awase," *Asahi Shimbun* (June 25, 1933), Tokyo morning edition, p. 2, column 1.

87 Low, "The Japanese Colonial Eye," 105.

88 Low, "The Japanese Colonial Eye," 105.

89 Tokunaga, *Dai'ichi ji ManMô gakujutsu chôsa kenkyû-dan hôkoku*, unpaginated plate.

90 Tokunaga, *Dai'ichi ji ManMô gakujutsu chôsa kenkyû-dan hôkoku*, 45.

91 Tokunaga, *Dai'ichi ji ManMô gakujutsu chôsa kenkyû-dan hôkoku*, 74.

92 Tokunaga, *Dai'ichi ji ManMô gakujutsu chôsa kenkyû-dan hôkoku*, 72.

93 "A Japanese Scientific Expedition to Manchoukuo," *Nature* (March 23, 1935), 479; https://www.nature.com/articles/135479b0.pdf, accessed September 24, 2020.

94 "A Japanese Scientific Expedition to Manchoukuo," *Nature*, 479.

95 "A Japanese Scientific Expedition to Manchoukuo," *Nature*, 479.

96 "A Japanese Scientific Expedition to Manchoukuo," *Nature*, 479.

97 "A Japanese Scientific Expedition to Manchoukuo," *Nature*, 480.

98 In Tokunaga, *Dai'ichi ji ManMô gakujutsu chôsa kenkyû-dan hôkoku*, insert.

99 "A Japanese Scientific Expedition to Jehol in 1933," *The Geographical Journal*, 82.

100 Tokunaga, *Dai'ichi ji ManMô gakujutsu chôsa kenkyû-dan hôkoku*, plates 379–84.

101 Tokunaga, *Dai'ichi ji ManMô gakujutsu chôsa kenkyû-dan hôkoku*, plates 385–90.

102 Tokunaga, *Dai'ichi ji ManMô gakujutsu chôsa kenkyû-dan hôkoku*, last plate.

Notes

103 "Iodine: Fact Sheet for Consumers," National Institutes of Health Office of Dietary Supplements, https://ods.od.nih.gov/factsheets/Iodine-Consumer/, accessed October 7, 2020.
104 "Research Party Reaches Mukden: Japanese Scientists Return from 70 Day Trip Into Jehol," *Manchuria Daily News* (October 12, 1933), 7.
105 "Research Party Reaches Mukden," *Manchuria Daily News*, 7.
106 "Research Party Reaches Mukden," *Manchuria Daily News*, 7.
107 "Personal and Topical," "Waseda Professor," *Manchuria Daily News* (October 16, 1933), 2.
108 "Personal and Topical."
109 Bob Hackett, "*Rikugun byoinsen*: IJA Hospital Ship/Transport BAIKAL MARU: Tabular Record of Movement," (2013–14), http://www.combinedfleet.com/Baikal_t.htm, accessed September 1, 2020.
110 Hackett, "*Rikugun byoinsen*."
111 Donald A. Bertke, Gordon Smith, and Don Kindell, *World War II Sea War—Volume 7: The Allies Strike Back (Day-to-Day Naval Actions September through November 1942* (Dayton, OH: Bertke Publications, 2014), 184–7.
112 Hackett, "*Rikugun byoinsen*."
113 Tokunaga, *Dai'ichi ji ManMô gakujutsu chôsa kenkyû-dan hôkoku*, 76.
114 Robert A. Mullen, "Speed of Racing Pigeons," *The Auk*, Vol. 50, No. 3 (July 1, 1933), 374–5.
115 Asahi Shimbun, English, "Corporate Overview," http://www.asahi.com/shimbun/honsya/e/e-history.html, accessed September 8, 2020.
116 Author's correspondence with Norman Smith, August 21, 2020; *Manchuria Daily News* (August 29, 1933).
117 "Biggest Pigeonry in Orient Proposed by Kwantung Army," *Manchuria Daily News* (October 18, 1933), 1.
118 "Biggest Pigeonry," *Manchuria Daily News*, 1.
119 "Carrier Pigeons Army," *Manchuria Daily News* (October 24, 1933), 2.
120 "52 Carrier-Pigeons Received by [sic] Army," *Manchuria Daily News* (October 24, 1933), 3.
121 Andrea DenHoed, "Photo Booth: The Turn-of-the-Century Pigeons That Photographed Earth from Above," *The New Yorker* (April 14, 2018), https://www.newyorker.com/culture/photo-booth/the-turn-of-the-century-pigeons-that-photographed-earth-from-above, accessed October 10, 2020.
122 DenHoed, "Photo Booth."
123 Quoted in Dan Schlenoff, "Anecdotes from the Archive: Aerial Spying, 100 Years Before Drones," *Scientific American* (October 10, 2014), https://blogs.scientificamerican.com/anecdotes-from-the-archive/aerial-spying-100-years-before-drones/, accessed September 1, 2020.
124 Mizuno, "On a Few Birds Peculiar to Manchuria," 3.
125 (US) War Department, "Chapter III: Field Organization," *Technical Manual: Handbook on Japanese Military Forces* (December 15, 1944), 51, https://www.ibiblio.org/hyperwar/Japan/IJA/HB/HB-3.html, accessed September 1, 2020.
126 (US) War Department, "Chapter III: Field Organization," 51
127 US War Department, *Technical Manual: Handbook on Japanese Military Forces*, 51.
128 About CIA, CIA Museum, Experience the Collection, "Aerial Reconnaissance: Pigeon Camera," https://www.cia.gov/about-cia/cia-museum/experience-the-collection/text-version/collection-by-subject/aerial-reconnaissance.html, accessed September 1, 2020.

129 CIA Museum, "Aerial Reconnaissance," https://www.cia.gov/about-cia/cia-museum/experience-the-collection/text-version/collection-by-subject/aerial-reconnaissance.html.
130 About CIA, CIA Museum, Experience the Collection, "Aerial Reconnaissance: Pigeon Camera."
131 Tokunaga, *Dai'ichi ji ManMô gakujutsu chôsa kenkyû-dan hôkoku*,
132 Tokunaga, *Dai'ichi ji ManMô gakujutsu chôsa kenkyû-dan hôkoku*, 72.
133 Witmer Stone, "Recent Literature: *Birds of Jehol*," *The Auk*, Vol. 52, No. 4 (October 1, 1935), 476.
134 Stone, "Recent Literature," 476.
135 J. C. Barlow, S. N. Leckie, P. Pyle, and M. A. Patten, "Eurasian Tree Sparrow (*Passer montanus*), version 1.0," in S. M. Billerman, ed., *Birds of the World* (Ithaca, NY: Cornell Lab of Ornithology, 2020), https://birdsoftheworld.org/bow/species/eutspa/cur/introduction, accessed September 20, 2020.
136 Amherst College Library, Archives and Special Collections, T. Soffer Ornithology Collection, Soffer Ornithology Collection Notes, https://www.amherst.edu/library/archives/holdings/soffer/t#takatsukase2, accessed September 4, 2020. J. Anker, *Bird Books and Bird Art: An Outline of the Literary History and Iconography of Descriptive Ornithology* (New York: Springer, 2014), 203.
137 Anker, *Bird Books and Bird Art*, 203.
138 Kingsberg Kadia, *Into the Field*, 8.
139 Prince N. Taka-Tsukasa, Marquis M. Hachisuka, N. Kuroda, Marquis Y. Yamashina, and S. Uchida, *Birds of Jehol*, Section V, Division II, Part III (April 1935) in *Report of the First Scientific Expedition to Manchoukuo under the Leadership of Shigeyasu Tokunaga, June-October 1933* (Tokyo: Office of the Scientific Expedition to Manchoukuo, Faculty of Science and Engineering, Waseda University, 1935). I thank Yale University's Library for sending me an electronic copy.
140 Bibliothèque Nationale Français, Catalogue Générale, Notice Bibliographique, "Birds of Jehol," https://catalogue.bnf.fr/ark:/12148/cb316744362, accessed September 23, 2020.
141 *Nekkashô san chôrui seigôhyô*, in Takatsukasa Nobusuke, Hachisuka Masauji, Kuroda Nagamichi, Yamashina Yoshimaro, and Uchida Seinosuke, "Birds of Jehol," in "Report of the First Scientific Expedition to Manchoukuo Under the Leadership of Shigeyasu Tokunaga. Section V, Zoology. Division 2, Vertebrata; 3" (Tokyo: Office of the Scientific Expedition to Manchoukuo, Faculty of Science and Engineering, Waseda University, April 1935), frontispiece insert.
142 See Kuroda's bibliographical entries in Yamashina Yoshimaro, "Note on the Specimens of Manchurian Birds Chiefly Made by Mr. Hyojiro Orii in 1935," *Tori* [Birds] Vol. 10, No. 49 (1939), 540, https://www.jstage.jst.go.jp/article/jjo1915/10/49/10_49_446/_pdf, accessed September 10, 2020.
143 Yamashina, "Note on the Specimens of Manchurian Birds," 537–44.
144 Yamashina, "Note on the Specimens of Manchurian Birds," 544.
145 Yamashina, "Note on the Specimens of Manchurian Birds," 446–544.
146 Loukashkin emigrated to San Francisco in 1941, presumably due to Soviet entrance into the Second World War, and worked for the California Academy of Sciences as a Pacific Sardine specialist. He also ran the Museum of Russian Culture from 1954 until 1965. Information on Loukashkin can be found in "Register of the Anatole S. Loukashkin Papers," Museum of Russian Culture Staff, Hoover Institution Archives,

Stanford University, https://oac.cdlib.org/findaid/ark:/13030/c81z48xd/entire_text/, accessed October 8, 2020.
147 *Heilongjiangsheng bowuguan* [Heilongjiang Provincial Museum] website, *Lishi yange* [Historical Legacy], http://www.hljmuseum.com/system/201510/102323.html, accessed October 10, 2020.
148 "Heilongjiang Provincial Museum," China Travel Website, https://www.chinatravel.com/harbin-attraction/heilongjiang-provincial-museum/, accessed October 10, 2020.
149 Kelly (last name unspecified), China Highlights website, "Heilongjiang Provincial Museum," https://www.chinahighlights.com/harbin/attraction/provincial-museum.htm, accessed October 10, 2020. When I visited the museum in early September 2001, one hall housed vendors of high-end souvenirs, like white jade bracelets of Afghan jade. This lighter and airier space resembled an Art Nouveau exhibition hall, while halls housing exhibits with poorer lighting lent antique aspects to dusty animal furs and faded specimen feathers. A moose display and other specimens possibly dated to Loukashkin's time.
150 Localities where Ori'i collected included locations around Lamagulusu near Lake Dalainor, the Jalamte station 220 km southeast from Manchouli, the city of Aihon near Blagoveshchensk, the Khingan station near the Khingan Range, and mountains in Jehol (Rehe). All were border regions or near key rivers and mountains. Yamashina, "Note on the Specimens of Manchurian Birds," 449–50.
151 Anatole S. Loukashkin, "With the Orochons in Manchuria," http://loukashkin.com/Orochon/index.htm, accessed October 8, 2020. First published in *Pacific Discovery*, Vol. 20, No. 4, (July–August 1967), California Academy of Sciences, San Francisco.
152 Loukashkin, "With the Orochons in Manchuria."
153 Loukashkin, "With the Orochons in Manchuria."
154 Matsuda Michio, "Yo no naka, igai to semai—Manshu chôrui genshoku dai zukan," *Hobo yachô rokuon no hanashi* (September 8, 2012), http://syrinxmm.cocolog-nifty.com/syrinx/2012/09/post-c212.html, accessed October 8, 2020.
155 Yamashina Yoshimaro, *Ornithology and Bird Protection in USSR: Summary and Acknowledgment*, Vol. 7, No. 1 (June 30, 1973), 1.
156 Yamashina, "Note on the Specimens of Manchurian Birds," 449–51.
157 Yamashina, "Note on the Specimens of Manchurian Birds," 450.
158 Yamashina, "Note on the Specimens of Manchurian Birds," 446.
159 In 1938, missionary and Peiping-based educator George D. Wilder (1869–1946) wrote in *The Auk*, that Shaw's 1936 "Birds of Hopei Province" "is noteworthy in the world of ornithology, not only in that it is a model of its kind but also because it is the first scientific monograph on birds to be prepared entirely by a Chinese scientist." George D. Wilder, *The Auk*, Vol. 55, No. 2 (April 1938), 291. After graduate school at University of California-Berkeley and then at Stanford University in the mid-1920s, Shaw was one of Chinese ornithology's founders and also initiated vertebrate zoology. See Fuwen Wei and Dehua Wang, "Tsen-hwang Shaw: Founder of Vertebrate Zoology in China," *Protein and Cell* (September 4, 2020), https://link.springer.com/article/10.1007/s13238-020-00780-0, accessed October 9, 2020.
160 Yamashina, "Note on the Specimens of Manchurian Birds," 449.
161 Recent advances in modern photography enable this less destructive method to determine species diversity.
162 Yamashina, "Note on the Specimens of Manchurian Birds," 449.

Chapter 6

1. "Bird-Banding in Japan," (Tokyo: Department of Animal Husbandry, Ministry of Agriculture and Forestry, 1928).
2. In 1948, he published "Studies on the Biting Lice (*Mallophaga*) of Japan and Adjacent Territories," *The Japanese Medical Journal*, Vol. 1, No. 6 (1948), 535–56. This possibly intersected with vermin-group research for the *Institute*. In 1932, he published *Nippon konchû zukan: Fu konchû saishû benran* (Tokyo: Hokuryûkan, 1932).
3. Oliver L. Austin, Jr., "Japanese Ornithology and Mammology During World War II (An Annotated Bibliography)," Natural Resources Section Report Number 102 (January 30, 1948).
4. Herbert Bix, *Hirohito and the Making of Modern Japan* (New York: Harper Perennial, 2001). See Chapter 14, "A Monarchy Reinvented."
5. Yoshimi Yoshiaki, *Grassroots Fascism: The War Experience of the Japanese People* (New York: Columbia University Press, 2015), translated by Ethan Mark, 42, 102. Published in Japanese as *Kusa no ne no fuashizumu: Nihon no minshû no sensô taiken* (Tokyo: University of Tokyo Press, 1987).
6. Report of the Government Section, Supreme Commander for the Allied Powers, *Political Reorientation of Japan: September 1945 to September 1948* (Washington, DC: US Government Printing Office, 1948).
7. Yuma Totani, *The Tokyo War Crimes Trial: The Pursuit of Justice in the Wake of World War II* (Cambridge, MA: Harvard University Press, 2009).
8. Austin, "Japanese Ornithology and Mammology," 3.
9. Sheldon Harris, *Factories of Death: Japanese Biological Warfare, 1932–1945, and the American Cover-Up* (New York: Routledge, 1994, 2002), 14.
10. Yamashina Yoshimaro, *Dai 18-kai sengô no seikatsu, Yamashina Yoshimaro: Watakushi no Rirekisho*, http://www.yamashina.or.jp/hp/yomimono/rirekisho/rirekisho18.html, accessed August 30, 2020.
11. Yamashina Yoshimaro, *Manshû-san chôrui no shokusei* (Shinkyô [Changchun] *Manshûkoku Rinnôkyoku* (January 1, 1939).
12. Tama Shinnosuke, "Manshû ringyô imin to eirin jitsumu jisshû-sei seido," in Tama Shinnosuke, ed., *Sôryokusen taisei-ka no Manshû nôgyô imin* (Tokyo: Yoshikawa Kôbunkan), 1–20.
13. Iida-shi rekishi Kenkyûsho-hen, ed., *Manshû imin: Iida Shimo Ina kara no messeji* (Tokyo: Gendai shiryô shuppan, 2007).
14. Ronald Suleski, "Reconstructing Life in the Youth Corps Camps of Manchuria, 1938–1945: Resistance to Conformity," *East Asian History*, Vol. 30 (2005), 67–90.
15. Author's field notes, Tieli City, People's Republic of China, January 2000.
16. Annika A. Culver, "Japanese Mothers and Rural Settlement in Wartime Manchukuo: Gendered Reflections of Labor and Productivity in *Manshû Gurafu* [Manchuria Graph], 1936–1943," in Dana Cooper and Claire Phelan, eds., *Motherhood and War: International Perspectives* (London: Palgrave Macmillan, 2014), 95–113.
17. Culver, "Japanese Mothers and Rural Settlement in Wartime Manchukuo," 95–113.
18. Cornell University College of Veterinary Medicine, "Basic Duck Care," https://www.vet.cornell.edu/animal-health-diagnostic-center/programs/duck-research-lab/basic-duck-care, accessed December 24, 2020.
19. "Japan—Bacteriological Warfare Summary" (September 30, 1944), 390/39/33/04 165 486 27 JICA/CBI/SEA WD/G-2 JICA 8631 246/11, at National Archives and Records

Administration (NARA). William H. Cunliffe, "Select Documents on Japanese War Crimes and Japanese Biological Warfare, 1934–2006" (Washington, DC: National Archives and Records Administration, November 2006) https://www.archives.gov/files/iwg/japanese-war-crimes/select-documents.pdf, accessed December 23, 2020.,

20 Hachisuka, "Contributions on the Birds of Hainan," xii.
21 Marquess Hachisuka [Masauji], "Contributions on the Birds of Hainan," Supplementary Publication No. XV, *The Ornithological Society of Japan* (October 30, 1939), i.
22 Hachisuka, "Contributions on the Birds of Hainan," vii.
23 "Mrs. Birckhead, Dies, Julia Ward Kin," *The Newport Mercury and Weekly News* (February 11, 1972), 3.
24 S. Dillon Ripley, "Page 2 of a Letter from S. Dillon Ripley to Herbert S. Friedmann," (December 13, 1943), Smithsonian Institution Archives, Record Unit 305, Box 899, Folder: 170221, https://ids.si.edu/ids/deliveryService?id=SIA-SIA2010-0584&max=1024, accessed December 28, 2020.
25 "Mrs. Birckhead, Dies, Julia Ward Kin," *The Newport Mercury*, 3.
26 Smithsonian archives, "Ripley," at https://ids.si.edu/ids/deliveryService?id=SIA-SIA2010-0584&max=1024, accessed December 28, 2020.
27 "XXIV—Notices of Recent Ornithological Publications: Birckhead on Szechwan Birds," *Ibis*, Vol. 80, No. 2 (April 1938), 352.
28 Hachisuka, "Contributions on the Birds of Hainan," v.
29 Kawada Shin'ichirô and Yasuda Masatoshi, Hyōhon o meguru saishū hito to bōeki-shō to shūshū-ka–Hainan mogura no raberu o yomitoku, *Honyûrui kagaku*, Vol. 52, No. 2 (2012), 260.
30 This appears as "1927" in Hachisuka's "Contributions on the Birds of Hainan," vi.
31 Hachisuka, "Contributions on the Birds of Hainan," vii.
32 Hachisuka, "Contributions on the Birds of Hainan," vii.
33 Hachisuka Masauji, "Notes on Hainan Mammals," *Bulletin of the Biogeographical Society of Japan*, Vol. 11, No. 3 (March 1941), 9–14.
34 Austin, "Japanese Ornithology and Mammology during World War II," 28. Austin lists Katsumata Zensaku as "Y. Katsumata."
35 Hachisuka Masauji, Kuroda Nagamichi, and Takatsukasa Nobusuke, "A List of the Birds of Micronesia Under Japanese Mandatory Rule," in Hachisuka, Kuroda, and Takatsukasa, eds., *Handlist of Japanese Birds* (Tokyo: 1942) 169–211.
36 Austin, "Japanese Ornithology and Mammology during World War II," 16.
37 Based on a search of the holdings in the National Diet Library, Tokyo, https://ndlonline.ndl.go.jp/, accessed December 28, 2020.
38 "Overview: Heibonsha (1878–1961)," Oxford Reference, https://www.oxfordreference.com/view/10.1093/oi/authority.20110803095928491, accessed December 28, 2020.
39 Kuroda Nagamichi, *A Bibliography of the Duck Tribe, Anatidae, Mostly from 1926 to 1940, Exclusive of That of Dr. Phillips's Work* (Tokyo: Herald Press, 1942).
40 Austin, "Japanese Ornithology and Mammology during World War II," 15.
41 Notes from author's interview with Nakamura Tsukasa, Shinjuku, Tokyo, Japan, November 17, 2015.
42 Two volumes advertised for sale on Amazon included a hand-written dedication by Yamashina to Austin, clearly gifted to him during the Occupation. After Tony Austin's 2019 death, his descendants likely sold some of his father's rare book collections. See Amazon.com sales site for "A Natural History [Ecology] of

Japanese Birds (*Nippon no Chôrui to Sono Seitai*), in 2 volumes (1933–1934–1941), complete. Hardcover—January 1, 1933 by Yoshimaro Yamashina (Author)," https://www.amazon.com/Natural-History-Japanese-1933-1934-1941-complete/dp/B00SQIDDAW, accessed November 30, 2020.
43 Amherst College, Archives and Special Collections, Soffer Ornithology Collection Notes, https://www.amherst.edu/library/archives/holdings/soffer/y#Yamashina, accessed November 30, 2020.
44 Alex Danchev and Daniel Todman, eds., *War Diaries, 1939–1945: Field Marshal Lord Alanbrooke* (London: Phoenix Press, 2002), 521.
45 Steffen Kowalsky, *Alanbrooke: A Life in Books* (Westport, Ireland: West Coast Rare Books, 2018), https://www.westcoastrarebooks.com/data/news/news_7_doc1.pdf, accessed December 3, 2020.
46 Kowalsky, *Alanbrooke: A Life in Books*, 112–13.
47 Danchev and Todman, *War Diaries, 1939–1945*, xxv.
48 Danchev and Todman, *War Diaries, 1939–1945*, xxv.
49 The Isseido Booksellers Website, Takehiko Sakai, "About Us," http://www.isseido-books.co.jp/about/en_index.html, accessed December 3, 2020.
50 Amazon.com sales site for "A Natural History of Japanese Birds."
51 Prince Nobusuke Takatukasa (sic), *Japanese Birds* (Tokyo: Board of Tourist Industry Japanese Government Railways, 1941).
52 Takatukasa, *Japanese Birds*, "Editorial note," n.p.
53 David Leheny, "'By Other Means': Tourism and Leisure as Politics in Pre-war Japan," *Social Science Japan Journal*, Vol. 3, No. 2 (October 2000), 184.
54 Leheny, "By Other Means," 172.
55 Leheny, "By Other Means," 174.
56 Leheny, "By Other Means," 182.
57 Takatukasa, *Japanese Birds*, 9.
58 Takatukasa, *Japanese Birds*, 9.
59 Prince Nobusuke Takatukasa (sic), *Japanese Birds*.
60 Takatukasa, *Japanese Birds*, 11–12, 14, 24–5.
61 Takatukasa, *Japanese Birds*, 59.
62 Takatukasa, *Japanese Birds*, 41–2.
63 Takatukasa, *Japanese Birds*, 73.
64 Takatukasa, *Japanese Birds*, 73.
65 Takatukasa, *Japanese Birds*, 73.
66 Takatukasa, *Japanese Birds*, 81.
67 Sheldon Garon, "Luxury is the Enemy: Mobilizing Savings and Popularizing Thrift in Wartime Japan," *The Journal of Japanese Studies*, Vol. 26, No. 1 (Winter 2000), 42.
68 Leheny, "By Other Means," 184.
69 Katarzyna J. Cwiertka with Yasuhara Miho, *Branding Japanese Food: From Meibutsu to Washoku* (Honolulu: University of Hawaii Press, 2020).
70 Hans Martin Krämer, "'Not Befitting Our Divine Country': 'Not Befitting Our Divine Country': Eating Meat in Japanese Discourses of Self and Other from the Seventeenth Century to the Present," *Food and Foodways*, Vol. 16, No. 1 (January 2008), 33–62.
71 Krämer, "Not Befitting Our Divine Country," 39.
72 Suga Yutaka, *Yô Shôgun to tsuru no miso shiru–Edo no chô no bishokugaku* (Tokyo: Kôdansha, 2021), 48–52, 78–84.
73 Austin, "Mist Netting for Birds in Japan," 13.

74 1943 data are entirely absent, while 1944 and 1945 statistics are only partial. Austin, "Mist Netting for Birds in Japan," 18.
75 Austin, "Mist Netting for Birds in Japan," 13.
76 Austin, "Waterfowl of Japan," 97.
77 Austin, "Waterfowl of Japan," 96–7.
78 Austin, "Waterfowl of Japan," 94–5.
79 Austin, "Waterfowl of Japan," 99.
80 Austin, "Waterfowl of Japan," 91.
81 Austin, "Waterfowl of Japan," 91.
82 Austin, "Waterfowl of Japan," 91.
83 Benjamin Uchiyama, *Japan's Carnival War: Mass Culture on the Home Front* (Cambridge, UK: Cambridge University Press, 2019), 5.
84 "Shigen kagaku renmei hakkai," *Asahi Shimbun*, morning edition, January 17, 1941, 7.
85 *Pan no tonosama—Tôki Akira*, Exhibition Information, Numata City website, http://www.city.numata.gunma.jp/kanko/bunka/1006877/1006878/1009550.html, accessed December 23, 2020.
86 "Shigen mitomete—kagaku kenkyu-jo hossokusuru," *Asahi Shimbun*, evening edition (July 10, 1941), 2.
87 Umberto Quattrocchi, *CRC World Dictionary of Plant Names: Common Names, Scientific Names, Eponyms, Synonyms, and Etymology—Volume IV R-Z* (Boca Raton, FL: CRC Press, 1999), 2474.
88 Andrew Benson, *Hiroshi Tamiya, 1903–1984: A Biographical Memoir—National Academy of Sciences Biographical Memoirs, Volume 86* (Washington, DC: National Academic Press, 2005), 4.
89 *Asahi Shimbun* (July 14, 1953).
90 Morning in Oahu, Hawaii on 7 December was equivalent to Tokyo time 3:00 a.m. the next day.
91 "Kyôeiken he kagaku tanken-tai," *Asahi Shimbun*, morning edition (December 9, 1941), 3.
92 "Kyôeiken he kagaku tanken-tai," *Asahi Shimbun*, 3.
93 "Kyôeiken he kagaku tanken-tai," *Asahi Shimbun*, 3.
94 David S. G. Goodman, "JinJiLuYu in the Sino-Japanese War: The Border Region and the Border Region Government," *The China Quarterly*, No. 140 (December 1994), 1014.
95 Tai Wei Lim, "The Historical Development of Shanxi's Coal Industry as a Case Study," in *Energy Transitions in Japan and China* (Singapore: Palgrave Macmillan, 2017), 77–97.
96 Tracy G. Miller, "Northern Song Architecture in Southern Shanxi Province," *Journal of Song-Yuan Studies*, No. 29 (1999), 135.
97 "Wakaki Toshihide o môra—Senseishô chôsadan seizoiroi," *Asahi Shimbun*, evening edition (March 21, 1942), 2.
98 "Senseishō gakujutsu chōsa kenkyū-dan no kyo (1) / kaihatsu e no danryoku-sei/ Tôki Akira," *Asahi Shimbun*, morning edition (April 19, 1942), 4.
99 "Tōa minzoku kenkyūjo o sōsetsu kyōei-ken kakuritsu no ki," *Asahi Shimbun*, morning edition (May 20, 1942), 1.
100 Yamashina Yoshimaro, "Kiyosu Yukiyasu hakase no omoide," *Nippon Chôgakkai-ihô*, Vol. 25, No. 99 (June 1976), 8.
101 Shimomura Kenji and Kiyosu Yukiyasu, *Yachô seitai shashinshû* (Tokyo: Unsôdô, 1940).

102 Uchida Seinosuke and Shimomura Kenji, *Chôrui seitai shashinshû* (Tokyo: Sanseidô, 1930).
103 Witmer Stone, "Uchida's Photographs of Bird Life in Japan," *The Auk*, Vol. 47, No. 3 (July 1, 1930), 433.
104 Kiyosu Yukiyasu, "Ôsuichin, Dankyoku hômen no chôrui / dôbutsu-ban," *Asahi Shimbun* (June 22, 1942), 4.
105 Kiyosu Yukiyasu, "Ôsuichin, Dankyoku hômen no chôrui," 4.
106 "Sensei no shigen kaihatsu he—dai'ichi-ji gakujutsu chôsa kenkyu shokuin no shuki," *Asahi Shimbun* (June 22, 1942), 4.
107 Yamashina Yoshimaro and Kiyosu Yukiyasu, "A New Race of a Sparrow from Shansi, China," *Bulletin of the Biogeographical Society of Japan*, Vol. 13, No. 5 (February 1943), 39–41. Also mentioned in Austin, "Japanese Mammology and Ornithology During World War II," 25
108 "Sensei chôsadan shushuhin-ten," *Asahi Shimbun*, evening edition (April 2, 1943), 2.
109 "Sensei chôsadan shushuhin-ten," 2.
110 Aso, *Public Properties*, 2.
111 Austin, "Japanese Mammology and Ornithology during World War II," 44. These were published in March 1943, April 1943, October 1943, November 1943, March 1944, August 1944, October 1944, and June 1945. The last June issue was published prior to 25 May, when the bombings devastated the Institute, but still retained the anticipated June date.
112 Austin, "Japanese Mammology and Ornithology during World War II," 44.
113 Austin, "Japanese Mammology and Ornithology during World War II," 3.
114 Austin, "Japanese Mammology and Ornithology during World War II," 3.
115 Austin, "Japanese Mammology and Ornithology during World War II," 3.
116 Austin, "Japanese Mammology and Ornithology during World War II," 3.
117 Austin, "Japanese Mammology and Ornithology during World War II," 3.
118 Kiyosu Yukiyasu, "Shokuryô shigen to naru chôrui ni tsuite," *Shigen kagaku kenkyûjo ihô*, Vol. 4 (Tokyo: Kasumigaseki shobô, November 1943), 1–119.
119 Austin, "Japanese Mammology and Ornithology during World War II," 9.
120 Austin, "Japanese Mammology and Ornithology during World War II," 9.
121 Austin, "Japanese Mammology and Ornithology during World War II," 35.
122 Austin, "Japanese Mammology and Ornithology during World War II," 36–7.
123 Austin, "Japanese Mammology and Ornithology during World War II," 36–7.
124 Austin, "Japanese Mammology and Ornithology during World War II," 37.
125 Austin, "Japanese Mammology and Ornithology during World War II," 38.
126 Austin, "Japanese Mammology and Ornithology during World War II," 38–9.
127 Harris, *Factories of Death* (and Hal Gold, *Unit 731 Testimony* (North Clarendon, VT: Charles E. Tuttle Co., 1996).
128 Gold, *Unit 731 Testimony*, 67–8.
129 Noriko Aso, *Public Properties: Museums in Imperial Japan* (Durham, NC: Duke University Press, 2013), 55.
130 Ian Jared Miller, *The Nature of the Beasts: Empire and Exhibition at the Tokyo Imperial Zoo* (Berkeley: University of California Press, 2013), 122.
131 Miller, *The Nature of the Beasts*, 130–1.
132 Miller, *The Nature of the Beasts*, 121–2.
133 Miller, *The Nature of the Beasts*, 158.
134 Miller, *The Nature of the Beasts*, 153.
135 Miller, *The Nature of the Beasts*, 125.

136 Miller, *The Nature of the Beasts*, 127–8.
137 Miller, *The Nature of the Beasts*, 120.
138 Miller, *The Nature of the Beasts*, 124.
139 Miller, *The Nature of the Beasts*, 125.
140 Miller, *The Nature of the Beasts*, 129.
141 National Diet Library of Japan, Portraits of Modern Japanese Historical Figures website, "Odachi Shigeo (1892–1955)," https://www.ndl.go.jp/portrait/e/datas/401.html, accessed December 28, 2020.
142 Klaus Antoni, "Yasukuni-Jinja and Folk Religion: The Problem of Vengeful Spirits," *Asian Folklore Studies*, Vol. 47, No. 1 (1988), 127–8. For more on Shintô and *onryô*, see John Breen and Mark Teeuwen, *A New History of Shintô* (Chichester, UK: John Wiley and Sons, 2010), 77–8.
143 Animal memorial statues appear on Shrine precincts. "Precinct Map," Yasukuni Jinja website, https://www.yasukuni.or.jp/english/precinct/map.html, accessed December 28, 2020.
144 Wesley Craven and James Cate, eds., *The Pacific: Matterhorn to Nagasaki. The Army Air Forces in World War II. Volume V* (Chicago: University of Chicago Press, 1953), 638–9.
145 Austin, "Japanese Mammology and Ornithology," 4.
146 J. Delacour, *The Auk*, Vol. 77, No. 1 (January 1960), 118.
147 Austin, "Japanese Mammology and Ornithology During World War II," 4.
148 Kuroda Nagamichi, "A List of the Birds of Paradise and the Bower-Birds," *Tori* [Birds], Vol. 11 (1944), 620.
149 Clifford B. Frith, "Ornithological Literature from the Papuan Subregion 1915 to 176: An Annotated Bibliography," *Bulletin of the Museum of American Natural History*, Vol. 164, Article 3 (1979), 414.
150 Oliver L. Austin, letter to James Cowen Greenway, November 25 1946, headed as "NRS, GHQ, etc. etc., Austin correspondence," Ikk Binder 5.
151 Jean Delacour and Ernst Mayr, "Notes on the Taxonomy of the Birds of the Philippines," *Zoologica: Scientific Contributions of the New York Zoological Society*, Vol. 30, Part III, No. 12–15 (November 15, 1945), 105–18.
152 Ernst Mayr, "In Memoriam: Jean (Theodore) Delacour," *The Auk*, Vol. 103 (July 1986), 605.
153 Delacour and Mayr, "Notes on the Taxonomy of the Birds of the Philippines," 105.
154 Delacour, "Obituaries: Masauji, 18th Marquess Hachisuka," 521.
155 Lt.-Col, W.P.C. Tenison, D.S.O., ed., *Bulletin of the British Ornithologists' Club*, Vol. 47, Session 1946–1947 (London: H. F. & G. Witherby, 1947), xi.
156 Tenison, *Bulletin of the British Ornithologists' Club*, xii.
157 Tenison, *Bulletin of the British Ornithologists' Club*, xii.
158 Tenison, *Bulletin of the British Ornithologists' Club*, xv.
159 Tenison, *Bulletin of the British Ornithologists' Club*, xvi.
160 Tenison, *Bulletin of the British Ornithologists' Club*, xvi.
161 Tenison, *Bulletin of the British Ornithologists' Club*, Iv.
162 Tenison, *Bulletin of the British Ornithologists' Club*, preface (unpaginated).
163 Nakamura Tsukasa, interview with the author, Shinjuku, Tokyo, Japan, November 17, 2015.
164 Nakamura, interview with the author.
165 Murakami, *Zetsumetsu dori dôdô wo oimotometa otoko*, 331.

166 "The Oliver L. Austin Photographic Collection" online archive, curated by Annika A. Culver, https://austin.as.fsu.edu/exhibits/show/occupation/item/1355, accessed August 4, 2020.
167 Murakami, *Zetsumetsu dori dôdô wo oimotometa otoko*, 332.
168 Murakami, *Zetsumetsu dori dôdô wo oimotometa otoko*, 302–3.
169 I thank Masako Hachisuka (Paris Mako Hachiska) for sharing her memories in mid-May 2021. Email correspondence with the author, May 14, 2021.
170 Murakami, *Zetsumetsu dori dôdô wo oimotometa otoko*, 302.
171 Murakami, *Zetsumetsu dori dôdô wo oimotometa otoko*, 304–5.
172 Report on the Commission on Wartime Relocation and Internment of Civilians, *Personal Justice Denied* (Washington, DC: US Government Printing Office, 1982, 1983), 118.
173 Murakami, *Zetsumetsu dori dôdô wo oimotometa otoko*, 302–3.
174 "Marquis Asks Divorce from Nisei Wife," *Pacific Citizen* (April 28, 1951), https://pacificcitizen.org/wp-content/uploads/archives-menu/Vol.032_%2316_Apr_28_1951.pdf, accessed August 6, 2020.
175 Austin, "Japanese Mammology and Ornithology During World War II," 4.
176 Elliot McClure, *Stories I Like to Tell: An Autobiography* (Camarillo, CA: Self-Published, 1995), 127.
177 Yamashina Yoshimaro, *Yamashina Yoshimaro hakase no ayumi* (Tokyo: Sôbunshinsha, 1984), 24.
178 My translation. Yamashina, *Dai 18-kai sengô no seikatsu*.
179 Nakamura, interview with the author.
180 My translation. Yamashina, *Yamashina Yoshimaro: Watakushi no Rirekisho*.
181 *Yamashina Yoshimaro: Watakushi no Rirekisho*.
182 *Yamashina Yoshimaro: Watakushi no Rirekisho*.
183 *Yamashina Yoshimaro: Watakushi no Rirekisho*.
184 *Yamashina Yoshimaro: Watakushi no Rirekisho*.
185 Nagahisa Kuroda, *The Ornithologists and Avifauna in [sic] Early History of Japanese Field Ornithology (Chiefly Until 1950)* (Tokyo: Maruzen, 2004), V.
186 Yamashina Yoshimaro, *Watakushi no rireki-sho*.
187 These are digitized and featured on the digital humanities website that I curate: https://austin.as.fsu.edu/.
188 The next chapter illuminates this concept, which I discussed in a past article. Annika A. Culver, "Saving the Birds: Oliver L. Austin's Collaboration with Japanese Scientists in Revising Wildlife Policies in US-Occupied Japan, 1946–1950," *Endeavour*, Vol. 41, No. 4 (December 2017), 151–65.
189 Augé, *Oblivion*, 3.
190 Ruiz, question and answer session, February 18, 2016.

Chapter 7

1 I thank *Endeavour* for permission to publish edited portions of "Saving the Birds: Oliver L. Austin's Collaboration with Japanese Scientists in Revising Wildlife Policies in US-Occupied Japan, 1946–1950," *Endeavour*, Vol. 41, No. 4 (December 2017), 151–65.
2 SCAP refers to the position held by General MacArthur during his leadership of Japan's postwar Allied Occupation.

3   A Fall 2015 Japan Foundation Research Fellowship at Waseda University, a 2015 D. Kim Foundation Summer Research Grant, and a 2015 Committee on Faculty Research Support summer grant from FSU's Council on Research and Creativity supported this chapter. In October 2015, I presented a draft at the Third Conference of East Asian Environmental History at Kagawa University in Takamatsu, Japan. I thank Professor Setoguchi Akihisa of Kyoto University for insightful comments. I am grateful to Dr. Tsurumi Miyako, YIO Collections Director and Division of Natural History Specialist, for generously sharing materials, showing specimen collections, and arranging bird-banding observation.
4   Austin, "Japanese Ornithology and Mammalogy during World War II."
5   Oliver L. Austin, Jr. and Nagahisa Kuroda, "The Birds of Japan, Their Status and Distribution," *The Bulletin of the Museum of Comparative Zoology at Harvard College* Vol. 109, No. 4 (1953), 277–637.
6   Matsuda Takeshi, *Soft Power and Its Perils: US Cultural Policy in Early Postwar Japan and Permanent Dependency* (Washington, DC: Woodrow Wilson Center Press, 2007); *Sengo Nihon ni okeru Amerika no sofuto pawā: han'eikyūteki izon no kigen* (Tokyo: Iwanami shoten, 2008); and *Taibei izon no kigen: Amerika no sofuto pawâ no senryaku* (Tokyo: Iwanami shoten, 2015).
7   The first student exchanges between Kyoto University of Foreign Studies and FSU began in spring 2015.
8   Dr. Oliver L. Austin, "Speech for Bird Day," April 10, 1947, in Oliver L. Austin, Sr.'s Ikk Binder #IV, Correspondence from Japan from Oliver L. Austin, Jr. and Family, University of Florida, Florida Museum of Natural History, 4.
9   "Pheasants are reported to have made a comeback in numbers during the last three years, the result of few winters, less shooting and the scarcity of ammunition, as the birds are difficult to net or trap. I found them not uncommon … ." Oliver L. Austin, Jr., Memorandum on "Field Trip to Investigate Wildlife in the Districts of Yamanashi and Shizuoka," NR 091.33, APO 500, General Headquarters Supreme Commander for the Allied Powers Natural Resources Section, October 31, 1946, 3.
10 Princess Takamado, "The Green Pheasant: The National Bird of Japan," Bird Life International-Asia Org, News (March 1, 2020), https://www.birdlife.org/asia/news/green-pheasant-national-bird-japan, accessed on April 7, 2020.
11 Takamado, "The Green Pheasant."
12 Morris Low, *Science and the Building of a New Japan* (London: Palgrave Macmillan, 2005), 197.
13 Joseph Nye, *Bound to Lead: The Changing Nature of American Power* (London: Basic Books, 1990).
14 Joseph Nye, *Soft Power: The Means to Success in World Politics* (New York: Public Affairs, 2004), X.
15 Matsuda, *Soft Power and Its Perils* (2007); *Sengo Nihon ni okeru Amerika no sofuto pawā* (2008); and *Taibei izon no kigen* (2015).
16 *Tori*, Vol. 12, No. 56 (November 1947), 1–2.
17 *Tori*, 1–2.
18 Shinobu Seizaburô, *Taishô demokurashii-shi* (Tokyo: Nippon hyôron-sha, 1954).
19 Max Ward, *Thought Crime: Ideology and State Power in Interwar Japan* (Durham, NC: Duke University Press, 2019).
20 Morris Fraser Low, "Japan's Secret War? 'Instant' Scientific Manpower and Japan's World War II Atomic Bomb Project," *Annals of Science*, Vol. 47, No. 4 (1990), 347–60.

21 Duck liming refers to a prewar Japanese practice continuing into the 1940s, which involved steeping a quarter-inch thick line of string in boiled bark of Mochi (*Ilex integra*) or Yamaguruma (*Trochodendron aralioides*) trees on the afternoon before it was set, usually in the evening. Then, it is rolled up on a reel holding 2000-metres of line. According to Austin, " … it catches birds with a sticky material, working like flypaper." The sticky line "is set after dark on the shallow waters where the ducks congregate to rest and feed. With the reel on a spindle in the stern of his punt, the hunter fastens one end of the lime-string to a bamboo marker and starts to pay it out loosely in loops and whorls as his assistant poles haphazardly over the area in pitch blackness. As soon as his mile and a half of lime-string is floating on the surface, he poles through the nearby marshes where the birds have gone to feed and scares them up so they will retire to the center of the pond where his string is set." Austin, "Waterfowl of Japan," 85–6.
22 Andrew J. Grad, "Land Reform in Japan," *Pacific Affairs*, Vol. 21, No. 2 (June, 1948), 115–35; and Tsutomu Ōchi, "The Japanese Land Reform: Its Efficacy and Limitations," in *The Developing Economies* 4/2 (1966), 129–50.
23 Lebra, *Above the Clouds*, 1.
24 Prime Minister of Japan and His Cabinet website, "The Constitution of Japan," https://japan.kantei.go.jp/constitution_and_government_of_japan/constitution_e.html, accessed November 8, 2020.
25 Dower, *Embracing Defeat*, 320–38.
26 Dower, *Embracing Defeat*, 23.
27 José Rabasa, *Tell Me the Story of How I Conquered You: Elsewheres and Ethnosuicide in the Colonial Mesoamerican World* (Austin: University of Texas Press, 2012).
28 Rabasa, *Tell Me the Story*, 2.
29 Rabasa, *Tell Me the Story*, 1.
30 Rabasa, *Tell Me the Story*, 1.
31 Rabasa, *Tell Me the Story*, 4.
32 Rabasa, *Tell Me the Story*, 5.
33 Figure 0.5, "Ornithological Society."
34 Austin, Jr. and Kuroda, "The Birds of Japan," 277–637.
35 Nagahisa Kuroda, "Preface," *The Ornithologists and Avifauna in [sic] Early History of Japanese Field Ornithology* (Tokyo: Maruzen Company, 2004), IV.
36 Erwin Stresemann, "*The Birds of Japan, Their Status and Distribution*. Oliver L. Austin, Jr., Nagahisa Kuroda," in the section "New Biological Books," of *The Quarterly Review of Biology*, Vol. 29, No. 3 (September 1954), 269.
37 Stresemann, "*The Birds of Japan*," 269.
38 The number eighty-eight, spoken in Mandarin Chinese as *bābā*, sounds like "wealth," with lucky connotations, so Kuroda likely chose to publish that number of copies for this reason. I thank Dr. Tsurumi who gifted me a copy. Kuroda, *The Ornithologists and Avifauna in Early History*, 562.
39 Kuroda, *The Ornithologists and Avifauna in Early History*, frontispiece.
40 Kuroda, *The Ornithologists and Avifauna in Early History*, frontispiece.
41 Kuroda, *The Ornithologists and Avifauna in Early History*, 1.
42 Kuroda, *The Ornithologists and Avifauna in Early History*, 53.
43 Kuroda, *The Ornithologists and Avifauna in Early History*, 51.
44 Bernard Brisé uses this term to describe his photos constructing interplays between real and imagined worlds issuing from ruined spaces. Bernard Brisé, *Lieux d'Ailleurs* (Paris: Bord de l'Eau, 2004).

45 For example, the *Kasumi Kaikan* [Kasumi (Misty) Meeting Hall], a long-standing private social club for former aristocrats in Tokyo's Kasumigaseki area, still exists, and was once directed by Kuroda Nagahisa's brother Nagahide, though its exclusivity is belied by inexpensive yearly fees. A 1997 *The Washington Post* interview described Nagahide as "an elegant, thin man who walks with perfect posture," who said "If we were interested in money, we could sell just one piece of art and buy a gorgeous house …. But then Japanese culture would deteriorate. It is our duty to keep it intact." Quoted in Mary Jordan, "The Last Retreat of Japan's Nobility," *The Washington Post* (May 21, 1997), https://www.washingtonpost.com/archive/lifestyle/1997/05/21/the-last-retreat-of-japans-nobility/8fa34c94-e808-43eb-916d-3806924759bd/, accessed November 9, 2020. Yet, in 2015, the Kuroda family's *yuteki tenmoku* [oil-spot glazed] Song Dynasty-era tea bowl was decommissioned as national treasure and later sold.

46 Revised portions are reprinted from *Endeavour*, Vol. 41, No. 4, Annika A. Culver, "Saving the Birds: Oliver L. Austin's Collaboration with Japanese Scientists in Revising Wildlife Policies in US-Occupied Japan, 1946–1950," 151–65 (December 2017), with permission from Elsevier, obtained October 29, 2020.

47 Ōchi, "The Japanese Land Reform," 129–50.

48 According to the Australian Energy Market Commission, "Embedded networks are private electricity networks which serve multiple premises and are located within, and connected to, a distribution or transmission system through a parent connection point in the National Electricity Market." http://www.aemc.gov.au/ Rule-Changes/ Embedded-Networks, accessed August 17, 2017. "Embedded networks," whose original definition came from the electrical energy transmission system pioneered recently in Australia and adopted in Canada, can be used in a postwar Allied occupation context, for circulation of individuals within already established private networks dedicated to promotion of certain activities (like focusing intellectual energies toward ornithology) connecting to larger political entities composed of occupiers, of which Austin, as SCAP ornithologist, was a representative (and could channel energies back to the central point.)

49 Section C, "Working for the Occupation" in III. "Biographical Introduction" at https://library.osu.edu/projects/bennett-in-japan/1c_intro.html; and Section D, "Colleagueship: PO&SR's Relationship to the Japanese" in "A History of the Public Opinion and Sociological Research Division, SCAP," at https://library.osu.edu/projects/bennett-in-japan/3a_docs.html). John W. Bennett and Iwao Ishino, *Paternalism and the Japanese Economy: Anthropological Studies of* Oyabun-Kobun *Patterns* (Minneapolis: University of Minnesota Press, 1963).

50 Section C "Working for the Occupation," in III. "Biographical Introduction" at https://library.osu.edu/projects/bennett-in-japan/1c_intro.html.

51 John W. Bennett, cited in Section D, "Colleagueship."

52 Oliver L. Austin, Jr. letter to Oliver L. Austin, Sr., February 19, 1947, Austin Correspondence, Ikk Binder #6.

53 Oliver L. Austin, Jr. letter to Oliver L. Austin, Sr., February 19, 1947.

54 Oliver L. Austin Junior and Edythe Austin, interview with Steve Kerber, August 30, 1978, University of Florida oral history project, http://ufdc.ufl.edu/ UF00005958/00001/pdf, 31–2, accessed August 18, 2017.

55 Joseph Stromberg, "What Is the Anthropocene and Are We in It?—Efforts to Label the Human Epoch Have Ignited a Scientific Debate Between Geologists and Environmentalists" Smithsonian Magazine (January 2013), http://www.smithsonianmag.com/science-nature/ what-is-the-anthropocene-and-are-we-in-it-164801414/, accessed October 1, 2015.

56 Tsutsui, "Landscapes in the Dark Valley."
57 Austin and Austin, interviewer with Kerber, 21.
58 Austin received his Ph.D. in Ornithology from Harvard University in 1931. Austin and Austin, interviewer with Kerber, 22.
59 Austin and Austin, interview with Kerber, 35–6.
60 Austin and Austin, interview with Kerber, 38.
61 Oliver L. Austin, Jr., letter to Oliver L. Austin, Sr. (February 22, 1946), 1–2, Florida State Museum of Natural History, Gainesville, Florida, Correspondence of Oliver L. Austin Sr. (hereafter Austin correspondence), Ikk Binder 5. "Ikk" is a nickname for Oliver L. Austin, Jr.
62 Bruce Cumings, *Korea's Place in the Sun: A Modern History* (New York: W. W. Norton and Company, 2005), 185–236.
63 Austin and Austin, interview with Kerber, 36.
64 Oliver L. Austin Jr., letter to Oliver L. Austin Sr. (March 2, 1946), Florida State Museum of Natural History, Gainesville, Florida, Correspondence of Oliver L. Austin Sr. (hereafter Austin correspondence), Ikk Binder 5.
65 According to Alastair Roberts, such ideas on race and public administration harboured long-standing histories, and derive from Americans administrating the Philippines: "It compels a reassessment of our understandings about their commitment to democracy, and about the supposed differences between American and European public administration at that time." Alastair Roberts, "Bearing the White Man's Burden: American Empire and the Origins of Public Administration," *Perspectives on Public Management and Governance*, Vol. 3, No. 3 (September 2020), 185–96.
66 Austin and Austin, interview with Kerber, 36.
67 Austin and Austin, interview with Kerber, 36.
68 Most of northern Korea's birds passed through Korea's southern portion in seasonal migrations, while Austin relied on printed colonial Japanese sources to determine systematics for birds common to the north.
69 Austin and Austin, interviewer with Kerber, 37.
70 Austin and Austin, interview with Kerber,
71 These were the only mist-nets available in the US until Austin introduced Japanese mist-nets in the early 1950s. He began using Italian mist nets at his Wellfleet home on Cape Cod between 1928 and 1929. They allowed researchers to catch birds en-masse for banding, and even target specific sizes, while avoiding bait. Prior to Japanese mist-nets, these represented the most efficient method of capturing birds for banding. Austin and Austin, interview with Kerber, 10–11.
72 Oliver Luther Austin, "The Birds of Korea," *Bulletin of the Museum of Comparative Zoology* Vol. 101, No. 1 (1948), 1–301.
73 Austin and Austin, interview with Kerber, 37.
74 Oliver L. Austin Jr., letter to Oliver L. Austin Sr. (April 4, 1946), Ikk Binder 5.
75 Oliver L. Austin Jr., letter to Oliver L. Austin Sr. (March 23, 1946), Ikk Binder 5.
76 Austin Jr., letter to Austin Sr. (April 9, 1946), Ikk Binder 5.
77 Conrad Totman, *Pre-Industrial Korea and Japan in Environmental Perspective* (Leiden: Brill Academic Publishing, 2004), 159.
78 Few studies of colonial Korea's environment exist besides Totman, *Pre-Industrial Korea and Japan in Environmental Perspective*.
79 Austin Jr., letter to Austin Sr. (March 23, 1946), Ikk Binder 5.

80 Claude M. Adams Photographs, finding aid, http://www.oac.cdlib.org/findaid/ark:/13030/c8280bwp/admin/#ref39, accessed August 22, 2017.
81 A. Myra Keen, "Memorial to Hubert Gregory Schenck, 1897–1960," Memorials—The Geological Society of America 10 (1960): 1–6, //rock.geosociety.org/pub/Memorials/v10/Schenck-HG.pdf.
82 Austin and Austin, interview with Kerber, 37, 39.
83 Colonel Hubert G. Schenk, "Natural Resources Problems in Japan," Science, Vol. 108, No. 2806 (1948), 367–72, on 367.
84 Schenk, "Natural Resources Problems in Japan," 367–72.
85 For the colonizers, "go-betweens' work was vital for the creative adaptation and imitation of technique in economic and commercial innovation. It also mattered a great deal for the establishment of the matters of fact in which the new sciences traded." Quoted in Simon Schaffer et al., editors, *The Brokered World: Go-Betweens and Global Intelligence, 1770–1820* (New York: Science History Publications, 2009), xxiv. Kapil Raj, "Mapping Knowledge Go-Betweens in Calcutta, 1770–1820," in Shaffer et al., *The Brokered World*, 105–50. Raj argues that, for the Portuguese in India, "interaction with the various communities and political authorities concerned was rendered possible only through the mediation of professional go-betweens with specific literary, technical, juridical, administrative and financial skills," 108.
86 Christopher Aldous, "A Dearth of Animal Protein: Reforming Nutrition in Occupied Japan (1945–1952)," in *Food and War in Mid-Twentieth Century East Asia*, ed. Katarzyna J. Cwiertka (Burlington, VT: Ashgate, 2013), 53–72. Bruce F Johnston, with Mosaburo Hosoda and Yoshio Kusumi, *Japanese Food Management in World War II* (Palo Alto: Stanford University Press, 1953). Austin headed the Foods Branch, Price and Rationing Division, Economic and Scientific Section, GHQ SCAP. Aldous, "Dearth of Animal Protein," 56.
87 Mary H. Clench and J. William Hardy, "In Memoriam: Oliver L. Austin Jr.," *The Auk* 106 (October 1989), 706–23, on 708.
88 Tokyo Port Wild Bird Park, http://www.wildbirdpark.jp/en/, accessed November 10, 2020.
89 Austin and Austin, interviewer with Kerber, 41–2.
90 Austin's description appears in a SCAP report: "The mist net is made of strong silk thread, dyed black to make it invisible when stretched against a dark background. It operates on a simple principle …. The mesh is mounted loosely on a taut frame of strong twine, crossed by horizontal braces called shelfstrings [sic] (tanaito) about two feet apart, with one-half again that amount of mesh between each shelf. The excess hangs in a loose bag or pocket below the lower shelf-string. A bird striking the net falls into the pocket and remains helpless and unharmed until removed by the hunter." Austin, "Mist Netting for Birds in Japan," 7.
91 Oliver L. Austin Jr., letter to Oliver L. Austin Sr., September 8, 1946, Austin correspondence, Ikk Binder 5.
92 Austin, "Japanese Ornithology and Mammalogy during World War II."
93 John M. MacKenzie coined "collecting imperialism" in *Museums and Empire*, 60.
94 Oliver L. Austin, Junior, letter to Jimmy (James L.) Peters, September 19, 1946, 1.
95 Oliver L. Austin, Jr., letter to Oliver L. Austin Sr., September 23, 1946, Austin correspondence, Ikk Binder 5.
96 Oliver L. Austin, Jr., letter to Oliver L. Austin, Sr., September 20, 1946, Austin correspondence, Ikk Binder 5.

97 Austin and Austin, interview with Kerber, 48; and Aldous, "Dearth of Animal Protein."
98 Austin, "Japanese Ornithology and Mammalogy during World War II."
99 Lieutenant Colonel Assistant Adjutant General W. W. McMillion, Order AGPO 268–1, letter to "Individuals Concerned," General Headquarters United States Army Forces, Pacific (September 25, 1946).
100 Austin, letter to Peters, September 19, 1946.
101 Oliver L. Austin, Jr., letter to Oliver L. Austin, Sr., October 23, 1946, Tokyo, Japan, 1.
102 Shibusawa, *America's Geisha Ally*, 5.
103 Nakamura, interview with the author, November 17, 2015.
104 Murakami, *Zetsumetsu dori dôdô wo oimotometa otoko*, 242, 301.
105 Austin, "Mist Netting for Birds in Japan" and "Waterfowl of Japan."
106 General Headquarters Supreme Commander for the Allied Powers Natural Resources Section, HGS/JFJ/OLA/mc, APO 500, October 31, 1946, ME 091.33 Memorandum for "Field Trip to Investigate Wildlife in the Districts of Yamanashi and Shizuoka."
107 Oliver L. Austin Jr., letter to Oliver L. Austin Sr. (November 3, 1946), Austin correspondence, Ikk Binder 5.
108 The Brookline Bird Club (BBC) was a wealthy birding group founded on June 18, 1913 in the fancy Boston suburb's Brookline Public Library. Brookline's inhabitants were fanatical about birds and appointed a town "bird warden"; this group initially boasted thirty members "to study, observe, and protect native song birds and to encourage their propagation." Now welcoming over 1,000 members, the BBC bills itself as "one of the country's oldest, largest, and most active bird clubs." John Nelson, "The Brookline Bird Club, 1913–1945," Brookline Bird Club website, https://www.brooklinebirdclub.org/wp-content/uploads/2020/04/BBC-History-1913-1945.pdf, accessed November 11, 2020.
109 Oliver L. Austin, Jr., letter to Oliver L. Austin Sr., December 5, 1946, Austin correspondence, Ikk Binder 5.
110 Austin, Jr., letter to Austin, Sr. December 15, 1946, Austin correspondence, Ikk Binder 5, 3–4.
111 Oliver Luther Austin, Supreme Commander for the Allied Powers, Natural Resources Section, "Japanese Wildlife Sanctuaries and Public Hunting Grounds" (Tokyo: General Headquarters, Supreme Commander for the Allied Powers, 1948), Preliminary Study, Supreme Commander for the Allied Powers, Natural Resources Section, No. 28.
112 Oliver L. Austin Jr., letter to Oliver L. Austin Sr. (October 23, 1946), Austin correspondence, Ikk Binder 5, 1.
113 Austin, "Mist Netting for Birds in Japan," 5.
114 Austin, "Mist Netting for Birds in Japan," 7.
115 Austin, "Mist Netting for Birds in Japan," 13.
116 Oliver L. Austin Photographic Collection, https://austin.as.fsu.edu/files/original/d19ff1a7b13e950d0a694e7e2995e98b.jpg, accessed August 9, 2021.
117 Oliver L. Austin, Jr., letter to Oliver L. Austin, Sr., April 18, 1947, Austin correspondence, Ikk Binder 6.
118 Oliver L. Austin, Jr., letter to Oliver L. Austin, Sr., February 1950, Austin correspondence, Ikk Binder 8.

119 Nishimatsu Hideki, Japan Foundation, Tokyo, email correspondence with author, October 9, 2015.
120 Oliver L. Austin, Daily Notes, November 16, 1946, Tajimi, Gifu-Ken, Austin correspondence, Ikk Binder 5.
121 Sadao Yoshida, "Isao Ijima: The Father of Parasitology in Japan (With Potrait [sic] Plate)," *The Journal of Parasitology*, Vol. 10, No. 3 (1924), 165–7.
122 Tôzawa Kôichi, *Rekidai kaichô no kotoba*, in *Nihon Chôgakkai—100 nen no rekishi*, Nihon Chôgakkai 100 shûnen kinen tokubetsu gô, *Nihon Chôgakkai-shi*, Vol. 61 (November 2012), 8–9.
123 Tôzawa, *Rekidai kaichô no kotoba*, 8–9.
124 Oliver L. Austin, Jr., letter to Oliver L. Austin, Sr., October 9, 1946, Tokyo, Austin Correspondence, Ikk Binder 5; Oliver L. Austin, Jr., letter to Oliver L. Austin, Sr., October 23, 1946, Tokyo, Austin Correspondence, Ikk Binder 5; and Oliver L. Austin, Jr., letter to Jimmy Peters, November 25, 2020. NRS, GHQ, etc., etc., Austin correspondence, Ikk Binder 5.
125 Austin interview with Culver, 3.
126 Oliver L. Austin, Jr., letter to Oliver L. Austin, Sr., November 26, 1946, Austin Correspondence, Ikk Binder 5.
127 "The Widow Ariyoshi," image in the Oliver L. Austin Photographic Collection, https://austin.as.fsu.edu/items/show/1363, accessed November 18, 2020.
128 This amount was calculated via this website. "Inflation: Calculate the Value," *Dollar Times*, https://www.dollartimes.com/inflation/inflation.php?amount=15&year=19 46#:~:text=Adjusted%20for%20inflation%2C%20%2415.00%20in,over%20this%20 period%20was%203.64%25., accessed November 10, 2020.
129 Oliver L. Austin, Jr., letter to Oliver L. Austin, Sr., October 23, 1946, Tokyo, Austin Correspondence, Ikk Binder 5.
130 Oliver L. Austin, Jr., letter to John Todd Zimmer, September 23, 1946, headed as NRS, GHQ, etc. etc., Austin correspondence, Ikk Binder 5, 3.
131 In contrast, Yamashina notes nine people in his 1982 biography: himself, Hachisuka Masauji, Kuroda Nagamichi, Takatsukasa Nobusuke, Uchida Seinosuke, Momiyama Tokutarô, Matsuyama Shirô, Kumatani Saburô, and Oda Haruo. Yamashina and Aoki, *Yamashina Yoshimaro no shôgai*, 207.
132 Oliver L. Austin, Jr., letter to Oliver L. Austin, Sr., April 17, 1947, Austin correspondence, Ikk Binder 6.
133 Oliver L. Austin, Jr., letter to Oliver L. Austin, Sr., 17 April 1947, Austin correspondence, Ikk Binder 6.
134 Oliver L. Austin, Jr., letter to Oliver L. Austin, Sr., 17 April 1947, Austin correspondence, Ikk Binder 6.
135 Oliver L. Austin, letter to John Todd Zimmer, September 23, 1946, Austin correspondence, Ikk Binder 5, 3.
136 Oliver L. Austin, Jr., letter to James Cowen Greenway, November 25, 1946, headed as NRS, GHQ, etc. etc., Austin correspondence, Ikk Binder 5.
137 Austin and Austin, interview with Kerber, 33.
138 Austin and Austin, interview with Kerber, 21.
139 Mizuno, *Science for the Empire*, 71.
140 Reported in a letter to General MacArthur given to his interpreters by Japanese witnesses. Austin and Austin, interview with Kerber, 50.
141 Aldous, "Dearth of Animal Protein."

142 Austin interview with Culver, 25.
143 Austin interview with Culver, 45, 102.
144 Kuroda, "The Ornithologists and Avifauna in Early History," V.
145 Austin and Austin, interview with Kerber, 49–50.
146 Austin and Austin, interview with Kerber, 49.
147 Austin and Austin, interview with Kerber, 49–50.
148 Oliver L. Austin, Jr., General Headquarters, Supreme Commander for the Allied Powers, Natural Resources Section, "Memorandum: Outline of Argument to be presented for consideration at Wildlife Conservation Conference," held at 13:30, Tuesday, December 3, 1946, Bureau of Scientific Education, Room 41, Ministry of Education Building, NR 337 (December 2, 1946) File, 3.
149 Austin, "Memorandum," 3.
150 In 1946, hunting guns were again allowed with a license. "Imperial Ordinance Concerning the Prohibition of the Possession of Guns and Other Arms," *Imperial Ordinance No. 300 of 1946* (June 3, 1946).
151 "Elizabeth Gray Vining, Japan's Royal Tutor, Died on November 27th, Aged 97," *The Economist* (December 9, 1999), accessed October 20, 2015, http://www.economist.com/node/267202.
152 Elizabeth Vining, *Windows for the Crown Prince* (Philadelphia: J. B. Lippincott and Company, 1952).
153 Oliver L. Austin Jr., letter to Oliver L. Austin Sr; 1 (February 25, 1947), Austin correspondence, Ikk Binder 6.
154 Austin Jr., letter to Austin Sr; 1 (February 25, 1947), Austin correspondence, Ikk Binder 6, 1.
155 Sydney B. Whipple, "Let's Protect Japan's Birds: Serious Economic Loss to a Nation Foreseen Unless Destructive Custom of Netting Wild Fowl Abandoned," *Nippon Times Magazine*, February 27, 1947, 3–4.
156 "Chapter 6: The Reverse Course, 1947–1952," in William R. Nester, *Power across the Pacific: A Diplomatic History of American Relations with Japan* (London: Palgrave Macmillan, 1996), 224–60.
157 The Rocky Mountain Elk Foundation (RMEF) provides one of the most succinct definitions for the North American Model of Conservation (NAMC) in stating that "fish and wildlife belong to all Americans" and that they need to be managed in a sustainable fashion. The RMEF notes that the basic principles of the NAMC can be summed up in the Seven Sisters for Conservation, whose principles include the following: 1) wildlife is held in the public trust, 2) prohibition in commerce of dead wildlife, 3) democratic rule of law, 4) hunting opportunity for all, 5) non-frivolous use, 6) international resources, and 7) scientific management. Rocky Mountain Elk Foundation, "The North American Wildlife Conservation Model," http://www.rmef.org/Conservation/HuntingIsConservation/NorthAmericanWildlifeConservationModel.aspx, accessed August 22, 2017.
158 Whipple, "Let's Protect Japan's Birds," 3.
159 J. F. Organ, V., Geist, S. P. Mahoney et al., "The North American Model of Conservation," *The Wildlife Society Technical Review*, Vol. 12, No. 4 (2012).
160 Oliver L. Austin Jr., "Speech for Bird Day," April 10, 1947, Austin correspondence, Ikk Binder 8.
161 Austin Jr., "Speech for Bird Day."
162 "Prime Minister of Japan and His Cabinet" website, "The Constitution of Japan."

163 See Beata Sirota Gordon, *The Only Woman in the Room: A Memoir of Japan, Human Rights, and the Arts* (Chicago: University of Chicago, 1997).
164 Austin, interview with Culver.
165 Nibe Tominosuke, with a foreword by Uchida Seinosuke, *Nô no tori no seitai* (Tokyo: Kôbunsha, 1951).
166 Yamashina Yoshimaro, *Nihon no chôrui no seitai to hogo* (Tokyo: Yamashina Ornithological Institute, 1951).
167 Yamashina Yoshimaro, *Birds in Japan: A Field Guide* (Tokyo: Tokyo News Service, 1961).
168 S. Dillon Ripley, "In Memoriam: Yoshimaro Yamashina, 1900–1989," *The Auk*, Vol. 108, No. 4 (1989), 721.
169 Mizuno, *Science for the Empire*, 183.
170 Seth H. Low, "Banding with Mist Nets," *Journal of Field Ornithology*, Vol. 28, No. 3 (1957), 115–28.
171 Austin and Kuroda, "The Birds of Japan: Their Status and Distribution," 284.
172 Iida, *Chôgaku no hyakunen—chô ni sudamaserareta hitobito*, 70–1.
173 Dong-Won Kim, Yoshio Nishina: Father of Modern Physics in Japan (Boca Raton, FL: CRC Press, 2007); Morris Low, ed., Building a Modern Japan: Science, Technology, and Medicine in the Meiji era and Beyond(New York: Palgrave Macmillan, 2005); and Morris Low, Science and the Building of a New Japan (New York: Palgrave Macmillan, 2005).
174 John Bowen, *The Gift of the Gods: The Impact of the Korean War on Japan* (Norfolk, VA: Bowen, 1984).

Chapter 8

1 Dower, *Embracing Defeat*.
2 Shibusawa, *America's Geisha Ally*.
3 Nakamura Tsukasa, interview with the author, Tokyo, November 17, 2015.
4 Francesca Di Marco, "Act or Disease? The Making of Modern Suicide in Early Twentieth-century Japan," *The Journal of Japanese Studies*, Vol. 39, No. 2 (Summer 2013), 325–58; and Alan Stephen Wolfe, *Suicidal Narrative in Modern Japan: The Case of Dazai Osamu* (Princeton, NJ: Princeton University Press, 1990).
5 Ivan Morris, *The Nobility of Failure: Tragic Heroes in the History of Japan* (Tokyo: Kurodahan Press, 1975, 2013).
6 H. Elliott Mcclure, *Stories I Like to Tell: An Autobiography* (Camarillo, CA: Self-Published, 1995), 126.
7 McClure, *Stories I Like to Tell*, 127.
8 McClure, *Stories I Like to Tell*, 127.
9 Norman Boyd Kinnear, "Obituary: The Marquess Hachisuka," *Ibis*`, Vol. 96, No. 1 (1954), 150.
10 Morinaka Sadaharu, Chairman of the Biogeographical Society of Japan, *Nihon seibutsu chiri gakkai sôsetsu-sha–Hachisuka Masauji ni tsuite*, http://biogeo.a.la9.jp/main/hachisuka.htm, accessed July 31, 2020.
11 "Nodal point" usually refers to urban space, like in Melanie U. Pooch's work, but I describe a person centered within a larger, dynamic network. Pooch emphasizes " … the cultural significance of urban centers … the global city's peculiarity as a *cultural*

nodal point of a global network of flows is of particular importance." "Global Cities as Cultural Nodal Points," in Melanie U. Pooch, *DiverCity—Global Cities as a Literary Phenomenon: Toronto, New York, and Los Angeles in a Globalizing Age* (Bielefeld, Germany: Transcript Verlag, 2016), 27.

12 Smithsonian Institution biologist David Challinor denied Ripley's CIA involvement. Michael Lewis, "Scientists or Spies? Ecology in a Climate of Cold War Suspicion," *Economic and Political Weekly* (2002), 37.
13 Oliver L. Austin, Jr.'s "Daily Notes," November 18, 1946, Atami, Shizuoka Prefecture, Japan. Archived in the Florida State Museum of Natural History, University of Florida, Gainesville.
14 Austin, "Daily Notes."
15 Austin, "Daily Notes."
16 MCZBase: The Database of Zoological Collections, Museum of Comparative Zoology-Harvard University, https://mczbase.mcz.harvard.edu/SpecimenResults.cfm, accessed December 9, 2020.
17 https://iiif.lib.harvard.edu/manifests/view/ids:17329067, accessed December 12, 2020.
18 Kerber, "Interview with Oliver L. Austin, Jr. and Edythe Austin," 22–4.
19 Kerber, "Interview with Oliver L. Austin, Jr. and Edythe Austin," 23.
20 Austin, "Daily Notes."
21 Austin, "Daily Notes."
22 Hellman, "Profiles: Curator Getting Around."
23 David Valley, *Gaijin Shogun: Gen. Douglas MacArthur Stepfather of Postwar Japan* (San Diego: Sektor, 2000).
24 Hellman, "Profiles: Curator Getting Around."
25 Nakamura Tsukasa, *Yue Kuroda Nagahisa Sensei wo shinonde, Bulletin of the Biogeographical Society of Japan*, Vol. 64 (20 December 2009), 1. AOU Committee on Memorials, "In Memoriam: Nagahisa Kuroda, 1916–2009," *The Auk*, Vol. 128, No. 3 (2011), 592.
26 Oliver L. Austin Photographic Collection, https://austin.as.fsu.edu/files/original/3956b617ada2d19f125aa0e099f6ebb4.jpg, accessed August 9, 2021.
27 National Institute of Infectious Diseases (NIID), "Outline-Organization: History," https://www.niid.go.jp/niid/en/aboutniid-2.html, accessed December 15, 2020.
28 Sheldon Harris, "Japanese Biomedical Experimentation During the World War II Era," in Thomas E. Beam and Linette R. Sparacino, eds., *Military Medical Ethics Volume 2* (Washington, DC: Borden Institute at Walter Reed Medical Center, 2003), 495, https://www.laguardia.edu/maus/files/Ethics-ch-16.pdf, accessed December 14, 2020.
29 Thompson Report, "Summary: Japanese Biological Warfare (BW) Activities," (January–March 1946), IWG Reference Collection of Select Documents on Japanese Biological Warfare and on Japanese War Crimes, Textual Research Room (Room 2000) of Archives II, College Park, MD, JWC 001a.
30 In April 2018, Shiga University of Medical Science Professor Emeritus Nishiyama Katsuo announced disclosure of a January 1, 1945 document requested from Japan's National Archives, listing 3,607 researchers connected to Unit 731. Many names appeared redacted, but negotiations with the archive allowed further declassification. Kyodo Staff Report, "Names of 3,607 members of Imperial Japanese Army's Notorious Unit 731 Released by National Archives," *The Japan Times* (April 16,

2018), https://www.japantimes.co.jp/news/2018/04/16/national/history/names-3607-members-imperial-japanese-armys-unit-731-released/, accessed December 14, 2020.
31 Neil R. Smith, "Infectious Disease Research Laboratory," Report of Investigation Division, Legal Section, GHQ, SCAP (December 6, 1946), JWC 258/18, https://www.archives.gov/files/iwg/japanese-war-crimes/select-documents.pdf, accessed December 15, 2020.
32 Harris, *Factories of Death*, 64–5.
33 Hal Gold, *Unit 731: Testimony*, 30–1, 189.
34 Brody et al., "United States Responses."
35 Hiroshi Yoshikura and Takeshi Kurata, "Institute Profile: National Institute of Infectious Diseases," *Trends in Microbiology*, Vol. 9, No. 10 (October 2001), 512.
36 Akemi Nakamura, "Officials Feel Remains Best Left Buried: Group Alleged Unit 731 victims' bones still mystery," *Japan Times* (August 4, 2004), https://www.japantimes.co.jp/news/2004/08/04/national/alleged-unit-731-victims-bones-still-mystery/, accessed December 14, 2020.
37 Nakamura, "Officials Feel Remains Best Left Buried."
38 Nakamura, "Officials Feel Remains Best Left Buried."
39 Nishitani Kôsuke, *Toyama Kyôkai no kenbutsu ga shûkan-shi no Koramu ni saizai saremashita*, in *Bokushi-shitsu no mado*, https://toyama-tokyo.com/church/週刊誌のコラム掲載/, accessed December 14, 2020.
40 Amidst scientific work, Hirohito even forget personal space. E. J. H. Corner, "His Majesty Emperor Hirohito of Japan, K. G.: April 29, 1901–January 7, 1989, Elected F.R.S. 1971," *Biographical Memoirs of Fellows of the Royal Society*, Vol. 36 (December 1990), 250.
41 Toyama Yôchien website, https://toyama-tokyo.com/kindy/accss/, accessed December 14, 2020.
42 Toyama Kyôkai website, https://toyama-tokyo.com/church/about-us/, accessed December 14, 2020.
43 Dr. Nishitani, *Bokushi-shitsu no mado*, https://toyama-tokyo.com/church/category/column/, accessed December 14, 2020.
44 Masashi Yoshii and Nagahisa Kuroda, "In Memoriam: H. Elliott McClure, 1910–1998," *The Auk*, Vol. 116, No. 4 (1999), 1126.
45 Imperial Ordinance Concerning the Prohibition of the Possession of Guns and Other Arms, Imperial Ordinance No. 300 of 1946 (June 3, 1946). Library of Congress, Sayuri Umeda, Foreign Law Specialist, "Firearms-Control Legislation and Policy: Japan," (February 2013), https://www.loc.gov/law//help/firearms-control/japan.php#t13, accessed November 25, 2020.
46 More specifically, it is called the Order Concerning Firearms and Swords, Cabinet Order No. 334 of 1950 (November 15, 1950). Umeda, "Firearms-Control Legislation and Policy: Japan."
47 McClure, *Stories I Like to Tell*, 127.
48 McClure, *Stories I Like to Tell*, 127.
49 McClure, *Stories I Like to Tell*, 127.
50 Yoshii and Kuroda, "In Memoriam: H. Elliott McClure, 1910–1998," 1126.
51 Department of the Army, "US Army Activity in the US Biological Warfare Programs: Vol. 1, Unclassified (February 24, 1977)," 27, https://nsarchive2.gwu.edu//NSAEBB/NSAEBB58/RNCBW_USABWP.pdf, accessed November 30, 2020.
52 Department of the Army, "US Army Activity," 36.

53 Fort Detrick website, "History," https://home.army.mil/detrick/index.php/about/history, accessed December 14, 2020.
54 Yoshii and Kuroda, "In Memoriam: H. Elliott McClure, 1910–1998," 1126.
55 US Army Academy of Health Sciences, Stimson Library Digital Collections, Medical Bulletin of the US Army Far East, "Functional Chart: 406th Medical General Laboratory," https://stimson.contentdm.oclc.org/digital/collection/p15290coll1/id/68, accessed December 14, 2020.
56 "Functional Chart: 406th Medical General Laboratory."
57 Brody et al., "United States Responses."
58 Brody et al., "United States Responses."
59 Eric Horowitz, "The Nazi Interrogator Who Revealed the Value of Kindness," *Pacific Standard* (July 3, 2014), https://psmag.com/social-justice/nazi-interrogator-revealed-value-kindness-84747, accessed December 13, 2020. Raymond F. Toliver, *The Interrogator: The Story of Hanns-Joachim Scharff, Master Interrogator of the Luftwaffe* (Atglen, PA: Schiffer Military History Books, 1997).
60 Peter Grier and Faye Bowers, "How Interrogation Tactics Have Changed: A Review of CIA and Military Manuals from Several Decades Show Techniques Ranging from Low-Pressure to Coercive," *The Christian Science Monitor* (May 27, 2004), https://www.csmonitor.com/2004/0527/p02s01-usmi.html, accessed December 13, 2020.
61 Michaela Nesvarova, "Secrets of the 'Master Interrogator,'" *UToday* (September 5, 2019), https://www.utoday.nl/spotlight/67300/secrets-of-the-master-interrogator, accessed December 13, 2020.
62 Brody et al., "United States Responses."
63 Kondo S., ed., *Unit 731 and Biological Warfare: a CD-ROM Collection* (Tokyo: Kashiwa shobô, 2003). Report by Norbert H. Fell, April 22, 1947, Interrogation of Masuda Tomosada, 1–2.
64 Compiled in November 2006 by NARA archivist William H. Cunliffe to compile declassified US documents authorized by the Japanese government in 1999, *Select Documents on Japanese War Crimes and Japanese Biological Warfare, 1934–2006*, indicates that in late 1946, "SCAP CinCFE implemented a policy that all national security matters would fall under Washington control of the JCS and the SWNCC, and release required prior approval of the JCS [JWC 296/03, 243/49]," 4; "'SCAP Legal Section, Investigative Division Case No. 330': SCAP Intelligence Division (G-2) classified the entire case as SECRET and stopped all further investigation by SCAP Legal Section, Investigative Division [JWC 261/4, JWC 285]," 5; and "Secretary of War and Joint Research & Development Board recommended that all information about the biological warfare program be held at the TOP SECRET level [JWC 305/01]," 5. A June 9, 1947 "Intelligence Information on Bacteriological Warfare Note from General Willoughby, SCAP G-2 to the SCAP Legal Section" relates the "desire of the Chief of the Chemical Warfare Service to keep all of the BW information in intelligence channels and out of any war crimes trials." 290/12/04/06 331 1294, 1434 SCAP/G-2 SCAP/Legal Willoughby, Gen. Case File #330 Folder #13 255/14, at the National Archives and Records Administration (NARA).
65 Sheldon Harris, *Factories of Death*, 302.
66 Brody et al., "United States Responses."
67 Peter Williams and David Wallace, *Unit 731: Japan's Secret Biological Warfare in World War II* (New York: Free Press, 1989), 178–9.
68 Brody et al., "United States Responses."

69 American medical scientists performed their own human experimentation research overseas and domestically on what they viewed as non-white marginal individuals. Kayte Specter-Bagdady and Paul A. Lombardo, "From In Vivo to In Vitro: How the Guatemala STD Experiments Transformed Bodies into Biospecimens," *Milbank Quarterly*, Vol 96, No. 2 (June 2018), 244–71.

70 Clarence G. Lasby, *Project Paperclip: German Scientists and the Cold War* (New York: Scribner, 1975), Annie Jacobsen, *Operation Paperclip: The Secret Intelligence Program That Brought Nazi Scientists to America* (New York: Little, Brown, and Company, 2014), and Brian E. Crim, *Our Germans: Project Paperclip and the National Security State* (Baltimore: Johns Hopkins University Press, 2018).

71 General Headquarters, United States Army Forces, Pacific, Military Section, General Staff, APO 500, "Memorandum for Record: Russian Request to Interrogate Japanese on Bacteriological Warfare, to Chief of Staff" (February 7, 1946), JWC 243/46, https://www.archives.gov/files/iwg/japanese-war-crimes/select-documents.pdf, accessed December 14, 2020.

72 McClure, *Stories I Like to Tell*, 283.

73 Cited as footnote 42 in John W. Powell, "Japan's Germ Warfare: The US Cover-up of a War Crime," *Bulletin of Concerned Asian Scholars*, Vol. 12, No. 4 (1980), 17.

74 Yoshii and Kuroda, "In Memoriam: H. Elliott McClure, 1910–1998," 1126.

75 AOU Committee on Memorials, "In Memoriam: Nagahisa Kuroda, 1916–2009," *The Auk*, Vol. 128, No. 3 (2011), 592.

76 Laura Hein, *Post-Fascist Japan: Political Culture in Kamakura After the Second World War* (London: Bloomsbury, 2019), 74.

77 In Hein, *Post-Fascist Japan*, 74. Originally quoted in Charles E. Martin, "Report of the United States Cultural Science Mission to Japan," to Supreme Commander for the Allied Powers: Civil Information and Education Section, Seattle: University of Washington, Institute of International Affairs, January 1949, 3.

78 Robert D. Eldridge, "How Eisenhower's June 1960 trip to Okinawa became a catalyst for reversion," *Japan Times* (June 17, 2020), https://www.japantimes.co.jp/opinion/2020/06/17/commentary/japan-commentary/eisenhowers-june-1960-trip-okinawa-became-catalyst-reversion/, accessed December 4, 2020.

79 Nick Kapur, *Japan at the Crossroads: Conflict and Compromise After Anpo* (Cambridge, MA: Harvard University Press, 2018), 1.

80 Yamashina Yoshimaro, *Yamashina Yoshimaro hakase no ayumi* (Tokyo: Sôbun-shinsha, 1984).

81 My italics. International House of Japan website, "History," https://www.i-house.or.jp/eng/history/, accessed November 29, 2020.

82 International House of Japan website, "History."

83 Suzuki Hiroyuki, *Landscape Gardener Ogawa Jihei and His Times: A Profile of Modern Japan* (Tokyo: JPIC, 2018).

84 International House of Japan Website, "Floor Plan," https://www.i-house.or.jp/eng/floormap.html, accessed December 4, 2020.

85 Yoshii et al., "Japanese Bird-Banding Now and Past," 310.

86 Lewis, "Scientists or Spies," 2323.

87 Quoted by Devesh Gadhavi, "Sagacious Salim Ali," The Corbett Foundation blog (December 2, 2015), http://corbettfoundation.org/articles/2015/12/02/sagacious-salim-ali/, accessed December 16, 2020.

88 Gadhavi, "Sagacious Salim Ali."

89 Warren King, "International Council for Bird Preservation: Historical Aspects and Current Projects of the ICBP," *American Birds*, Vol. 40, No. 1 (Spring 1986), 49.
90 S. Dillon Ripley, "In Memoriam: Yamashina Yoshimaro, 1900–1989," *The Auk*, Vol. 106, No. 4 (October 1989), 721.
91 Delacour, *The Living Air*, 137.
92 Delacour, *The Living Air*, 137.
93 Delacour, *The Living Air*, 137.
94 Delacour, *The Living Air*, 137–8.
95 Roger Stone, *The Lives of Dillon Ripley: Natural Scientist, Wartime Spy, and Pioneering Leader of the Smithsonian Institution* (Lebanon, NH: ForEdge, 2017).
96 Geoffrey T. Hellman, "Profiles: Curator Getting Around," *The New Yorker* (August 26, 1950), https://www.newyorker.com/magazine/1950/08/26/curator-getting-around, accessed September 6, 2020.
97 Geoffrey T. Hellman, "Delacour Reobserved," in the "Talk of the Town" section, *The New Yorker* (December 17, 1960), 29, https://www.newyorker.com/magazine/1960/12/17/delacour-reobserved, accessed September 4, 2020.
98 Hellman, "Delacour Reobserved," 28.
99 Hellman, "Delacour Reobserved," 28.
100 Geoffrey T. Hellman, *The Smithsonian: Octopus on the Mall* (Philadelphia: J. B. Lippincott Publishing Company, 1966).
101 Geoffrey T. Hellman, *Bankers, Bones & Beetles: The First Century of the American Museum of Natural History* (Garden City, New York: Natural History Press, 1969).
102 Delacour, *The Living Air*, 138.
103 Hellman, "Delacour Reobserved," 29.
104 Hellman, "Delacour Reobserved," 28.
105 Hellman, "Delacour Reobserved," 29.
106 Mikhail V. Kalyakin and Pavel S. Tomkovich, "History of the Bird Collections at the Zoological Museum at Moscow State University and Their Role for Russian Ornithology," *Bonner Zoologische Beiträge*, Vol. 51 (2002), No. 2/3 (September 2003), 168.
107 Kalyakin and Tomkovich, "History of the Bird Collections," 171.
108 Kalyakin and Tomkovich, "History of the Bird Collections," 168.
109 Kalyakin and Tomkovich, "History of the Bird Collections," 168.
110 Kalyakin and Tomkovich, "History of the Bird Collections," 171.
111 Yoshii and Kuroda, "In Memoriam," 1126.
112 Yoshii and Kuroda, "In Memoriam," 1126.
113 Michael Lewis, "Scientists or Spies? Ecology in a Climate of Cold War Suspicion," *Economic and Political Weekly*, Vol. 27, No. 24 (15–21 June 2002), 2326.
114 Yoshii and Kuroda, "In Memoriam," 1126.
115 Masashi Yoshii, Fumio Sato, Kiyoaki Ozaki, Yoshimitsu Shigeta, Shigemoto Komeda, Keiko Yoshiyasu, and Amane Mitamura, "Japanese Bird Banding, Now and Past," *Journal of the Yamashina Institute of Ornithology*, Vol. 21 (1989), 310, https://www.jstage.jst.go.jp/article/jyio1952/21/2/21_2_309/_pdf, accessed December 5, 2020.
116 Lewis, "Scientists or Spies," 2325.
117 Lewis, "Scientists or Spies," 2326.
118 Roy MacLeod, "'Strictly for the Birds': Science, the Military and the Smithsonian's Pacific Ocean Biological Survey Program, 1963–1970," *Journal of the History of Biology*, Vol. 34, No. 2 (June 2001), 318, https://citeseerx.ist.psu.edu/viewdoc/download?doi=10.1.1.23.4835&rep=rep1&type=pdf, accessed December 15, 2020.

119 Ted Gup, "The Smithsonian's Secret Contract: The Link between Birds and Biological Warfare," *The Washington Post Magazine* (May 12, 1985), 15.
120 Lewis, "Scientists or Spies?," 2327.
121 "[A]ccusations made to the effect that the Smithsonian is conducting germ warfare research or even that we are serving as a cover for the Department of the Army in such research are unwarranted." Quoted in MacLeod, "Strictly for the Birds," 337.
122 MacLeod, "Strictly for the Birds," 345.
123 MacLeod, "Strictly for the Birds," 317, 319.
124 John F. Kennedy Presidential Library and Museum website, "Nuclear Test Ban Treaty," https://www.jfklibrary.org/learn/about-jfk/jfk-in-history/nuclear-test-ban-treaty, accessed December 15, 2020.
125 MacLeod, "Strictly for the Birds," 323.
126 Lewis, "Scientists or Spies?," 2323.
127 AOU Committee on Memorials, "In Memoriam: Nagahisa Kuroda, 1916–2009," 592.

Conclusion

1 Eric Johnston, "Hunting in Japan: Gun Control Is the Norm and Discipline Is Rigid," *The Japan Times* (January 29, 2013), https://www.japantimes.co.jp/news/2013/01/29/reference/in-japan-gun-control-is-the-norm-and-discipline-is-rigid/, accessed December 17, 2020; and Karl Denzer, "Behold the Four-Month Process of Buying a Gun in Japan," *Washington Post* (October 5, 2017), https://www.washingtonpost.com/opinions/behold-the-four-month-process-of-buying-a-gun-in-japan/2017/10/05/72283fea-2375-11e7-b503-9d616bd5a305_story.html, accessed December 17, 2020.
2 Johnston, "Hunting in Japan."
3 Masashi Yoshii, Fumio Sato, Kiyoaki Ozaki, Yoshimitsu Shigeta, Shigemoto Komeda, Keiko Yoshiyasu and Amane Mitamura, "Japanese Bird-Banding Now and Past," *Journal of the Yamashina Ornithological Institute*, Vol. 21, No. 1 (August 1989), 309, https://www.jstage.jst.go.jp/article/jyio1952/21/2/21_2_309/_pdf, accessed December 7, 2020.
4 Yoshii et al., "Japanese Bird-Banding Now and Past," 310.
5 Yoshii et al., "Japanese Bird-Banding Now and Past," 310.
6 "In Memoriam," 592.
7 "In Memoriam: Nagahisa Kuroda, 1916–2009," *The Auk*, Vol. 123, No. 3 (July 1, 2011), 592.
8 Kuroda Nagahisa, "A Bird Census in the Imperial and Akasaka Palaces for 1965," *Journal of the Yamashina Institute for Ornithology*, Vol. 4, No. 5 (1966), 269–79.
9 Kuroda Nagahisa, "A Bird Census in the Imperial and Akasaka Palaces for 1966," *Journal of the Yamashina Institute for Ornithology*, Vol. 5, No. 1 (1967), 1–12.
10 Kuroda Nagahisa, "Analysis of Banding Data (1923–'43) of the Tree Sparrow in Japan," *Journal of the Yamashina Institute for Ornithology*, Vol. 4, No. 5 (1966), 397–402.
11 Kuroda Nagahisa, *Chôrui no kenkyû: Seitai* (Tokyo: Shin-shicho-sha, 1967).
12 Kerber, Interview with Austin and Austin, 66.
13 See Chapter Six.

14 AbeBooks.com website, Kuroda Nagahisa—Study of Birds: The Ecology, about This Item, https://www.abebooks.com/Study-Birds-Ecology-Kuroda-Nagahisa-Tokyo/14991783546/bd, accessed December 7, 2020.

15 IUCN website, "IUCN: A Brief History," https://www.iucn.org/about/iucn-a-brief-history, accessed December 7, 2020.

16 Bird Life International website, "Our History," https://www.birdlife.org/worldwide/partnership/our-history, accessed June 26, 2020.

17 Bird Life International, "Data Zone: Japan," http://datazone.birdlife.org/country/japan, accessed May 28, 2020.

18 Bird Life International, "Data Zone: Japan," "Crested Shelduck," http://datazone.birdlife.org/species/factsheet/crested-shelduck-tadorna-cristata, accessed May 28, 2020.

19 Bird Life International website, "Data Zone: Japan," http://datazone.birdlife.org/country/japan, accessed June 26, 2020.

20 Nakamura Tsukasa, *Yamanashi Daigaku Meshô Kyôju, Kenkyû. Kyôiku. Shizen hôgô. Shakai katsudô: 50-nen no ayumi (gaiyo)* (Kôfu, Japan: Yamanashi University, 2005), 1–2.

21 Hiroyoshi Higuchi, "Tsukasa Nakamura, 1926–2018," *The Auk* (December 31, 2019), https://academic-oup-com.proxy.lib.fsu.edu/auk/advance-article/doi/10.1093/auk/ukz076/5691185, accessed March 26, 2020.

22 Nakamura, *Kenkyû. Kyôiku. Shizen hôgô. Shakai katsudô*, 10.

23 Morinaka Sadaharu, *Hachisuka Masauji-shi seitan hyaku-nen kinen shinpojiumu wo oete, Nihon seibutsu chiri gakkai,* Bulletin 58 (2003), 115–19; and *Hachisuka Masauji-shi seitan hyaku-nen kinen ni yosete, Nihon seibutsu chiri gakkai,* Bulletin 58 (2003), 95–119.

24 Murakami, *Zetsumetsu dori dôdô wo oimotometa otoko.*

25 Yamashina Institute for Ornithology online website, "About Us: Chronological Table of Our History," http://www.yamashina.or.jp/hp/english/about_us/history.html, accessed March 24, 2020.

26 Yamashina Institute for Ornithology website, "About Us: President, His Imperial Highness Crown Prince Akishino (Akishinonomiya Fumihito), PhD, President, Yamashina Institute for Ornithology," http://www.yamashina.or.jp/hp/english/about_us/president.html, accessed March 24, 2020.

27 A. Fumihito, T. Miyake, S. Sumi, M. Takada, S. Ohno, and N. Kondo, "One Subspecies of the Red Junglefowl (*Gallus gallus gallus*) Suffices as the Matriarchic Ancestor of All Domestic Breeds," *Proceedings of the National Academy of Sciences*, Vol. 91 (1994), 12505–9; A. Fumihito, T. Miyake, M. Takada, S. Ohno, "The Genetic Link between the Chinese Bamboo Partridge (*Bambusicola thoracica*) and the Chicken and Junglefowl of the Genus Gallus," *Proceedings of the National Academy of Sciences*, Vol. 92 (1995), 11053–6, A. Fumihito, T. Miyake, R. Shinju, T. Endo, T. Gojobori, N. Kondo, and S. Ohno, "Monophyletic Origin and Unique Dispersal Patterns of Domestic Fowl," *Proceedings of the National Academy of Sciences*, Vol. 93 (1996), 6702–5.

28 Akishinomiya Fumihito, Ryôzô Kakizawa, Michael Roberts, and Victoria Roberts, *Ôshû no kakin zukan* (Tokyo: Tankobon, 1994).

29 Akishinomiya Fumihito, *Niwatori to hito–minzoku seibutsugaku no shiten kara* (Tokyo: Shôgakukan, 2000).

30 Akishinomiya Fumihito and Nishino Yoshiaki, *Chôgaku daizen (Chô no biosofuia Yamashina korekushon he no sasoi)* (Tokyo: University of Tokyo Press, 2008).

31 Oliver L. Austin, Jr., Arthur Zinger, and Herbert S. Zim, eds., *Birds of the World: A Survey of the Twenty-Seven Orders and One Hundred and Fifty-Five Families* (Racine, WI: Western Publishing Company, Inc., 1961).
32 Akishinomiya Naruhito, *Temuzu to tomoni: Eikoku no ninenkan* (Tōkyō: Gakushūin Sōmubu Kōhōka, 1993). Akishinomiya Naruhito, with Hugh Cortazzi, *The Thames and I: A Memoir of Two Years at Oxford* (Folkestone, UK: Global Oriental Press, 2006).
33 Tim Fernholz, "Gobies of Fun: Emperor Akihito of Japan Is a Published Expert on Tiny Fish," *Quartz* (October 2, 2018), https://qz.com/1410538/emperor-akihito-of-japan-is-an-expert-marine-biologist/, accessed March 26, 2020.
34 Daniel Sneider, "Celebrating Emperor Hirohito, Marine Biologist Extraordinaire," *Christian Science Monitor* (October 29, 1986), https://www.csmonitor.com/1986/0429/ohiro.html, accessed March 26, 2020.
35 Môri Hideo, *Tennô-ke to seibutsugaku* (Tokyo: Asahi Shimbun Publications, 2015).
36 Hideo Mohri, *Imperial Biologists: The Imperial Family of Japan and Their Contributions to Biological Research* (Singapore: Springer Nature, 2019).
37 "BirdLife's Honorary President: Her Imperial Highness Princess Takamado of Japan," https://www.birdlife.org/worldwide/programme-additional-info/birdlifes-honorary-president, accessed May 28, 2020.
38 BirdLife International Tokyo Annual Report 2019 (January 1–December 31, 2019), 11, https://tokyo.birdlife.org/sites/wp-content/uploads/2020/01/BL2019_AR_eng_final.pdf, accessed December 7, 2020.
39 BirdLife International Tokyo Annual Report 2019, 1.
40 "BirdLife's Honorary President."
41 *Nippon yachô no kai* website, *Meshô. Mokuteki nado*, https://www.wbsj.org/about-us/summary/about/, accessed December 7, 2020.
42 *Yachô no kai* Bird Shop 2020 *tsûhan katarogu baado shoppu* (Autumn/Winter 2020 edition), 20, https://www.wbsj.org/webcatalog/img/2020aw.pdf, accessed December 7, 2020.
43 Bird Shop 2020, 20.
44 These include Izumo Oyashiro [Izumo Grand Shrine], Izumo Oyashiro [Izumo Grand Shrine] Cultural Foundation, Fushimi Inari Taisha [Great Shrine], Hokkaidô Jingu [Shrine], Masumida Shrine, Samukawa Shrine, and Daihonzan Sojiji Temple. BirdLife International Tokyo Annual Report 2019,13.
45 BirdLife International Tokyo Annual Report 2019, 13.
46 *Santorî no aichô katsudô*, Nippon no chô hyakka, https://www.suntory.co.jp/eco/birds/encyclopedia/, accessed December 7, 2020.
47 Suntory Hakushû Distillery website, "How to Enjoy the Distillery," https://www.suntory.com/factory/hakushu/info/, accessed 7 December 2020.
48 Suntory Corporate Website, *Santori no aichô no katsudô*, https://www.suntory.com/csr/activity/environment/eco/birds/, and Suntory corporate website, "Water Sustainability: Bird Conservation Activities," https://www.suntory.co.jp/eco/birds/fund/, accessed 7 December 2020.
49 Personal interview with Dr. Tsurumi Miyako, Chief of the Division of Natural History, Yamashina Ornithological Institute, November 11, 2015.

Works Cited

Archival Collections

Dean C. Worcester papers, Bentley Historical Library, University of Michigan.
Dean C. Worcester Photographic Collection, University of Michigan, Museum of Anthropology.
Florida State Museum of Natural History, University of Florida, Gainesville.
Harvard University, Museum of Comparative Zoology.
Museum of Russian Culture Collection, Hoover Institution Archives, Stanford University.
National Archives and Records Administration.
The Oliver L. Austin Photographic Collection, Annika A. Culver, curator
Rare Books Section, University of Cambridge Library.
Smithsonian Institution archives.
Soffer Ornithology Collection, Amherst College Library, Archives and Special Collections.
Tokyo University, Natural History Collections.
Waseda University, Natural History Collections.
Yale University, Natural History Collections.

Interviews

Oliver L. Austin Junior and Edythe Austin. Interview with Steve Kerber. University of Florida oral history project, August 30, 1978. http://ufdc.ufl.edu/UF00005958/00001/pdf.
Tony Austin. Interview with author (transcript). Florida State University History Department, November 12–14, 2014.
Masako Hachisuka. Phone interview with author, May 12, 2021.
Hasegawa Michiko. Interview with author. Association for Asian Studies Annual Meeting, Washington, DC, March 23, 2018.
Nakamura Tsukasa. Interview with author. Shinjuku, Tokyo, Japan, November 17, 2015.
Tsurumi Miyako, Chief, Division of Natural History. Interview with author. Yamashina Ornithological Institute, November 11, 2015.

Newspapers

Asahi Shimbun
The Japan Times
Manchuria Daily News
New York Times
Washington Post

Japanese-Language Periodicals

Nihon seibutsu chiri gakkai-kaihô
Ryôyû
Shigengaku kenkyûjo-ihô
Tori

English-Language Periodicals

The Auk
The China Journal
Ibis
Nature
The New York Times
The New Yorker
Washington Post

Select Primary and Secondary Sources

Abel, Emily K. *Tuberculosis and the Politics of Exclusion: A History of Public Health and Migration to Los Angeles.* New Brunswick: Rutgers University Press, 2007.

Adam, Hajo and Adam D. Galinsky. "Enclothed Cognition." *Journal of Experimental Social Psychology* 46, no. 4 (2012): 918–25.

Akishinomiya, F. and Nishino Yoshiaki. *Chôgaku daizen (Chô no biosofuia Yamashina korekushon he no sasoi).* Tokyo: University of Tokyo Press, 2008.

Akishinomiya, Naruhito. *Temuzu to tomoni: Eikoku no ninenkan.* Tôkyô: Gakushūin Sōmubu Kōhōka, 1993.

Alberti, Samuel J. M. M. "Placing Nature: Natural History Collections and Their Owners in Nineteenth-Century England." *The British Journal for the History of Science* 35, no. 3 (September 2002): 291–311.

Aldous, Christopher. "A Dearth of Animal Protein: Reforming Nutrition in Occupied Japan (1945–1952)." In *Food and War in Mid-Twentieth Century East Asia*, edited by Katarzyna J. Cwiertka, 53–72. Burlington, VT: Ashgate, 2013.

Aldrich, Robert. *Colonialism and Homosexuality.* London: Routledge, 2002.

Anker, J. *Bird Books and Bird Art: An Outline of the Literary History and Iconography of Descriptive Ornithology.* New York: Springer, 2014.

Aso, Noriko. *Public Properties: Museums in Imperial Japan.* Durham, NC: Duke University Press, 2013.

Austin, Jr., Oliver L. "Japanese Ornithology and Mammalogy during World War II." *Wildlife Leaflet* 305 (March 1948), reproduction of Natural Resources Section Report Number 102, GHQ, SCAP.

Austin, Jr., Oliver L. "Mist Netting for Birds in Japan." *Supreme Commander for the Allied Powers, Natural Resources Section, Report 88,* August 1947.

Austin, Jr., Oliver L. "Waterfowl of Japan." *Natural Resources Section Report No. 118.* Tokyo: General Headquarters, Supreme Commander for the Allied Powers, 1949.

Austin, Jr., Oliver L. and Kuroda Nagahisa. *Birds of Japan: Their Status and Distribution.* Cambridge: Bulletin of the Museum of Comparative Zoology, Harvard University, 1953.

Australian information office. *Teien-Oosutoraria taishikan*. Tokyo: Shunhôsha, possibly 1988.

Aylmer, Charles. "The Memoirs of H. A. Giles." *East Asian History*, no. 13-14 (June/December 1997): 1-90.

Barrow, Mark. *A Passion for Birds: American Ornithology after Audubon*. Princeton, NJ: Princeton University Press, 1998.

Beebe, William. *A Monograph of the Pheasants: Four Volumes*. London: George Witherby and Company, 1918-22.

Benjamin, Walter. *The Arcades Project*. Translated by Howard Eiland and Kevin McLaughlin. Cambridge, MA: Harvard University Press, 2002.

Bibliothèque Nationale Français. "Birds of Jehol." Catalogue Générale, Notice Bibliographique. https://catalogue.bnf.fr/ark:/12148/cb316744362.

Blakiston, Thomas Wright. Henry James Stovin Pryer, Leonhard Stejneger, and Alexander Wetmore, *Birds of Japan (Revised to 1882)*. Yokohama: Asia Society of Japan, 1882.

Bowring, Richard. "The Selwyn Swastika." In *Calendar: Selwyn College Cambridge Vol. 125*, edited by Peter Fox et al. (2017-18), 63-4. Cambridge, UK: Selwyn College, 2018.

Browne, Janet. "A Science of Empire: British Biogeography before Darwin." *Revue d'Histoire de Sciences* [Review of the History of Science] 45, no. 5 (1992): 453-75.

Butler, Judith. *Gender Trouble: Feminism and the Subversion of Identity*. New York: Routledge, 1990.

Charbonneau, Oliver. "'A New West in Mindanao': Settler Fantasies on the U.S. Imperial Fringe." *The Journal of the Gilded Age and Progressive Era* 18, no. 3 (2019): 304-25.

Cook, Matt. ed. *Queer Domesticities: Homosexuality and Home Life in Twentieth Century London*. New York: Palgrave Macmillan, 2014.

Corner, E. J. H. "His Majesty Emperor Hirohito of Japan, K. G.: 29 April 1901-7 January 1989, Elected F.R.S. 1971." *Biographical Memoirs of Fellows of the Royal Society* 36 (December 1990): 242-72.

Cortezzi, Sir Hugh. *Japan in Late Victorian London: The Japanese Native Village in Knightsbridge and "The Mikado", 1885*. London: Sainsbury Institute, 2009.

Craven, Wesley and James Cate, eds. *The Pacific: Matterhorn to Nagasaki. The Army Air Forces in World War II. Volume V*. Chicago: University of Chicago Press, 1953.

Cross, Sherrie. "Prestige and Comfort: The Development of Social Darwinism in Early Meiji Japan, and the Role of Edward Sylvester Morse." *Annals of Science* 53, no. 4 (January 1996): 323-44.

Culver, Annika A. "Saving the Birds: Oliver L. Austin's Collaboration with Japanese Scientists in Revising Wildlife Policies in US-Occupied Japan, 1946-1950." *Endeavour* 41, no 4 (December 2017): 151-65.

Cunliffe, William H. "Select Documents on Japanese War Crimes and Japanese Biological Warfare, 1934-2006." Washington, DC: National Archives and Records Administration, November 2006.

Dacudao, Patricia Irene. "ABACA: The Socio-Economic and Cultural Transformation of Frontier Davao, 1898-1941." Master's Thesis, Department of History, Ateneo de Manila University, 2017.

Delacour, Jean. *The Living Air: The Memoirs of an Ornithologist*. London: Country Life, 1966.

Delacour, Jean and Ernst Mayr. "Notes on the Taxonomy of the Birds of the Philippines." *Zoologica: Scientific Contributions of the New York Zoological Society* 30 Part III, no. 12-15 (November 15, 1945): 105-18.

Demetrio, Francisco R. "Shamans, Witches and Philippine Society." *Philippine Studies* 36, no. 3 (Third Quarter 1988): 372-80.

Department of the Army. "US Army Activity in the US Biological Warfare Programs: Vol. 1, Unclassified." February 24, 1977. https://nsarchive2.gwu.edu//NSAEBB/NSAEBB58/RNCBW_USABWP.pdf

Dower, John. *Embracing Defeat: Japan in the Wake of World War II*. New York: W. W. Norton and Company, 1999.

Duara, Prasenjit. *Sovereignty and Authenticity: Manchukuo and the East Asian Modern*. Lanham, MD: Rowman and Littlefield, 2003.

"Elizabeth Gray Vining, Japan's Royal Tutor, Died on November 27th, Aged 97." *The Economist*, December 9, 1999. http://www.economist.com/node/267202.

Flugfelder, Greg. *Cartographies of Desire: Male-Male Sexuality in Japanese Discourse, 1600–1950*. Berkeley: University of California Press, 1999.

Fuchs, Barbara. *Mimesis and Empire: The New World, Islam, and European Identities*. Cambridge, UK: Cambridge University Press, 2001.

Gailbraith, Patrick W. *Otaku Spaces*. Seattle: Chin Music Press, 2012.

Gates, Barbara T. "Introduction: Why Victorian Natural History?" *Victorian Literature and Culture* 35, no. 2 (2007): 539–49.

General Headquarters, United States Army Forces, Pacific, Military Section, General Staff. APO 500, "Memorandum for Record: Russian Request to Interrogate Japanese on Bacteriological Warfare, to Chief of Staff." February 7, 1946, JWC 243/46. https://www.archives.gov/files/iwg/japanese-war-crimes/select-documents.pdf.

Giles, Herbert A., and Masa U. Hachisuka. *Record of Strange Nations: From the Chinese of 1392 A.D.–Publishers' Announcement*. London: Percy Lund, Humphries and Co. Ltd., early 1927.

Gillespie, Greg. *Hunting for Empire: Narratives of Sport in Rupert's Land, 1840–70*. Vancouver: University of British Columbia Press, 2008.

Glaser, Barney G. "The Local-Cosmopolitan Scientist." *American Journal of Sociology* 69, no. 3 (November 1963): 249–59.

Gold, Hal. *Unit 731: Testimony*. Rutland, VT: Tuttle Publishing, 1996.

Gopinath, Praseeda. "Imperial Masculinities." *Oxford Bibliographies* In "Literary and Critical Theory." New York and Oxford: Oxford University Press, June 2017. https://www.oxfordbibliographies.com/view/document/obo-9780190221911/obo-9780190221911-0032.xml

Gordon, Andrew. ed. "Consumption, Leisure, and the Middle Class in Transwar Japan." *Social Science Japan Journal* 10, no. 1 (2007): 1–21.

Gould, Carol G. *The Remarkable Life of William Beebe: Explorer and Naturalist*. Washington, DC: Island Press, 2004.

Grad, Andrew J. "Land Reform in Japan." *Pacific Affairs* 21, no. 2 (June 1948): 115–35.

Gup, Ted. "The Smithsonian's Secret Contract: The Link between Birds and Biological Warfare." *The Washington Post Magazine*, May 12, 1985.

Hachisuka [Masauji], Marquess. *The Birds of the Philippine Islands: With Notes on the Mammal Fauna, Volume 1 [Parts I & II, Galliformes to Pelecaniformes]*. London: H. F. & G. Witherby, 1931.

Hachisuka [Masauji], Marquess. *A Comparative Hand List of the Birds of Japan and the British Isles*. Cambridge, UK: Cambridge University Press, 1925.

Hachisuka [Masauji], Marquess. "Contributions on the Birds of Hainan." *The Ornithological Society of Japan*: Supplementary Publication no. XV, October 1939.

Hachisuka [Masauji], Marquess. *The Dodo and Kindred Birds: The Extinct Birds of the Mascarene Islands*. London: H. F. & G. Witherby, 1953.

Hachisuka, Toshiko. *Daimyô kazoku*. Tokyo: Mikasa shobô, 1957.

Hama-rikyu Gardens. *Special Place of Scenic Beauty and Special Historic Site*. Tokyo: Pamphlet, Hamarikyu Garden Office, 2015.

Hanami, Kaoru. *Tennô no takajô*. Tokyo: Sôshi-sha, 2002.
Harris, Sheldon H. *Factories of Death: Japanese Biological Warfare, 1932–1945 and the American Cover-Up*. London: Routledge, 2002.
Harris, Sheldon H. "Japanese Biomedical Experimentation during the World War II Era." In *Military Medical Ethics Volume 2*, edited by Thomas E. Beam and Linette R. Sparacino, 463–506. Washington, DC: Borden Institute at Walter Reed Medical Center, 2003.
Hartert, Ernst, F. C. R. Ticehurst, and H. F. Witherby. *A Hand-List of British Birds: With an Account of the Distribution of Each Species in the British Isles and Abroad*. London: Witherby & Co, 1912.
Hellman, Geoffrey T. "Delacour Reobserved." *The New Yorker*, December 17, 1960 "Talk of the Town" section. https://www.newyorker.com/magazine/1960/12/17/delacour-reobserved.
Hellman, Geoffrey T. "Profiles: Curator Getting Around." *The New Yorker*, August 26, 1950. https://www.newyorker.com/magazine/1950/08/26/curator-getting-around.
Holthuis, L. B., and Tsune Sakai. *Ph. F. von Siebold and Fauna Japonica: A History of Early Japanese Zoology*. Tokyo: Academic Press of Japan, 1970.
Horiuchi, Sanmi. *Nippon chôrui shuryo-hô: shashin kiroku*. Tokyo: Sansho-dô, 1939.
Horiuchi, Sanmi. *Nippon chôrui shuryo-hô: shashin kiroku*. Tokyo: Sansho-dô, 1942.
Horiuchi, Sanmi. *Nippon dentô shuryô-hô: shashin kiroku*. Tokyo: Shuppan kagaku kenkyû-jo, 1984.
Howard, L. O. "Biographical Memoir of Edward Sylvester Morse, 1838–1925." National Academy of the Sciences of the United States of America Biographical Memoirs, Vol. XVII, First Memoir (1935), presented to the Academy at the Annual Meeting, 1–29.
Howell, David L. "The Social Life of Firearms in Tokugawa Japan." *Japanese Studies* 29, no. 1 (2009): 65–80.
Iida, Tetsuji. *Chôgaku no hyakunen–chô ni sudamasereta hitobito*. Tokyo: Heibonsha, 2012.
Iijima, Isao., and Sasaki Chûjirô. *Okadaira Shell Mound at Hitachi, Being an Appendix to Memoir, Vol. 1, Part 1 of the Science Department*. Tokyo: University of Tokio [sic], 1882.
Ikuta, Shigeru. *Shuryôkai no kashitsu bôshi to ryôjû sôsahô teigi ichimei shuryô jôshiki sôjû dokuhon*. Tokyo: Ikuta Shigeru shôten, 1939.
Isono, Naohide. "Contributions of Edward S. Morse to Developing Japan." In *Foreign Employees in Nineteenth Century Japan*, edited by Edward R. Beauchamp and Akira Iriye, 193–212. New York: Routledge, 1990.
Jacobsen, Annie. *Operation Paperclip: The Secret Intelligence Program That Brought Nazi Scientists to America*. New York: Little, Brown, and Company, 2014.
"A Japanese Scientific Expedition to Jehol in 1933." "*The Monthly Record*" of *The Geographical Journal* 86, no. 1 (July 1935): 82–5.
Johnston, Bruce F., Mosaburo Hosoda, and Yoshio Kusumi. *Japanese Food Management in World War II*. Palo Alto: Stanford University Press, 1953.
Johnston, Eric. "Hunting in Japan: Gun Control Is the Norm and Discipline Is Rigid." *The Japan Times*, January 29, 2013. https://www.japantimes.co.jp/news/2013/01/29/reference/in-japan-gun-control-is-the-norm-and-discipline-is-rigid/.
King, Warren. "International Council for Bird Preservation: Historical Aspects and Current Projects of the ICBP." *American Birds* 40, no. 1 (Spring 1986): 49–50.
Kingsberg Kadia, Miriam. *Into the Field: Human Scientists of Transwar Japan*. Palo Alto, CA: Stanford University Press, 2020.
Kiyosu, Yukiyasu. *Shokuryô shigen to naru chôrui ni tsuite*. *Shigen kagaku kenkyûjo ihô*, Issue 4. Tokyo: Kasumigaseki shobô, November 1943.

Kopel, David B. "Japanese Gun Control." *Asia Pacific Law Review* 2, no. 2 (1993): 26–52.
Koyama, Noboru. *Japanese Students at Cambridge University in the Meiji Era, 1868–1912: Pioneers for the Modernization of Japan*. Translated by Ian Ruxton. Morrisville, NC: Lulu Press, 2004.
Koyama, Sachiko. "History of Bird-Keeping and the Teaching of Tricks Using *Cyanistes Varius* (Varied Tit) in Japan." *Archives of Natural History* 42, no. 2 (Edinburgh University Press, 2015): 211–25.
Koyama, Sachiko. *Yamagara no gei: bunkashi to kôdôgaku no shiten kara*. Tokyo: Hôsei Daigaku shuppankyoku, 1999.
Krämer, Hans Martin. "'Not Befitting Our Divine Country': Eating Meat in Japanese Discourses of Self and Other from the Seventeenth Century to the Present." *Food and Foodways* 16, no. 1 (January 2008): 33–62.
Kuroda, Nagahisa. *Chôrui no kenkyû: Seitai*. Tokyo: Shin-shicho-sha, 1967.
Kuroda, Nagahisa. *The Ornithologists and Avifauna in [sic] Early History of Japanese Field Ornithology: Chiefly Until 1950—the Draft Prepared for "The Birds of Japan: Their Status and Distribution" by Oliver L. Austin and Kuroda Nagahisa, 1953*. Tokyo: Self-Published Memorial Publication of 88 Copies, 2004.
Kuroda, Nagamichi. *A Bibliography of the Duck Tribe, Anatidae, mostly from 1926 to 1940, exclusive of that of Dr. Phillips's Work*. Tokyo: Herald Press, 1942.
Kuroda, Nagamichi. *Birds of the Island of Java, Volume One: Passeres*. Tokyo: Self-Published, 1933.
Kuroda, Nagamichi. *Birds of the Island of Java, Volume One: Non-Passeres*. Tokyo: Self-Published, 1936.
Kuroda, Nagamichi. *The Pheasants of Japan Including Korea and Formosa*. Tokyo: Self-Published, 1926.
Lebra, Takie Sugiyama. *Above the Clouds: The Status Culture of the Modern Japanese Nobility*. Berkeley: University of California Press, 1995.
Leheny, David. "'By Other Means': Tourism and Leisure as Politics in Pre-war Japan." *Social Science Japan Journal* 3, no. 2 (October 2000): 171–86.
Lewis, Michael. "Scientists or Spies? Ecology in a Climate of Cold War Suspicion." *Economic and Political Weekly* 37, no. 24 (June 15–21, 2002): 2323–32.
Library of Congress. Sayuri Umeda, Foreign Law Specialist, "Firearms-Control Legislation and Policy: Japan." February 2013. https://www.loc.gov/law//help/firearms-control/japan.php#t13.
Lovett, Irby and John W. Fitzpatrick. *Cornell Laboratory of Ornithology Handbook of Bird Biology: Third Edition*. Hoboken, NJ: Wiley Blackwell, 2016.
Low, Morris, ed. *Building a Modern Japan: Science, Technology, and Medicine in the Meiji Era and Beyond*. New York: Palgrave Macmillan, 2005.
Low, Morris. "The Japanese Colonial Eye: Science, Exploration, and Empire." In *Photography's Other Histories*, edited by Christopher Pinney et al., 105–15. Durham, NC: Duke University Press, 2003.
Low, Morris. *Science and the Building of a New Japan*. London: Palgrave Macmillan, 2005.
Lubar, Steven. *Inside the Lost Museum: Curating, Past and Present*. Cambridge, MA: Harvard University Press, 2017.
MacGillivray, William. *Manual of British Ornithology: Being a Short Description of the Birds of Great Britain and Ireland, Including the Essential Characters of the Species, Genera, Families, and Orders: Part I: The Land Birds*. London: Scott, Webster, and Geary, 1840.
MacKenzie, John M. *The Empire of Nature: Hunting, Conservation, and British Imperialism*. Manchester: Manchester University Press, 1997.

MacKenzie, John M. *Museums and Empire: Natural History, Human Cultures, and Colonial Identities*. Manchester: Manchester University Press, 2009.

MacLeod, Roy. "'Strictly for the Birds': Science, the Military and the Smithsonian's Pacific Ocean Biological Survey Program, 1963–1970." *Journal of the History of Biology* 34 (2001): 315–52.

Mangan, J. A. *Manufactured' Masculinity: Making Imperial Manliness, Morality and Militarism*. London: Routledge, 2012.

Marcon, Frederico. "Honzôgaku after Seibutsugaku: Traditional Pharmacology as Antiquarianism After the Institutionalization of Modern Biology in Early Meiji Japan." In *Antiquarianism, Language, and Medical Philology: From Early Modern to Modern Sino-Japanese Medical Discourses*, edited by Benjamin Elman, 148–62. Leiden: Brill, 2015.

Marcon, Frederico. *The Knowledge of Nature and the Nature of Knowledge in Early Modern Japan*. Chicago: University of Chicago Press, 2015.

Martin, Charles E. *Report of the United States Cultural Science Mission to Japan. to Supreme Commander for the Allied Powers: Civil Information and Education Section*. Seattle: University of Washington, Institute of International Affairs, January 1949.

Massey, Doreen. "Power-geometry and a Progressive Sense of Place." In *Mapping the Futures: Local Cultures, Global Change*, edited by J. Bird, B. Curtis, T. Putnam, G. Robertson and L. Tickner, 59–69. London: Routledge, 1993.

Massey, Doreen. *Space, Place, and Gender*. Minneapolis: University of Minnesota Press, 1994.

Matsuda, Takeshi. *Soft Power and Its Perils: US Cultural Policy in Early Postwar Japan and Permanent Dependency*. Washington, DC: Woodrow Wilson Center Press, 2007.

Mauss, Marcel. *The Gift: The Form and Reason for Exchange in Archaic Societies*. London: Routledge, 1990.

McClure, Elliot. *Stories I Like to Tell: An Autobiography*. Camarillo, CA: Self-Published, 1995.

McGregor, Richard Crittenden, and Dean Conant Worcester. *A Hand-list of the Birds of the Philippine Islands*. Manila: Bureau of Printing, 1906.

Merrill, Lyn L. *The Romance of Victorian Natural History*. New York: Oxford University Press, 1989.

Miller, Ian Jared. *The Nature of the Beasts: Empire and Exhibition at the Tokyo Imperial Zoo*. Berkeley: University of California Press, 2013.

Mizuno, Hiromi. *Science for the Empire: Scientific Nationalism in Modern Japan*. Stanford: Stanford University Press, 2008.

Mizuno, Kaoru. *Manshû chôrui bunpu mokuroku*. Tokyo: privately published by Kuroda Nagamichi, October 1934.

Mizuno, Kaoru. *Manshû chôrui genshoku dai-zukan*. Tokyo: Purosesu-sha, 1940.

Mohri, Hideo. *Imperial Biologists: The Imperial Family of Japan and Their Contributions to Biological Research*. Singapore: Springer Nature, 2019.

Morse, Edward S. *Japan Day by Day, 1877, 1878–1879, 1882–1883*. Boston: Houghton Mifflin, 1917.

Morse, Edward S, Iijima Isao, and Sasaki Chûjirô. *Shell Mounds of Omori: Volume 1, Part 1 of Memoirs of the Science Department*. Tokyo: Tôkyô Daigaku, 1879.

Murakami, Kimio. *Zetsumetsu dori dôdô wo oimotometa otoko: soratobu kôshaku Hachisuka 1902–1953*. Tokyo: Fujiwara shoten, 2016.

Nakamura, Masanori. *Sengoshi*. Tokyo: Iwanami shoten, 2005.

Nakamura, Tsukasa. *Kenkyû. Kyôiku. Shizen hôgo. Shakai katsudô: 50-nen no ayumi (gaiyo)*. Kôfu, Japan: Yamanashi University, 2005.

Nakayama, Thomas K., and Judith N. Martin, ed. *Whiteness: The Communication of Social Identity*. Thousand Oaks, CA: Sage Publications, 1998.

National Diet Library online archives, Nihon hôrei sakuin, Jûhô torishimari kisoku: Daijôkan fukoku dai 28 gô (January 29, 1872), https://hourei.ndl.go.jp/simple/detail?laWId=0000000013&current=-1, accessed October 28, 2021.

Nester, William R. *Power across the Pacific: A Diplomatic History of American Relations with Japan*. London: Palgrave Macmillan, 1996.

Niibe, Tominosuke. *Nô no tori no seitai*. Tokyo: Kôbunsha, 1951.

Nye, Joseph. *Soft Power: The Means to Success in World Politics*. New York: Public Affairs, 2004.

Ôchi, Tsutomu. "The Japanese Land Reform: Its Efficacy and Limitations." *The Developing Economies* 4, no. 2 (1966): 129–50.

Organ, J. F., V., Geist, S. P. Mahoney et al. "The North American Model of Conservation." *The Wildlife Society Technical Review* 12, no. 4 (2012).

Otsuka, Noriko. "Falconry: Tradition and Acculturation," *International Journal of Sport and Health Science* 4 (2006): 198–207.

Ôtsuka, Takeyuki. *Chôjû gyôsei no ayumi*. Tokyo: Forestry Agency, 1969.

Pandey, Rajyashree. "Gender in Pre-Modern Japan." In *The Routledge Companion to Gender and Japanese Culture*, edited by Jennifer Coates, Lucy Fraser, and Mark Pendleton New York: Routledge, 2019.

Parley, Norman to Herbert G. Giles. "*Record of Strange Nations*" (letter). On behalf of Percy Lund Humphries & Co Ltd. London, January 14, 1929.

Perraton, Hilary. *A History of Foreign Students in Britain*. New York: Springer, 2014.

Pitelka, Morgan. *Spectacular Accumulation: Material Culture, Tokugawa Ieyasu, and Samurai Sociability*. Honolulu: University of Hawaii Press, 2015.

Report on the Commission on Wartime Relocation and Internment of Civilians. *Personal Justice Denied*. Washington, DC: US Government Printing Office, 1982, 1983.

Rice, Mark. *Dean Worcester's Fantasy Islands: Photography, Film, and the Colonial Philippines*. Ann Arbor: University of Michigan Press, 2014.

Ritvo, Harriet. *The Animal Estate: The English and Other Creatures in the Victorian Age*. Cambridge, MA: Harvard University Press, 1987.

Roden, Donald. *Schooldays in Imperial Japan: A Study in the Culture of a Student Elite*. Berkeley: University of California Press, 1980.

Roden, Donald. "Thoughts on the Early Meiji Gentleman." In *Gendering Modern Japanese History*, edited by Barbara Moloney, 493–519. Cambridge, MA: Harvard University Asia Center, Harvard University Press, 2005.

Rothschild, Miriam. *Walter Rothschild: The Man, the Museum, and the Menagerie*. London: Natural History Museum, 2008.

Rudlin, Pernille. *The History of Mitsubishi Corporation in London: 1915 to Present Day*. London: Routledge, 2014.

Ruoff, Kenneth J. *The People's Emperor: Democracy and the Japanese Monarchy, 1945–1995*. Cambridge, MA: Harvard University Asia Center, 2001.

Sand, Jordan. *House and Home in Modern Japan: Architecture, Domestic Space, and Bourgeois Culture*. Cambridge, MA: Harvard University Press, 2005.

Schaffer, Simon et al., ed. *The Brokered World: Go-Betweens and Global Intelligence, 1770–1820*. New York: Science History Publications, 2009.

Schenk, Colonel Hubert G. "Natural Resources Problems in Japan." *Science* 108, no. 2806 (1948): 367–72.

Scott, Joan Wallach. "Gender: A Useful Category of Historical Analysis." *The American Historical Review* 91, no. 5 (December 1986): 1053–75.

Setoguchi, Akihisa. *Shuryô to dôbutsugaku no kindai–Tennôsei to "shizen" no poriteikkusu*. *Seibutsugaku-shi kenkyû*, no. 84 (2010): 73–83.

Setoguchi, Akihisa. Shuryô to kôzoku–zasshi "Ryôyû" ni miru dôbutsu wo meguru seiji. kagaku. gendaa. *Dôbutsu-kan kenkyû*. Vol. 13 (December 2009), 39.

Shibusawa, Naoko. *America's Geisha Ally: Reimagining the Japanese Enemy*. Cambridge, MA: Harvard University Press, 2010.

Shimomura, Kenji, and Kiyosu Yukiyasu. *Yachô seitai shashinshû*. Tokyo: Unsôdô, 1940.

Skabelund, Aaron. *Empire of Dogs: Canines, Japan, and the Making of the Modern Imperial World*. Ithaca, NY: Cornell University Press, 2011.

Smith, Neil R. "Infectious Disease Research Laboratory." *Report of Investigation Division, Legal Section, GHQ, SCAP, JWC 258/18*. December 6, 1946. https://www.archives.gov/files/iwg/japanese-war-crimes/select-documents.pdf.

Stoler, Ann Laura. *Along the Archival Grain: Epistemic Anxieties and Colonial Common Sense*. Princeton: Princeton University Press, 2010.

Stone, Roger. *The Lives of Dillon Ripley: Natural Scientist, Wartime Spy, and Pioneering Leader of the Smithsonian Institution*. Lebanon, NH: ForEdge, 2017.

Suga, Yutaka. *Yô no Shôgun to tsuru no miso shiru—Edo no chô no bishokugaku*. Tokyo: Kôdansha, 2021.

Suginami Ward Regional Museum staff. *Nakanishi Gotô seitan 120-nen: Yachô no chichi, Nakanishi Gotô wo meguru hito bito–tenji zuryoku*. Tokyo: Suginami Ward Regional Museum, 2015.

Taka-tsukasa, N., M. Hachisuka, and N. Kuroda et al. *Birds of Jehol: Report of the first scientific expedition to Manchoukuo under the leadership of Shigeyasu Tokunaga. Section V, Zoology. Division 2, Vertebrata*. Tokyo: Office of the scientific expedition to Manchoukuo, Faculty of science and engineering, Waseda University, 1935.

Takanami, Machiko. "Kôun no kenchiku. Asaka gutei–kenbutsu no chikara." *Kagu dogu shitsunai-shi gakkai, tokushû Asaka gutei to Aaru.Deko. Kagu dôgu shitsunai-shi* 6 (May 2014): 4–38.

Takatsukasa, Nobusuke. *Japanese Birds*. Tokyo: Board of Tourist Industry Japanese Government Railways, 1941.

Takatsukasa, Nobusuke. *Chakushoku zuhen kaidori shusei*. Tokyo: Yokendô, 1930.

Takatsukasa, Nobusuke, Marquis M. Hachisuka, N. Kuroda, Marquis Y. Yamashina, and S. Uchida. "*Birds of Jehol*, Section V, Division II, Part III (April 1935)." *Report of the First Scientific Expedition to Manchoukuo under the Leadership of Shigeyasu Tokunaga, June–October 1933*. Tokyo: Office of the Scientific Expedition to Manchoukuo, Faculty of Science and Engineering, Waseda University, 1935.

Tama, Shinnosuke. "Manshû ringyô imin to eirin jitsumu jisshû-sei seido." In *Sôryokusen taisei-ka no Manshû nôgyô imin*, edited by Tama Shinnosuke, 1–20. Tokyo: Yoshikawa Kôbunkan, 2016.

Tenison, W.P.C., Lt.-Col, D.S.O., ed. *Bulletin of the British Ornithologists' Club* Vol. XLVII, Session 1946-7. London: H. F. & G. Witherby, 1947.

Tiu, Macario C. *Davao: Reconstructing History from Text and Memory*. Davao, Philippines: Ateneo de Davao University Research and Publication Office for the Mindanao Coalition of Development NGOs, 2005.

Tokunaga, Shigemoto. "Tokunaga Shigeyasu shoden." *Chigaku zasshi* 94, no. 3 (1985): 54–6. https://www.jstage.jst.go.jp/article/jgeography1889/94/3/94_3_194/_pdf.

Totman, Conrad. *Japan's Imperial Forest Goryôrin, 1889–1946*. Folkestone, UK: Global Oriental, 2007.

Tôzawa, Kôichi. *Rekidai kaichô no kotoba*, in *Nihon Chôgakkai—100 nen no rekishi*, Nihon Chôgakkai 100 shûnen kinen tokubetsu gô, Nihon Chôgakkai-shi 61 (November 2012): 8–9.

Uchida, Seinosuke. "Bird-Banding in Japan." Tokyo: Department of Animal Husbandry, Ministry of Agriculture and Forestry, 1928.

Uchida, Seinosuke, and Shimomura Kenji. *Chôrui seitai shashinshû*. Tokyo: Sanseidô, 1930.

Uchiyama, Benjamin. *Japan's Carnival War: Mass Culture on the Home Front*. Cambridge, UK: Cambridge University Press, 2019.

US Army Academy of Health Sciences, Stimson Library Digital Collections, Medical Bulletin of the US Army Far East, "Functional Chart: 406th Medical General Laboratory," https://stimson.contentdm.oclc.org/digital/collection/p15290coll1/id/68.

Valley, David. *Gaijin Shogun: Gen. Douglas MacArthur Stepfather of Postwar Japan*. Sektor: San Diego, 2000.

Vandello, Joseph A. Jennifer K. Bosson, Dov Cohen, Rochelle M. Burnaford and Jonathan R. Weaver, "Precarious Manhood." *Journal of Personality and Social Psychology*, American Psychological Association 95, no. 6 (2008): 1325–39.

Venn, J. A. *Alumni Cantabrigienses: A Biographical List of All Known Students, Graduates, and Holders of Office at the University of Cambridge, from the Earliest Times to 1900*. Cambridge: Cambridge University Press, 1951.

Vining, Elizabeth. *Windows for the Crown Prince*. Philadelphia: J. B. Lippincott and Company, 1952.

Wain, Louis. "The Hon. Walter Rothschild's Pets: A Visit to Tring Museum." *The Windsor Magazine: An Illustrated Monthly for Men and Women*, Vol. II, July to December (December 1895): 661–70.

Wallace D., and Williams P. *Unit 731: Japan's Secret Biological Warfare in World War II*. New York: Free Press, 1989.

Whipple, Sydney B. "Let's Protect Japan's Birds: Serious Economic Loss to a Nation Foreseen Unless Destructive Custom of Netting Wild Fowl Abandoned." *Nippon Times Magazine*, February 27, 1947.

Wollaston, A. F. R. *The Life of Alfred Newton: Late Professor of Comparative Anatomy, 1866-1907, with a Preface by Sir Archibald Geikie*. New York: Dutton, 1921.

Worcester, Dean C. *The Philippine Islands and Their People: A Record of Personal Observation and Experience, with a Short Summary of the More Important Facts in the History of the Archipelago*. New York: The Macmillan Company, 1898.

Yamashina, Yoshimaro. *Birds in Japan: A Field Guide*. Tokyo: Tokyo News Service, 1961.

Yamashina, Yoshimaro. *Manshû-san chôrui no shokusei*. Shinkyô [Changchun]: Manshûkoku Rinnôkyoku, January 1939.

Yamashina, Yoshimaro. *Nihon no chôrui no seitai to hogo*. Tokyo: Yamashina Ornithological Institute, 1951.

Yamashina, Yoshimaro. *Yamashina Yoshimaro hakase no ayumi*. Tokyo: Sôbunshinsha, 1984.

Yamashina, Yoshimaro, Inukai Tetsuo, and Natori Bukô. *A List of Bird Skins Presented by Captain Blakiston in the University Museum of Natural History of Sapporo with a Brief Account of His Life in Hokkaidô*. Sapporo: Sapporo University Museum, December 1932.

Yoshida, Sadao. "Isao Ijima: The Father of Parasitology in Japan (With Potrait [sic] Plate)." *The Journal of Parasitology* 10, no. 3 (1924): 165–7.

Yoshii, Masashi, Fumio Sato, Kiyoaki Ozaki, Yoshimitsu Shigeta, Shigemoto Komeda, Keiko Yoshiyasu, and Amane Mitamura. "Japanese Bird Banding, Now and Past." *Journal of the Yamashina Institute of Ornithology* 21, no. 1 (1989): 309–25.

Young, Louise. *Japan's Total Empire: Manchuria and the Culture of Wartime Imperialism.* Berkeley: University of California Press, 1999.

Yu-Jose, Lydia N. *Japan Views the Philippines, 1900–1944.* Davao, Philippines: Ateneo de Manila University Press, 1999.

Yu-Jose, Lydia N., and Patricia Irene Dacudao. "Visible Japanese and Invisible Filipino: Narratives of the Development of Davao, 1900s to 1930s." *Philippine Studies: Historical and Ethnographic Viewpoints* 63, no. 1 (2015): 101–29.

Index

Abel, Emily K. 90
Abiko City Bird Museum 211
Adams, Claude M. (1907–74) 178
aesthetics/aestheticism 28, 55
Africa/Africans 9, 48, 56, 65, 82–3, 96, 246 n.16
　North Africa 39, 54, 70–1, 78, 92
Agassiz, Louis (1807–1873) 29
Aiwa Ong, *Neoliberalism as Exception: Mutations in Citizenship and Sovereignty* 226 n.56
Akishino Fumihito 218–19
Alberti, Samuel J. M. M. 50–1
Aldous, Christopher 187
Aldrich, Robert 91–2
Ali, Salim (1896–1987) 207, 209, 211
Allen, Joel Asaph (1838–1921) 36
　Christmas Bird Count 37
Allied Occupation 5, 10, 13, 17, 46, 52, 96, 136, 163, 168–9, 172, 178, 192, 195, 205, 209, 212, 266 n.2
American Indians 96
American Museum of Natural History (AMNH) 29, 36–7, 71, 74, 139, 157, 181, 208
American Ornithological Society (AOS) 32, 36–7, 41, 52, 54, 86–7, 114, 232 n.141
American Ornithological Union (AOU). *See* American Ornithological Society (AOS)
American Wigeon (*Mareca americana*) 57
Anglo-American world 2, 5–10, 12–17, 21–5, 32, 38–41, 43, 45, 51, 53, 59–60, 64–5, 71, 79, 81, 84, 86–8, 90, 93–4, 96, 99, 109, 113, 116–17, 119–20, 129, 135, 140, 142–3, 149, 158, 161, 168, 170, 181, 189, 192, 195, 205, 207, 209, 211, 213, 221–2
　imperial masculinity 87
　scientific study of ornithology in 32–7
　tourism 143–7

Anglo-European 14, 97, 141
Anglo-Japanese Alliance 38, 79, 84
Anglo-Saxon 5, 37, 66, 79
animals 27–8, 35, 50, 76, 83, 128–9, 132, 140, 153–6, 168, 218, 227 n.13, 249 n.90. *See also* plants
Anpô Treaty 205, 213
Anson, George (1697–1762) 246 n.15
anthropology/anthropologists 2, 6, 13, 15, 76, 104, 121, 162, 226 n.54, 226 n.56
anti-authoritarian 29, 61
aristocracy/aristocrats 4–6, 8, 10–11, 17, 20, 30, 35, 39, 44–5, 51–2, 55, 59, 65–8, 72–3, 78, 83, 89–93, 133, 137, 159, 161–2, 179, 185, 198, 206–7, 217
　deposed Japanese 168–71
　as imperial agent/s 37–40
　Kyoto 38–9, 62
　requisitioned house (late 1949) 6
　status culture of 6, 44
Ariyoshi Chûichi (1873–1947) 46
Ariyoshi Yoshiya (1901–84) 167
Asahi gurafu (Morning News Graph) 20
Asahi Shimbun newspaper 53, 59, 61, 73, 111–12, 117–19, 121, 127, 149, 151–2, 252 n.2
　League of Academic Associations for the Study of Natural Resources 150
Asano Nagatake (1895–1969) 62–3
Asano Sugako (1905–66) 62–3, 142
Asia/Asian 13, 21, 39, 54–5, 57, 73, 78, 83, 105, 192, 206–7, 222
　East 39, 45, 66, 78, 84, 116, 153, 158, 174, 179, 193, 200, 214
　Northeast 39, 126, 176
　and race 224 n.20
　Southeast 13, 56, 96–7, 99, 139, 149, 158, 197, 209, 218
Asian Exclusion Act (Immigration Act of 1924) 79

Asia-Pacific War (1937–45) 11, 40, 136–44, 166, 175
Aso, Noriko 152
Astley, Hubert Delaval (1860–1925) 91
Audubon, John James (1785–1851) 93
　The Birds of America 33
　Ornithological Biographies 33
Augé, Marc 2, 62
The Auk journal 15, 36, 54, 57, 71, 75, 115, 127, 157–8, 178, 181, 202, 215, 259 n.159
Austin, Oliver L., Jr. (1903–1988) 3–4, 6–7, 14–15, 19–20, 23, 46, 49, 136–7, 141, 143, 148–9, 153, 165–7, 170–7, 179–82, 184–5, 193, 195–6, 198, 208–11, 215–16, 223 n.13, 224 n.23, 225 n.30
　The Birds of Japan 7, 161
　Birds of the World 219
　Canon 35 mm Camera "Made in Occupied Japan" used by (1946) 3
　contributions in wildlife and conservation 187–92
　and "Jimmy" Peters 197
　"Mist Netting for Birds in Japan" 148, 182
　The Quarterly Review of Biology 170
　SCAP report 152, 161, 166–7, 179, 181–3, 187, 193, 198, 271 n.90
　"Waterfowl of Japan" 182
　and Yamashina 143
Austin, Oliver L., Sr. (1871–1957) 174
Austin, Tony 3
Australia 35, 47, 179, 198
　Australian Embassy 46, 159
　Australian Energy Market Commission 269 n.48
authentic/authenticity 9, 71, 133
authoritarian/authoritarianism 136, 187
avian imperialism 17, 21, 32–7, 39, 110–17, 133, 169, 221
　and Asia-Pacific War (1937–45) 136–43
avian species 18, 39, 49, 167, 221
avian surveillance 126–9, 133
aviaries 43, 51, 53–4, 57–8, 60, 63, 68, 71, 157, 160, 207
The Avicultural Magazine 54, 59
aviculture/aviculturists 54, 56, 59, 137–8

avifauna 39, 43, 50, 53, 65, 73, 82, 84–5, 93, 109–10, 113, 115, 125, 131, 140
Axis Alliance 146

bābā (wealth) 268 n.38
Backhouse, Edmund (1873–1944) 76
banbutsu (myriad things) 28
Barlow Sanitarium, LA 90
Barnes, C. M. 209
Barrow, Mark 37
bats 153, 174, 186
Beebe, William (1877–1962) 93
　A Monograph of the Pheasants 57–8
Belote, Theodore T. 51
Benjamin, Walter, *The Arcades Project* 49
Bennett, John W. (1915–2005) 172–3
Benson, Andrew (1917–2015) 150
Beveridge, Albert Jeremiah (1862–1927) 95
Binjiang Province 137
biogeography 92
Biological Survey Unit 36
biology/biologist 27, 40, 82, 150, 186, 193, 207, 216. *See also* zoology
　biological warfare (research) 11, 137, 199, 202–3, 278 n.64
　biological weapons 7, 136–8, 154, 202, 204, 209–11
　marine 29–31, 65, 136, 219, 240 n.1
Birckhead, Hugh (1913–1943) 139
bird-banding/bird-banders 15, 19, 192, 207, 221
　and conservation efforts for peaceful postwar nation 214–16
　and migration projects 209–11
"Bird Day," Japan's official 166, 186, 190
bird-harvesting 20, 149, 183, 185, 198
birding 1–2, 95, 107, 219–20, 272 n.108
bird-keeping 17–18, 41, 54–9
Bird Life International (BLI) organization 12, 53, 216, 219
Bird Migration Research Center 209
birds. *See also specific birds*
　caged/caging 56, 60, 62
　Cold War cooperation for 205–9
　fortune telling by 19
　Javanese 115–16
　killing methods of 182–3, 189
　meat 19–20, 148

migration (migratory birds) 12, 19, 21, 113, 207, 211, 213–14, 217, 270 n.68
  native 56
  observing 17, 62
  predatory 50, 52
  seed-eating 55, 58
  specimens (see specimen(s))
  study of (see ornithology)
  training for popular entertainment 18–21
  wild birds as food resources (1937–45) 147–9
  winter travels in Asia 54
Birds journal 17, 32, 41, 57, 59, 114–15, 131, 137, 167, 179, 186
bird skinning techniques 7, 175
birdsong 18–19, 32. *See also* songbirds
bird warden 272 n.108
bird-watching/ bird-watchers 11, 27, 32, 37, 44, 113, 142, 144, 219–22
  and homosociality 60–3
Black Birders Week 1
Black Grouse/Ussurian Black Cock (*Lyrurus tetrix*) 113
Black Lives Matter (BLM) 1
Blakiston, Thomas Wright (1832–1891) *Hand List* 39
  survey of Hokkaidô's birds 40
blue-and-white flycatcher (*Cyanoptila cyanomelana*) 18
Bolshevik Revolution (1918–21) 79
Bombay Natural History Society (BNHS) project 209–10
Bosson, Jennifer K. 8
Böttcher, Claudius 15
Bourns, Frank Swift (1866–1936) 85
  Distribution List of Philippine birds 86
Brazil 14, 82
breeding 15, 54, 56, 71, 135, 141, 144, 216
Brewster, William 36
Brisé, Bernard 268 n.44
British Empire 8, 14, 179
British Museum 34–5, 74, 106, 232 n.130
British Ornithologists' Club (BOC) 36, 70, 86
  members 158
British Ornithologists' Union (BOU) 32, 36, 38, 41, 67, 70, 86, 158

Bronx Zoo 58
Brookline Bird Club (BBC) 182, 272 n.108
Browne, Janet 92
Browning Auto-5 (A-5) shotgun 25, 175, 229 n.62
Browning, John 25
Bulletin of the British Ornithologists' Club journal 36
Bulletin of the Nuttall Ornithological Club journal 36
Bulwer-Lytton, Victor (1876–1947) 87
bunka jûtaku (culture house) 47
bureaucrats 3, 41, 58, 61, 83, 85–6, 94, 97, 155
Burnaford, Rochelle M. 8
The Butcher of Cuba. *See* Weyler, Valeriano
Butler, Judith 8–9
Buturlin, Sergey A. (1872–1938) 208

Canada 71, 73, 269 n.48
canine imperialism 35, 228 n.34
cannibalism 30, 230 n.101
Canon 35 mm Camera (used by Austin, 1946) 3
Carmichael, Leonard (1898–1973) 210
Carpenter, Philip Pearsall (1819–1877) 29
Carter, T. Donald (1893–1972) 139
Cassowaries 35
"The Castle" 34
Central Intelligence Agency (CIA) 128, 197, 210
Chapman, Frank (1864–1945) 37
Charbonneau, Olivier 96–7
China/Chinese 8, 11, 13, 18, 20–1, 27, 40, 50, 56, 76–7, 84, 87, 93–4, 111–12, 119, 124–7, 135, 139, 147, 182, 194, 204, 214
  artistic representation 28
  China Conflict 138, 144
  Classical Chinese 65–6, 78–9
  Communist China 136
  Hainan Island 39, 138–40, 153
  Mainland China 75
  Shanghai 48, 93, 121
  Taiwan 40, 56–7, 75, 81, 83–4, 110
The China Journal 120
China Society of Science and Arts 120
The Chinese Student journal 77

Chiyeko Nagamine (1909–1997) 47, 159–60
Chôgakkai (Ornithological Society) 4
Chôrui hyôshiki chôsa (Investigation of Bird Species Through Banding) 214
Churchill, Winston (1874–1965) 142
class 2–3, 5–6, 10, 14, 43. *See also* gender; race; social status
  bourgeoisie 22–6, 50, 60, 111, 172
  social 12–13, 16, 18, 215
  upper-class 17, 26, 41, 78, 84, 172, 179, 198
  wealthy ruling 23–4, 45–7, 50, 75, 89–90, 110, 163, 171, 181, 189, 222
Cohen, Dov 8
Cold War (1947–91) 10, 13, 128, 192, 194–205, 208, 215, 221
  bioweapons 209–11
  cooperation for birds 205–9
collecting imperialism 5, 81–4, 179, 224 n.24
colonialism 2, 34
colonization 5, 96, 130–1, 134, 152
commodity 28, 43, 49–50, 75, 99, 138
Communism 193–4
Comte, Auguste 27
Condor, Josiah (1852–1920), constructions/designs of 44–5
The Condor journal 37, 85, 249 n.96
Confucian 76
conservation/conservationists 8, 10–12, 20, 36, 53, 60–3, 83, 165–7, 171–3, 176–80, 187–94, 207, 213–16, 219–21
Convention on the Conservation of Migratory Species 216
Cook, Matt, *Queer Domesticities: Homosexuality and Home Life in Twentieth Century London* 63, 240 n.149
Cooper Ornithological Club (COC). *See* Cooper Ornithological Society (COS)
Cooper Ornithological Society (COS) 37, 41
Cornell Ornithology Lab 1, 107
cosmopolitan/cosmopolitanism 6, 10, 12, 14, 25, 28, 38, 44, 59, 65, 67, 89–90, 116, 183, 195, 213
  Cosmopolitan American 84
  internationalist 221

Coues, Elliott Ladd (1842–1899) 36
Cowen, Steven 33
cranes 56, 58, 71, 167, 180
creative non-fiction 2, 162
Crested Shelduck (*Tadorna cristata*) 216
Croutzen, Paul 173
cultural anthropology 15
cultural identity 3
cultural mimesis 12, 43–6, 64–5, 84
cultural science 205
cytology 40, 213, 218

Dacudao, Patricia Irene N. 96, 98
Dai hiroba (Great Plaza) 120
Daijôkan (Central Government Administration) 23
daimyô (feudral lords) 19, 38, 46, 50, 65, 91, 159, 183
Darnton, Robert 155
Darwin, Charles (1809–1882) 217
  theory of evolution 29–30, 36, 40
Datu Apang 103–4
Davao-kuo (Little Japan) 98
decoys 18–20, 148, 182, 214
Delacour, Jean Théodore (1890–1985) 2, 12, 53–4, 58–9, 62, 71, 73, 87–9, 92, 141, 157, 207–9
  "Delacour Reobserved" 208
  and Hachisuka 88, 90
  "Japanese Aviculture" 55
  The Living Air 54
  Notes on the Taxonomy of the Birds of the Philippines 158
  The Pheasants of the World 58
Dementiev, Georgy (1898–1969) 208
  Complete Key to the Birds of the USSR 209
democracy 13, 105, 168, 170, 187, 192, 194
  democratization 5, 11, 24, 44, 167
  SCAP 168–71
  imperial 136
DiMoia, John 13
dinosaur-like birds. *See* Cassowaries
diversity 33–4, 81, 88, 93, 104, 123, 139, 175, 215–16, 259 n.161
doves 15, 127, 141, 188
Dower, John 169
  Embracing Defeat 13, 195
Duara, Prasenjit 110, 252 n.1

Index

duck 57, 141, 182
   duck-hunt 20, 57, 148, 180, 184
   duck-liming 179, 183, 185, 188–9, 268 n.21
   duck production (meat and feathers) 138
   killing method of 183
Duke Takatsukasa Nobusuke (1890–1959) 7, 10–11, 27, 29, 32, 35, 37–8, 52–61, 68, 73, 79, 83–4, 90, 110, 115, 119, 130, 135, 139–40, 148–9, 162, 173, 178–9, 182, 201–2, 207, 218
   Birds of Japan project 59
   Birds of Jehol 39, 109, 114, 129–31
   The Birds of Nippon 115
   Collected Works on Bird-Keeping 59
   and Great Zoo Massacre 154–6
   Illustrated, Colored, Bird Collection, with Care and Feeding 59
   Japanese Birds booklet 59, 143–7
   A List of the Birds of Micronesia under Japanese Mandatory Rule 39, 140
   Monograph of the Birds of Greater East Asia 140
   Studies on the Galli of Nippon 141
Dutch 4, 12, 20, 27–8, 33

ecology 151, 191, 205, 211, 221, 231 n.112
Economic Miracle (1955–91) 205
Edo era (1603–1868) 18, 145, 148
Edwards, Scott V. 1
Edwards's Pheasant (*Lophura edwardsi*) 56
elite(s)
   American military 11
   business 44
   Chinese 82
   cosmopolitan 6, 44, 47, 221
   daimyô (feudral lords) 19, 38, 46, 50, 65, 91, 159, 183
   European 89
   French 55–6, 216
   hunting 83
   imperial 27, 30, 49, 65, 73, 83, 112, 150, 178
   Japanese 3, 5, 18, 21–4, 47, 49, 81, 94, 113, 167, 187, 191, 196, 213
   masculine 50
   political 9, 14, 111
   ruling 11, 20, 64, 116
   samurai 18, 30, 79, 91, 105, 110
   schoolboys 91
   scientific 10, 30, 191, 213
   Westernized 6, 17
   Yankee 37, 68, 165, 171, 179, 198
Emerson, Samuel D. I. 22–3
Emperor Akihito (b. 1933) 219
Emperor Hirohito (1900–89) 20, 28–9, 39, 45, 65–6, 68–9, 89, 91, 136, 150, 201, 208, 219, 277 n.40
Emperor Jimmu 103, 116
Emperor Monmu (r. 697–707) 19
Emperor Mutsuhito (1852–1912) 22, 29, 44, 57
Emperor Naruhito (b. 1960) 185
   A Study of Navigation and Traffic on the Upper Thames in the 18th Century 219
Emperor Nintoku (r. 319–399) 19
Emperor Yoshihito (1879–1926) 68, 208
Empress Shôken (1849–1914) 44
enclothed cognition 5, 8–9, 66, 224 n.22
England 11, 13, 15, 33, 36, 38–9, 53, 66, 68, 70, 72, 89, 93
Enlightenment era 14, 32
Ethiopia 86
ethnicity 21, 56, 105
Euro-American 13, 44
Europe/European 2, 12, 21, 25, 28, 31, 39, 41, 53–7, 68, 78, 88, 97, 113, 206–7
   Western 44, 53, 82, 152
Evans, Arthur Humble (A. H.) (1855–1943) 70

Fabrique Nationale de Herstal (FN) company 25
falcons/falconry 18–21, 41, 50, 110
   falconers (*takajo*/raptor masters) 20
   takagari (hunting with falcons) 18
Fascist/Fascism 146, 149, 167, 205
Fedman, David 83
femininity 8. *See also* gender; masculinity
Fenellosa, Ernest 27
firearms 9, 17, 21, 23–6, 94, 148, 168, 188, 201. *See also* shotguns and rifles
First Sino-Japanese War 57. *See also* Second Sino-Japanese War

First World War 38, 40, 53, 127. *See also* Second World War
Fitzpatrick, John W. 34
Fleming, James Henry (1872–1940) 68, 93
Fletcher, A. S. 101–3, 106
Florida Museum of Natural History at University of Florida 15
Flower, Stanley Smyth (1871–1946) 70, 243 n.42
foreign hired hands (*oyatoi-gaikokujin*) 10, 44
Forest Reserve for the Philippines on Mount Makiling 83
Fredenson-Walters, Maria Barbara 73
Friars, Dominican 169
Fuchs, Barbara
  cultural mimesis 12, 44–6, 84
  Mimesis and Empire 246 n.16
Fuller, George 26
Funabashi Yoichi 111

Gaijin shôgun (foreign Shôgun) 198
game birds 20, 22–3, 57, 113, 130, 141, 179–80, 183, 187
gender 5, 10, 12, 60–3, 181. *See also* class; race
  performative 8–9
  and precarious subaltern manhood 7–10
  Scott on 8
genetics 12, 40, 82, 219
The Geographical Journal 124
King George V (1865–1936) 66
Giles, Herbert A. (1845–1935) 65–6, 75–9
  conflict with fellow students 75–6, 78
  "Record of Strange Creatures" 75
  Yiyu tuzhi (Map and Chronicle of Exotic Places) 76
Gillespie, Gregory, hunting for empire 21, 82–4
Gladkov, Nikolai A. (1905–75) 209
Glaser, Barney G. 233 n.152
global ornithological networks 216–18
Gondô Yôkichi (1895–1970) 45
Goodfellow, Walter (1866–1953) 99
Gopinath, Praseeda 7–8
  on imperial masculinities 223 n.6
Gordon, Andrew 7, 23
goryôba (private-access royal hunting grounds) 22

Goryôrin (imperial forests) 83
Gotô Shimpei (1857–1927) 120
Gould, Carol G. 21
Great Britain 5, 32, 34, 37, 39, 41, 51, 57, 66, 79, 83, 91, 115, 117, 211, 242 n.28
Great Depression (1929–39) 78, 81, 106, 198
Greater East Asian Co-Prosperity Sphere 116, 150
Greater East Asian Travel Public Corporation 147
Great Indian Bustard (*Ardeotic nigriceps*) 207
Greco-Roman 66, 76
green pheasant (*Phasianus versicolor*) 58, 166, 190, 207
Greenway, James Cowen (1903–89) 141, 186
Grinnell, Joseph (1877–1939) 85, 91
  and McGregor 86, 246 n.27
Guillemard, Francis Henry Hill (F. H. H) (1852–1934) 70, 79
gun hunting culture 21–3. *See also* hunting/hunters (methods) of birds
licensed hunters 24, 148

Hachisuka Fueko (1898–1937) 47, 90
Hachisuka Masa'aki (1871–1932) 46–7, 67–8, 73, 88–9, 101
Hachisuka Masauji, Marquis (1903–53) 7, 9–12, 15, 18, 24, 32, 35, 37–8, 40, 43, 51, 59, 72, 74–5, 77–9, 81, 83–4, 103–4, 110, 114, 130, 135–6, 138–40, 149, 158–9, 161, 165, 173, 179–82, 188, 193, 196–7, 207, 216–18, 237 n.72, 246 n.27
  Atami *Bessô* (Second Home) 46–9
  "The Author at Lake Faggamb" 101
  Birds of Egypt 38
  Birds of Jehol 39, 129–31
  The Birds of the Philippine Islands 70, 86–93, 98, 102
  'The Birds of the Sage West China Expedition' 139
  British-Style Estate in Tokyo (Mita District) 46–9
  at Cambridge (1921–7) 66–72
  A Comparative Hand List of the Birds of Japan and the British Isles 38, 65, 72–3

"Contributions on the Birds of Hainan" 138–9
   death of 67
   and Delacour 88, 90
   and divorce proceedings 159–60
   explorations of 251 n.169
   game hunting in Africa 39
   gentleman of science 65
   Giles's conflict with 78
   A Handbook of the Birds of Iceland 71
   "Handlist of Japanese Birds" 186
   "Inscription on a Rock at the Summit of the Nameless Peak in the Apo Range" 102
   A List of the Birds of Micronesia under Japanese Mandatory Rule 39, 140
   matriculation record of (1924) 69
   Mount Apo Expedition (1929 & 1930) 99–107
   "Notes on Hainan Mammals" 140
   stopovers and voyage to Philippines 92–5
Haddon, Alfred Cort (1855–1940) 76–7
half-wild birds 62
Han Chinese 123
Hane Kôgyô (Feather Industry) Company 185
harai (exorcism) 156
Harris, Sheldon 137
Hartert, Ernst (1859–1933) 70–1, 79, 92, 133, 140
Harvard Museum of Comparative Zoology (MCZ) 15, 52, 81, 141, 175, 186
Hattori Hirotarô (1875–1965) 68, 201
Haussman, Baron 120
Hawaii 71, 73, 84, 92, 136, 210
Hawking Guild. *See* Hawk-Keeping Officers
Hawk-Keeping Officers 19
hawks 19–20, 52, 146, 161
Heian period (794–1185) 18–19
Hein, Laura 205
Heisei (1989–2019) 219
Hellman, Geoffrey T. (1907–77) 116, 198
   The Smithsonian: Octopus on the Mall 208
heteronormativity 91
heterosexuality 9, 91. *See also* homosexuality

H. F. & G. Witherby 93, 115
Hidefumi Imura 61
Hideki Yukawa 13, 193
Hideo Mohri 219
Hilgendorf, Franz Martin 29
Hiratsuka Raichô 61
Hiroi Tadakazu 214
Hiromi Mizuno 13, 37, 186, 192
historical ethnography 15
Holland 9, 12, 24
homosexuality 90–1, 240 n.149. *See also* heterosexuality
homosocial/homosociality 25, 43, 53, 91, 219
   and bird watching 60–3
honorary whiteness 2, 38, 43, 64, 66, 79, 81, 113, 181, 226 n.56, 241 n.4
honzôgaku (study of fundamental herbs) 28–9, 31, 38, 41
Honzô kômoku (Systematic *Materia Medica*) 28
Horiuchi Sanmi 20
   Nippon chôrui shuryô-hô (*Bird Hunting Methods in Japan*) 21
Hoshino Naoki (1892–1978) 155
House of Peers, National Diet's 11, 17, 27, 38–9, 41, 52, 64–5, 68, 73, 100, 110, 135, 144, 146, 165, 191
Howell, David 24
Hunter's Association, Tokyo 23, 27, 135, 147, 155, 188
hunting for empire 21, 82–4
hunting/hunters (methods) of birds 17, 20–1, 168, 175, 183, 185, 187, 190, 214. *See also specific methods*
   avian tradition 20
   duck-hunt 20, 57, 148, 180, 184
   game hunters 25, 39
   gun (*see* gun hunting culture)
   Indigenous 131
   license 23–4, 148, 214–15, 274 n.150
   in premodern era 18–21
   prewar hunting practices in Japan 25–6
   seal 191
   sport 21–3, 27, 113
   women hunters 25
Huxley, Julian (1887–1975) 216
   Evolution: The Modern Synthesis 40
   The New Systematics 40
hybrid cultural identity 13, 48

hybridization 40, 44
Hygiene Society of Japan 199

Ibis journal 15, 36
Iijima Isao (1861–1921) 22, 27, 30–2, 38, 41, 53, 119, 135, 150, 185
  Okadaira Shell Mound at Hitachi 30
  Shell Mounds of Omori: Volume 1, Part 1 of Memoirs of the Science Department 30
Ikuta Shigeru 25–6
imperial designs 14, 40, 158
Imperial Household Agency 20, 45, 48, 58, 90
imperialism 81, 83, 92, 113
  American 82
  avian (*see* avian imperialism)
  canine 35, 228 n.34
  collecting imperialism 5, 81–4, 179, 224 n.24
  imperialists 2, 5, 11, 38, 82–3, 97, 131, 221
  Japanese 16, 51, 65, 84, 97, 106, 111, 119, 140, 143, 158, 221
imperial masculinities 2, 7–10, 12, 17, 41, 44, 65, 81, 83, 92, 140, 181, 195, 202, 211, 219. *See also* postimperial masculinities
  Anglo-American 87
  Gopinath on 223 n.6
  in hunting and ornithology 21–7
Imperial Navy 138
Imperial Ordinance 201
Imperial Preserve 57–8
India/Indians 7–8, 58, 66, 96, 179, 207, 209–11
Indigenous peoples 2, 8, 50, 81, 85, 96–7, 99, 103–5
Indochina 39–40, 54, 56, 88, 126, 141, 149
insect-borne diseases 195, 202, 204
Institute for Infectious Diseases 199
International Council for Bird Preservation (ICBP) 11–12, 53, 206–8, 213, 215
International House of Japan 48, 206
International Military Tribunal for the Far East (IMTFE). *See* Tokyo, Tokyo Trial
International Ornithological Congress (IOC) 68, 217

International Ornithologists' Union (IOU) 217
International Union for Conservation of Nature (IUCN) 207, 216
Inukai Tetsuo (1897–1989) 40
Ishii Shirô (1892–1959) 154, 200
Ishii Unit (*Ishii butai*)/Unit 731 137–8, 200, 203
Ishikawa Tomomatsu 30
Ishiwara Kanji (1889–1949) 110
Isseidô Booksellers 142
Itagaki Seishirô (1885–1948) 110
Itô Keisuke (1803–1902) 31, 154
Itô S. 92, 100
Iwakawa Tomotarô 30
Iwasaki Koyata (1879–1945) 45, 48, 206, 207
Iwata Hisayoshi (unknown) 117
Izumi Sei'ichi 13–14

Jahn, Hermann (1911–87), *Zur Oekologie und Biologie der Vögel Japans* 170
Japan-Bacteriological Warfare Summary (1944) 138
Japan Carrier-Pigeon Association 127
Japanese-American Evacuation Claims Act 160
Japanese-Americans 47, 73, 90, 140–1, 160, 174, 180
Japanese Association for the Preservation of Birds 215
Japanese Association of Zoological Gardens and Aquariums 135, 155
Japanese blue flycatcher 146
Japanese encephalitis 196, 200–2, 209
Japanese robin (*Erithacus akahige*) 18, 54, 56, 58
Japan/Japanese 2, 9, 14–15, 31, 39, 43, 68, 83–4, 89, 173–7
  domestic 49, 66, 83, 112, 115, 117, 126, 137, 146
  elites 3, 5, 18, 21–4, 47, 49, 81, 94, 113, 167, 187, 191, 196, 213
  emigration 97, 112
  imperial Japan 3, 7, 9–12, 16, 19–21, 24, 27–32, 38, 44, 50–1, 53, 56, 62, 64–5, 79, 87, 110–12, 114–15, 117, 119, 129–30, 135–6, 140–1, 149, 152–3, 156, 165, 171, 182, 203

Ministry of the Environment 19, 213, 215
national bird of 166-7, 190
Occupied Japan 3, 86
official Bird Day of 166, 186, 190
postwar 5, 11-12, 18, 20, 148, 168, 193, 205, 219
   renewed ornithological collaboration in 177-87
   wildlife and conservation in 187-92
premodern 18-19, 21
   animal-based entertainments in 227 n.13
prewar 10-11, 13-14, 17, 21, 29, 38, 41, 44, 47, 59, 61, 65-6, 91, 125, 256 n.79, 268 n.21
and proper clothing 9
reformers 9, 44, 91
scientists 2, 5-13, 16-17, 20, 28, 36, 49, 79, 81-2, 84, 92, 96, 107, 111, 117, 120-1, 123, 130, 134, 136, 154, 157, 161, 165-6, 168-9, 178-80, 187-8, 192-5, 197, 204, 211, 216-17
Taiwan under rule of (1894-1945) 7-8
transwar 6-7, 10-16, 38, 60, 81, 136, 143, 169-70, 194-5, 213, 216-17, 221-2
Japan Travel Bureau (JTB) 143
Jehol 120-1, 123, 125, 140, 152
   map of (1933) 124
Jômon era (14,000-300 B.C.E.) 30
J. Purdey and Sons 16-gauge shotgun 25
Julius Neubronner (1852-1932) 127

kachôga (bird and flower paintings) 28
Kadia, Miriam Kingsberg 60
   Into the Field 130
Kaitôkaku (Opening Up the East Pavilion) 45
Kalyakin, Mikhail 209
Kamakura era (1185-1333) 18-19
Kamei Kan'ichirô 203
kamoba (duck-hunting ground) 57, 183, 185
kamoshaku-ami (spoon-shaped duck net) 185
Kano Tadao (1906-45) 115
Kantô Army 84, 87, 106-7, 109-12, 114, 119, 121, 130, 132-4, 137, 150
   carrier pigeons 126-9
   in final days of Tokunaga expedition 125-6
Kantô Earthquake (1923) 31, 45
Kaplan, Ann 95
Kapur, Nick 206
Karp, Matthew 82
kasumi-ami (mist-nets) 18, 183. *See also* mist-netting/netters
Kasumi Kaikan (Kasumi (Misty) Meeting Hall) 269 n.45
katakana 101, 103
Katô Hiroyuki (1836-1913) 29-30
Katsumata 139-40
Kawamura Tamiji (1883-1964) 151
kazoku (aristocrat) 6. *See also* aristocracy/aristocrats
Keene, Donald 253 n.21
Keh Ung Sang 173
Keio University 235 n.8
Kenkoku no hi (National Foundation Day) 103
Kigensetsu (Festival of the Accession of the First Emperor and the Foundation of the Empire) 103
Kikuchi Dairoku (1855-1917) 67
Kim Sung Jang 175
Kingsberg Kadia, Miriam L. 14
   Into the Field 13
Kinnear, Norman Boyd (1882-1957) 67, 70, 90, 196
Kipling, Rudyard (1865-1936), "The White Man's Burden: The United States and the Philippine Islands" 95-7
Kishida Kyûkichi (1888-1968) 129
Kishida Ryûsei (1891-1929) 61
Kitagawa Masao (1910-95) 123
Kiyosu Yukiyasu (1901-75) 60, 149, 151, 186
   "Bird types in the Hengshuizhen and Yuanqu Areas/Zoology Team" 151
   "On Birds That Can Become Food Resources" 153, 186
Kobayashi Shigekazu (1887-1975), *Birds of Japan in Natural Colors* 115-16, 130, 140, 142
Koch, Otto 99
koji (rice bran lees) 19
koku (unit of measure) 159

Kokusai bunka kaikan (International Culture Hall). *See* International House of Japan
Kokusai Kwankô (International Tourism) 144
Kon Wajirô (1888–1973) 47
Korea/Korean 50, 73, 82–3, 89, 110–11, 115, 126–7, 129–30, 153, 165, 172–9, 181, 183, 214
Korean Mandarin Duck 50
Korean Peninsula 82, 174, 176, 194, 204, 214
Korean War 193–4, 202, 204–5
Koyama, Sachiko 18–19, 70, 76, 78
Krämer, Hans Martin 147
Krizek, Robert L. 5
Kuroda Nagahisa (1916–2009) 7, 9, 11, 49, 60, 116, 136, 157, 161, 165, 170, 193–4, 198–9, 206, 211, 215–16, 269 n.45
  "The Birds of Japan, Their Status and Distribution" 170
  "The Ornithologists and Avifauna in (*sic*) Early History of Japanese Field Ornithology" 170
Kuroda Nagamichi (1889–1978) 11, 32, 37–8, 43, 49–50, 55–60, 62, 73, 84, 113–17, 119, 129–32, 135, 139–40, 149, 157, 166, 170–1, 173, 179, 185, 191, 193, 198, 200–5, 207–10, 268 n.38
  A Bibliography of the Duck Tribe 141
  Birds of Java 115
  Birds of Jehol 39, 109, 114, 129–31
  Birds of the Island of Java 115
  A List of the Birds of Micronesia under Japanese Mandatory Rule 39, 140
  memberships 116
  Monograph of the Birds of Greater East Asia 140
  Non-passeres 115
  Passeres 115, 130
  The Pheasants of Japan Including Korea and Formosa 57
  Pseudotadorna cristata (1917) 50
  The Quarterly Review of Biology 170
  The Study of Birds: Ecology 215
  study on Javanese birds 116
Kuroda Nagatoshi (1881–1944) 57
Kuroda Sayoko (b. 1969) 219

Kurosawa Shôju 175, 177, 179
Kuser, Anthony R. (1862–1929) 58
  Blood Partridge 58
kuwayaki (cooking style) 20
Kwantung Army 127

Lalique, René (1860–1945) 45
language 2, 9, 16, 35, 165, 172–3, 190
  classical 66, 76
  Oriental 76–7
  of science 11–12, 16, 40, 119
  as status markers 44
Latin America 82, 89, 169, 178
League of Nations 114, 119
Lebra, Takie Sugiyama 226 n.54
  status culture 6, 44
Le Corbusier (1887–1965) 206
Lee, Bruce (1940–73) 105
legitimacy 2, 41, 60, 65, 77–8, 110, 114, 131, 154–5, 186
Leheny, David 144
Lemay, Curtis (1906–90) 156
Lemon, Frank E. 12
Leopold, Aldo (1887–1948) 190
Leuckart, Rudolf (1822–98) 31
Lewis, Michael 210
Liaodong Peninsula 113, 132, 253 n.26
Liao Dynasty (916–1125) 118
Liberal Democratic Party 205
Li Bo (701–762) 77
Limited Nuclear Test Ban Treaty 211
Li Shizhen (1518–93) 28
Liszt, Franz von (1851–1919) 82
Lord Alanbrooke (1883–1963) 142
Lord Cromer (Evelyn Baring) (1841–1917) 75
Loukashkin, Anatole S. (1902–1988) 131–2, 258 n.146
Lovette, Irby J. 34
Lowell, Percival 31
Low, Morris 13, 119, 167
Lubar, Steven 34, 49
Lytton Commission (1932) 87, 114

MacArthur, Arthur (b. 1938) 187–8, 203
MacArthur, Douglas (1880–1964) 5, 142, 161, 168, 178, 198, 200, 273 n.140
MacGillivray, William, *Ornithological Biographies* 33

MacGregor, Richard (1871–1936) 37
MacKenzie, John 83
    collecting imperialism 5, 81–4, 179, 224 n.24
MacLeod, Roy 210
Maekawa Kunio (1905–1986) 206
Makigari (traditional hunt) 22–3
Makiling Center for Mountain Ecosystems (MCME) 245 n.10
mallophaga (biting lice) 31
Maloney, Barbara 9
mammology 137
Manchukuo 39–40, 87, 107, 109–11, 113–14, 117–19, 122–3, 127, 131, 135, 137, 182, 200, 252 n.1
    as laboratory of modernity 110–12
    Manchukuo expedition 119
    Manchukuo Forestry and Agricultural Bureau 137
    Manshû gurafu (Manchuria Graph) 138
    natural world 125, 129–31
Manchuria Daily News 112–13, 119–20, 125, 127
Manchuria/Manchurians 11, 13–14, 20, 84, 86–7, 98, 107, 109–10, 112–13, 117, 124, 126, 131, 133–4, 137–8
Manchurian Incident (1931) 110–17, 168
Mangan, J. A. 8
Manshû nichi nichi Shimbun. See *Manchuria Daily News*
Marcon, Frederico 28–9. See also *honzôgaku*
Marquis Kuroda Nagashige (1867–1939) 50, 57
maruta (logs) 200
masculinity 60. See also femininity; gender
    British 66
    imperial (see imperial masculinities)
    Japanese 81
    precarious manhood 7–10
Massey, Doreen 6
    power geometries theory 14
materia medica 28
Matsuda Masayuki (1892–1976) 89
Matsuda Michio 116
Matsuda Takeshi 166
Matsunoi Kanji (1895–1982) 48
    design of Toyô Eiwa Girls' Academy 48

Matsura Sayonhiko 30
Matsuyama Shiro 21
Mauss, Marcel 52
Mayr, Ernst (1904–2005) 88, 157, 186, 197
    Notes on the Taxonomy of the Birds of the Philippines 158
    Systematics and the Origin of Species from the Point of View of a Zoologist 40
McClure, H. Elliott (1910–98) 195–6, 198, 201–2, 204, 209–11
McGregor, Richard Crittenden (1871–1936) 88, 91–2, 94–5, 97, 99, 106, 247 n.35
    and Grinnell 86, 246 n.27
    Hand-list of the Birds of the Philippine Islands 85–7
    A Manual of Philippine Birds 85
McKinley, William (1843–1901) 84–5
Mearns, Edgar Alexander (1856–1916) 99
meat, bird 19–20, 148
megafauna 83, 154–5
Meiji Japan 5–6, 10, 19, 22–3, 27–8, 30, 32, 40, 50, 67, 90, 92, 114, 148, 166, 176, 184, 189, 201, 206, 214
    genrô (principle elder/oligarch) 29
    Meiji Civil Code (1898) 9, 91
    national project 44
Meiji Shrine 29, 135, 156–7, 201–2
Meiji Social Darwinism 245 n.2
Meise, Wilhelm (1901–2002) 133
memory 2, 15, 70, 162, 170. See also oblivion
Menage, Louis F. (1850–1924) 85
Merton, Robert K. 233 n.154
mice 153, 202. See also rats
Micronesia 39–40, 115
migration, bird (migratory birds) 12, 19, 21, 113, 207, 211, 213–14, 217, 270 n.68
Migratory Animals Pathological Survey (MAPS) project 209–10, 215
Mikado Pheasant 57
militarism 21, 111–14, 121, 127, 134, 168, 216
Miller, Ian 50
    The Nature of the Beasts 155
Miller, Tracy G. 150
Milne-Edwards, Alphonse 56

Mindanao 105–7, 110
  colonial history of 95–9
Ming dynasty (1368–1644) 28, 76
Ministry of Agriculture and Forestry 21, 24, 147–8, 151, 185, 188–9, 214–15
Ministry of Health and Welfare 200
Miroku Company 229 n.62
Mission Revival (1890–1915) 48
mist-netting/netters 18–19, 21, 24, 189, 192, 202, 210, 214, 270 n.71, 271 n.90
Mita Ward Museum 46
Mitsubishi Corporation 38, 44–5, 234 n.7, 235 nn.9–10
Mitsui Company 221
Mizuno, Hiromi 192
Mizuno Kaoru (1895–1980) 112–14, 117, 127–8, 131–2
  A Distributional List of the Manchurian Birds 116, 130
  The Great Illustrated Encyclopaedia of Manchurian Birds in Original Colours 116
MKULTRA program 210
Mochizuki Michito 214
modernity 18, 25, 38, 40, 44, 47, 66, 90–1, 121, 124, 134, 206
Momiyama Tokutarô (1895–1962) 31, 60, 73, 139, 198
Mongolia/Mongolians 14, 113, 117–18, 120, 124, 126, 153
Monkey-eating eagle" (*Pithecophaga jefferyi*)/Philippine Eagle 100
Monti, Martin J. (1921–2001) 203
Moore, Aaron S. 13
Mori Tamezô (1884–1962) 123
Morita Sohei (1881–1949) 61
Morse, Edward S. (1838–1925) 19, 27, 29–30
  Catalogue of the Morse Collection of Japanese Pottery 31
  donations of collections 31
  Japan Day by Day memoir 30
  Japanese Homes and Their Surroundings 30
  Shell Mounds of Omori: Volume 1, Part 1 of Memoirs of the Science Department 30
  special students of 30

motodamari (duck ponds) 20
Mount Apo Expedition 99–107
Murakami Kimio 217
Murray, John Owen Farquhar (1858–1944) 68
Museum of Modern Art 218
Museum of Russian Culture 258 n.146

Nadin Hée 82
Nagura Otohiko (unknown) 153
Nakai Takenoshin (1882–1952) 118, 123
Nakamori Akio 49
Nakamura Masanori 7
  kansen-shi 225 n.33
Nakamura Tsukasa (1926–2018) 11, 47–8, 89, 158, 196, 213, 217–18
Nakamura Yoshio (1890–1974) 83, 87, 109, 181–2
Nakanishi Gotô (1895–1984) 60–2, 216
  Hakua no yakata (Limestone Mansion/White House) 48, 62
  Yachô to tomo ni ((Living) Together with Wild Birds) 62
Nakayama, Thomas K. 5
Nanbara Shigeru (1889–1974) 205
Naoko Shibusawa 181
  America's Geisha Ally 195
Nara Period (710–794) 18
narratives 16, 35, 43
  colonial 5
  historical 2, 5, 9
  visual 5, 9
Nashimoto Morimasa (1874–1951) 159–60
National Audubon Society 36
national bird of Japan 166–7, 190
National Diet Library 228 n.46
national identity 10, 13, 49, 143
National Institute of Health and Nutrition 200
National Institute of Infectious Diseases 200–1
National Rifle Association (NRA) Hunter's Safety Course 25
Natori Bûko (1905–87) 40
Natsume Sôseki (1867–1916) 111, 196
  Kokoro (Sincerity) 91
Natural History Books 228 n.28

Natural History Museum, Tring 34, 39, 72, 87, 93
naturalists 28–9, 31, 33, 36, 39–40, 99, 106–7, 129, 136, 154, 190
Nature journal 15, 71, 87, 123
Nazi regime 78, 146, 157, 204
Negritos 105
netting harvest of birds 19
Newton, Alfred (1829–1907) 36
New York Zoological Park. *See* Bronx Zoo
Nicholl, Michael John (1880–1925) 70
Nihon shoki (Chronicles of Japan) 19
Niibe Tominosuke (1882–1947), *The Ecology of Birds in the Field* 191
Nippon Times Magazine 189
Nishina Yoshio (1890–1951) 13
Nishiyama Katsuo 276 n.30
Nixon, Richard (1913–94) 202
N. M. Rothschild and Sons 72
Noboru Koyama 69, 241 n.3
Noguchi Woodpecker (*Dendrocopus noguchii*) 51
Nomonhan Incident (1939) 121
non-Europeans 66. *See also* Europe/European
non-white 2, 7–8, 10, 13, 78, 86, 88, 158, 195, 279 n.69. *See also* white people
Norman Parley publisher 77
North American Model of Conservation (NAMC) 274 n.157
North American Wildlife Conservation Model 190
North-China Daily News and Herald 120
Notô Hisashi (unknown) 118
NRS Wildlife Bureau 19, 148–9, 177–8, 191, 197, 201
Nye, Joseph 167

Obata Shigeyoshi (1888–1971) 77
oblivion 2, 162. *See also* memory
Occupation 3–5, 7, 19–20, 81, 158, 161, 165–6, 168–9, 178, 183, 187, 196, 198, 201, 203, 205–6, 261 n.42. *See also* Allied Occupation
Odachi Shigeo (1892–1955) 155
Oda Nobunaga (1534–82) 159, 196
Odaya Yoshiya 214–15, 225 n.31
odeshi 53
Ogasawara Nagayoshi (1885–1935) 48

Ogasawara-Tei (Ogasawara Estate) 48
Ogawa Jihei VII (1860–1933) 206
Ogilvie-Grant, W. R. (1863–1924) 140
Okada Hisashi 200
Okamura Yasutsugu 127
Okinawa 40, 51, 56, 153, 237 n.59
Okinawa woodpecker 35, 51
Omori Tadashi 155
Operation Paperclip 204
Order Concerning Firearms and Swords 201
Oriental/Orientalist 30, 55, 66, 76–7, 193
Ori'i Hyojiri 109–10, 130, 137, 165, 182, 259 n.150
    collection of Manchurian birds 131–3
Ornithological Society 4–5, 11, 17, 22, 31–2, 37–8, 41, 51, 53–4, 57, 61, 68, 86, 114, 129, 135, 159, 166–7, 170, 173, 185, 187–8, 190, 193, 207, 215
    Hand-List of Japanese Birds 112–17, 140–1, 157
    lifetime member 254 n.46
    postwar rebirth 166–8
ornithology 11–12, 15–18, 21–2, 27–9, 31–2, 41, 44, 62, 64, 68, 73, 92–3, 95, 107, 109–10, 113, 117, 119, 135, 139, 144, 162, 173, 178, 192, 221–2
    homosociality 43
    as imperial agent/s 37–40
    in late 1940s 195–205
    ornithologists 4–8, 10, 12–16, 21, 27, 35, 37, 40, 43, 49–51, 53, 58, 60–1, 63, 65, 83, 97, 106, 109–10, 112, 114, 116–17, 135, 137, 141, 143, 162, 165, 167, 169, 186, 194, 206, 211, 214 (*see also* zoology, zoologists)
    and politics 1
    popularization of 218–21
    professionalization of 27, 32
    scientific study in Anglo-American world 32–7
Osaka Shosen K. K. (OSK Line) 126
Osgood, Wilfred (1875–1947) 94
otaku 49
Otsuka, Naoko 19
Oyanguren, José (1800–59) 99

Pacific Citizen 160
Pacific Ocean Biological Survey Program (POBSP) 210
Pacific War (1941–5) 10–11, 40, 51, 107, 110, 126, 128, 133, 135–6, 141, 143, 147, 149, 160, 162, 168, 173–4, 186, 192, 197
pangolin 94, 249 n.90
Pan-Pacific areas 207
Papua New Guinea 14, 198
parasitology 31, 41, 135, 137, 185
Paris International Exposition for Decorative Arts (1925) 45
peacocks 15, 116
Pearson, Thomas Gilbert (1873–1943) 12
Peers Club 150
Peer's School (*Gakushûin*) 30, 40, 67, 76, 173
People's Republic of China 204, 211
performative ethnography 15
Perraton, Hilary 66
Peru 14
Peters, "Jimmy" (James Lee) (1889–1952) 51–2, 174, 179, 197
Phasianinae 58
pheasants 19–27, 55–8, 115, 141, 166–7, 267 n.9
Philippine Insurrection (1899–1902) 105
The Philippines 11, 13, 39–40, 48, 73, 78–9, 81–2, 84, 89, 92–3, 96, 98–9, 104, 110–11, 136, 149, 157, 182, 188, 196
    abacá (Manila hemp) production in 97–8
    exploration in 84–6
    Filipinos 82, 98, 101, 103, 105–7
    Filippinization campaign 96
    Hachisuka
        on Philippine birds 86–92
        voyage to 92–5
    Japanese invasion of 52
    Manila 81, 85, 93–5, 106
    Mount Apo expedition 99–107
    Philippine Commissions 85
    Subic Bay 223 n.4
photographs/photography 5, 9–10, 25
    amateur 3–4
    decoding 5
Pitelka, Morgan 19

plants 28, 33, 90, 92, 94, 97, 104, 116, 121, 123, 144, 182, 194. *See also* animals
postimperial masculinities 10. *See also* imperial masculinities
poultry 135, 141, 148
power geometries theory 6, 14
precarious manhood/masculinity 7–10
Prince Asaka Yasuhiko (1887–1981)
    Art Deco villa/*Teien* (Estate Park) 45–6
    and Second Sino-Japanese War (1937–45) 45–6
Prince Nobosuke Taka-Tsukasa 54, 56, 115, 237 n.72
Princess Kitashirakawa Mineko (1911–70) 46, 73, 88–9
Princess Takamado 12
    'Through the Lens' 220
Princess Tokugawa Kikuko (1911–2004) 91
Prince Takamado Norihito (1954–2002) 219
Prince Takamatsu (1905–87) 66, 248 n.70
Prince Tokugawa Yoshinobu (1837–1913) 67, 88
private museums 43, 51, 139
Proceedings of the National Academy of Sciences 218
Proclamation on Gun Control Regulation 23
public literacy 33

Queen Noor of Jordan (b. 1951) 219

Rabasa, José 170, 192
    Tell Me the Story of How I Conquered You 169
race/racism 2–3, 5, 10, 13, 16, 56, 79, 177, 180. *See also* class; gender
    and Asian 224 n.20
    racial superiority 174
Raichô (thunderbird) 61
Rajal, Don Joaquin 99
Raj, Kapil 271 n.85
Rangaku (Dutch Learning) 28
Rangaku-sha (adherents of new Dutch Learning) 28
Rapin, Henri (1873–1939) 45
raptors 19–21, 50, 52, 81, 100, 116, 128, 161, 166, 188

rats 153–4. *See also* mice
Raymond, Antonin (1888–1976) 206
Red List of Threatened Species, IUCN 216
The Report of the National Museum 51
Research Institute for Natural Resources (1941–5) 7, 11, 147–8, 149–54, 178, 186
revived *Chôgakkai* (Ornithological Society) (1947) 4
Rice, Mary Blair (Blair Niles (1880–1959)) 58
Ridgeway, Robert, *The Birds of North and Middle America* 36, 232 n.142
Ridgeway, William (1853–1926) 76
Ripley, S. Dillon (1913–2001) 52, 116, 139, 197–8, 207–8, 210–11
Ritchie, Lizzie (unknown) 73
Ritvo, Harriet 32
Roberts, Alastair 270 n.65
Rocky Mountain Elk Foundation (RMEF) 274 n.157
rodents, zoological research on 7, 31. *See also* mice; rats
Roosevelt, Theodore 190
Rosenstone, Robert 4–5
Rothschild, Lord Lionel Walter (1868–1937) 34–5, 38, 43, 48, 51, 53, 70–4, 78–9, 86–7, 93, 139–40, 198, 242 n.40
  Hand-List of British Birds 72
Rothschild, Miriam (1908–2005)
  Dear Lord Rothschild 73
Rothschild, Nathan Mayer Freiherr von (1777–1836) 72
The Royal Geographical Society 124
Ruiz, Vicki L., creative non-fiction 2, 162
Russo-Japanese War (1904–5) 110, 112, 126, 134
Ryôyu (Hunter's Companion) magazine 17, 22–3, 25, 41

sadeami (scoop net) 20
Sage, Alida (Ann) M. (1915–44) 139
Sage, Dean, Jr. (1909–63) 139
Sage Expedition 139–40
Saitama *kamoba* 20
Sakakura Junzô (1901–69) 206
Sakenokimi 19
samurai 18, 30, 79, 91, 105, 110

sandaimeichô 18
Sand, Jordan 44, 47
Sanford, Leonard Cutler (1868–1950) 74
Sasaki Chûjirô
  Okadaira Shell Mound at Hitachi 30
  Shell Mounds of Omori: Volume 1, Part 1 of Memoirs of the Science Department 30
Satow, Ernest (1843–1929) 75
Sawada Kiyoshi (1903–94) 78
Schandenberg, Alexander (1852–96) 99
Scharff, Hanns-Joachim (1907–92) 203
Schenck, Hubert G. (1897–1960) 178
Schrader, Stuart 96
Schreurs, Miranda Alice 61
scientific exchange 11, 13, 17, 32, 44–6, 51, 192–3, 213
scientific knowledge 30, 41, 46, 51, 53, 66, 82, 84, 109, 132, 155, 168
scientific nationalism 13, 186, 192
Sclater, Philip L. (1829–1913) 36
Scott, Joan 8
Second Sino-Japanese War (1937–45) 90, 138, 168. *See also* First Sino-Japanese War
Second World War 19, 40, 136, 139, 149, 160, 165, 178, 181, 186, 203, 258 n.146. *See also* First World War
Seebohm, Henry (1832–95), "The Birds of the Japanese Empire" 170
Seikadô Bunko ("Hall of Tranquil Praise" Library) 45
Seikanron (Rectify Korea Discourse) 82
Selywn College Library 78, 242 n.29
Sengoku period (1467–1603) 24
Setoguchi Akihisa 22, 31
Seven Sisters for Conservation, principles of 274 n.157
sexual mentorship (*shudô*) 91
Seys, P. G. 133
Sheldon, William G. (1912–87) 139
sheldrakes (*Tadorna cristata*) 50, 59
shelfstrings (*tanaito*) 271 n.90
Shiba-inu (traditional Japanese hunting dog) 25
Shibata Keita (1877–1949) 150
Shibusawa, Naoko, *America's Geisha Ally* 13
Shichi-go-san Festival 227 n.13

Shigen kagaku kenkyû-jo. *See* Research Institute for Natural Resources
Shimizu Mitsuo (unknown) 153
Shimizu Saburô (1918–2004) 188
Shimoizumi Jûkichi (1901–75) 153
Shimomura Kenji (1903–67)
  Photo Collection on the Ecology of Bird Varieties 151
  Photo Collection on the Ecology of Wild Birds 151
Shimonaka Yasaburô (1878–1961) 140
Shinobu Seizaburô 167
Shintô belief system 156
Shirasu Jirô (1902–85) 66
Shiseidô Gurafu (Shiseido Graph) magazine 25
shosei (disciple/apprentice) 182, 217
shotguns and rifles 21–4, 25–6, 201–2. *See also* firearms; *specific guns*
Siberian Intervention 67, 79, 113
Sino-Japanese War (1894–5) 82
Sittiparus varius (varied tit) 19
Skabelund, Aaron, *Empire of Dogs* 35
Sloane, Hans (1660–1753) 34
Smith, Gordon 23
Smith, Norman 256 n.77
Smithsonian Institution, Washington, DC 34–5, 139, 210, 232 n.129, 281 n.121
  National Museum of Natural History 34, 36
Smithson, James (1765–1829) 34
sociability 19, 22, 24, 27, 45, 53, 59–60, 84, 100, 211, 219
Social Darwinism 27, 30
socialization 50–1, 64, 125
social status 10, 15, 49–50, 54, 111, 130, 166. *See also* class
Soffer, Richard L. 142
Sone, Tatsuzô (1853–1937) 44–5
songbirds 15, 18–19, 21, 58, 115, 130, 135, 148, 153, 183, 189, 214. *See also* birdsong
Song dynasty (960–1279) 50, 150
  jian tea bowl (Important Art Object) 50
Sôtô Zen Academy (now Setagaya Gakuen) 61
Southeast Asian Treaty Organization (SEATO) 209

South Manchuria Railway Company (SMRC) 110, 112, 116–17, 119–21, 127–8, 130, 132–4, 137
Soviet Union 126, 132, 136, 204, 209, 211
space(s) 1–2, 6, 8, 16, 43–4
  affective 5
  domestic 27, 43–4, 47, 49, 64, 162, 196
  private 47–9, 62
  social 5, 10–11, 15, 81, 169–70
Spain 44, 84, 99, 104, 169
Spaniards 104, 170, 178
Spanish Colonial Revival (1915–31) 48
sparrows 19, 133, 148, 153, 215
specimen(s) 15, 20, 31, 35–6, 49–51, 59, 63, 70, 74, 81, 106, 109–10, 114, 117, 121, 129–32, 139–40, 151, 154, 157, 174, 183, 201, 210, 218, 221
  avifaunal 43, 50
  collecting/hunting 12, 15, 21, 31, 33–4, 39–40, 81, 113, 137, 168, 197
  as politicized sociality 49–53
  exchange 43, 49–53, 68, 115, 213, 219
  Great Auk 71
  live 56, 107, 202
  natural history 49–50
Spencer, Herbert (1820–1903) 27
Spivak, Gayatri C. 35
sport hunting (*yûryô*) 21–3, 27, 113
Stalin, Josef (1878–1953) 121
State-War-Navy Coordinating Committee (SWNCC) 203
Steere, Joseph Beal (1842–1940) 85
Steinbacher, Joachim (1911–2005) 133
Stoler, Ann Laura 2
  historical ethnography 15
Stone, Witmer (1866–1939) 57, 71, 114–16, 151
Stresemann, Erwin (1889–1972) 170
subalterns 2, 5, 35, 79, 96, 105, 178, 196, 198
  non-white 8, 10
  precarious subaltern manhood 7–10
Sub-Saharan Africans 105
Sugawara no Michizane (845–903) 156
Suga Yutaka 148
sukiyaki (meat fondue) 20, 184
Suntory corporation 220

Suntory Fund for Bird Preservation trust 220
Swing, Joseph ("Jumping Joe") May (1894–1984) 188
Swinhoe, Robert (1836–77) 140
Syrmaticus mikado 57

Tachibana Tanekatsu (b. 1912–unknown) 73
tairiku yôme (continental brides) 137
Taishô era (1912–26) 22–3, 41, 91, 168, 189
takagari (hunting with falcons) 18
Takahashi Hishikari 127
Takama Sôshichi (1903–86) 26
Takashima Haruo (1907–62) 136, 152
Tamura Kaîchirô (unknown) 112
Tanaka Ryô (unknown) 153
Tanaka Yoshio (1838–1916) 154
Tateishi Shinkichi (1894–1977), "Advice to Rat Catchers" 153
taxidermic objects 35
taxonomy 28, 38, 40, 49, 68, 82, 86, 110, 123, 132, 140–1, 153, 158, 213, 218
technology 13, 32–3, 37–8, 122, 125, 194, 217
Temminck, M. J. C. (Jacob Coenraad), *Manuel d'ornithologie, ou Tableau systématique des oiseaux qui se trouvent en Europe* (Manual of Ornithology, or Systematic Classification of Birds Found in Europe) 33
Thomas, Julia Adeney 30
tlacuiloque (scribes) 169–70
Todd, Vivian Edmiston (1912–82) 188
Tôjo Hideki (1884–1948) 136, 191
Tôki Akira (1892–1979) 150
Tokuda Mitoshi (1906–75) 153–4
Tokugawa era (1603–1868) 4, 18–20, 27, 32, 59, 62, 91, 110, 146, 166, 211
  daimyô 46, 50
Tokugawa Ienari (1773–1841) 67
Tokugawa Yorisada (1892–1955) 69, 78
Tokunaga Expedition 109, 111–12, 117–25, 131, 133, 137, 140, 152
  Kantô Army in 125–6, 150
  carrier pigeons 126–9
  members (1933) 122, 129–30
  route of 124

Tokunaga Shigeyasu (1874–1940) 109, 118, 122–3, 125–6, 128
Tokyo 11, 13, 17, 19, 24, 29, 43, 48, 56, 60, 76, 114–15, 117, 126, 142–3, 177, 207
  Tokyo Army Medical College 154
  Tokyo Imperial University 15, 17–18, 22, 27, 29–31, 38, 41, 44
  Tokyo Trial 136, 191
Tokyo Asahi Shimbun newspaper 127
Tomkovich, Pavel 209
Tori (Birds) 11, 15. *See also* birds
torisha (cages) 19
Toshiko (Hachisuka) (1896–1970) 73
Totman, Conrad 176
tourism 83, 113, 120, 144–7
toyaba (bird hut place) 18
Toyama Yôchien 201
Toyotomi Hideyoshi (1537–98) 159, 196
  Sword Hunt 24
Tôzawa Kôichi 185
transnational/transnationality 5, 7, 12, 14, 37–8, 41, 129, 154, 172, 191–3, 210
  migrations 173, 213
  trade 176
Treaty of Berlin (1884–5) 82
Treaty of Kanghwa (1876) 176
Treaty of San Francisco (1951) 10
Tripartite Pact/Berlin Pact. *See* Axis Alliance
Tsuneishi Kei'ichi 200
Tsurumi Miyako 218, 220, 231 n.112, 267 n.3
Tsutsui, William 173
Turnix sylvaticus masaaki 88

Uchida Seinosuke (1884–1975) 11, 21, 31–2, 38, 60, 73, 109, 119, 135, 147–9, 157, 185, 189, 191
  Birds of Jehol 39, 114, 129–31
  Monograph of the Birds of Greater East Asia 140
Uchiyama, Benjamin 149
Ueno Zoo 154–6
United Nations Educational, Scientific, and Cultural Organisation (UNESCO) 216
The United States 1–2, 5, 12, 15, 32, 34, 36–7, 39, 41, 44, 53–4, 58, 63, 71,

84, 87, 89–90, 96, 99, 105, 117, 136, 141, 144, 146, 158–60, 165, 167, 172, 192, 202–3, 205, 211
Americanization 45
Immigration Act of 1924 79
The United States National Museum. *See* Smithsonian Institution, Washington, DC
The University of Cambridge, Preliminary Examination 76
The University of Cambridge Library 75
Uno, Kathleen 9
upland hunting method 15, 25–7
USAMRIID (United States Army Medical Research Institute of Infectious Diseases) 202
US invasion of Mexico (1846) 82
US-Japan Alliance 10
US-Japan Conference on Cultural and Educational Interchange (CULCON) 213, 217
US-Japan Mutual Security Treaty (*Sôgo kyôryoku oyobi anzen hoshô jôyaku*). *See* Anpô Treaty
Usui Mitsue (unknown) 60, 159, 180

Vandello, Joseph A., precarious manhood/masculinity 8–9
Van Tienhoven, P. G. (1875–1953) 12
Verbeck, Guido (1830–98) 4
Victorian Era (1837–1901) 33
Vincent, J. Keith 91
Vining, Elizabeth Gray (1902–99) 188–9
violence 21, 82, 171, 173
virology 210
Volkov, Michael 131
von Siebold, Philipp Franz (1796–1866) 31
  Fauna Japonica (Animal Life of Japan) 33
Vories, William Merrell (1880–1964) 48

Wada Hiroʼo (1903–67) 188
Wade-Giles system 75
Wade, Thomas (1818–95) 75
Wain, Louis (1860–1939) 72
Wakulla Springs State Park, Florida 15
Walker, Walton Harris (1889–1950) 188
Walter Rothschild Zoological Museum. *See* Natural History Museum, Tring

Warblers 18, 56
War Memorial Fund 78
war/wartime. *See also specific wars*
  contingencies 157–62
  impact on avian populations (Korea and Japan) 173–7
  Japanese civilians for sacrifice 154–6
  violence 173
  war crimes 136, 191, 203
  wartime exigency 200
  wild birds as food resources (1937–45) 147–9
Waseda University 15, 17
washoku (traditional cuisine of Japan) 147
Watase Shôzaburo (1862–1929) 92
Waterstradt, Johannes (1869–1944) 74, 99
Waxbills (Estrilda rhodopyga) 58
Weaver, Jonathan R. 8
West Indies 8, 37
Weyler, Valeriano (1838–1930) 85
Whipple, Sydney B. (1888–1975) 189–90
white Anglo-Saxon Protestant (WASP) 37
White, Hayden 5
Whitehead, Neil, performative ethnography 15
white people. *See also* non-white
  domination of white men 2
  white/honorary white settlers 2
Whitman, Charles Otis (1842–1910) 31
Whitney, Gertrude Vanderbilt (1875–1972) 73
Wild Birds 132
Wild Bird Society 11, 17, 29, 61, 191, 215–16, 221
Wilder, George D. (1869–1946) 259 n.159
wild pheasant 57
Williams, Raymond, structure of feeling 5
Wilson, S. S. 93
The Windsor Magazine 72
W. M. Vories Company 48
Wolfe, Kenneth Bonner (1896–1971) 237 n.59
Wolfe, Lloyd R. 81, 92, 94, 99, 106, 187–8, 237 n.59
The Birds of Japan 161
women hunters 25
Worcester, Dean Conant (1866–1924) 84–5, 97, 104

Menage Expedition 85
The Philippine Islands and Their People 85
Worcester, Frederick L. 97, 104
World Wildlife Fund (WWF) 207
Wright, Frank Lloyd (1867–1959) 197
Wulf, Andrea 14
   Founding Gardeners 13
   The Invention of Nature 13

Yachô (Wild Bird) journal 60
Yachô no kai. See Wild Bird Society
yakitori (roasted bird) 19, 183
Yamagara 18–19
Yamagata Aritomo (1838–1922) 82
Yamashina Institute for Ornithology (YIO) 15, 51, 209, 211, 213–15
   aristocratic lineages 218–21
Yamashina Yoshimaro 7, 9–12, 15, 21, 32, 38, 43, 52–3, 56, 58–9, 61–4, 81, 83, 109–10, 129–33, 135, 138–41, 160, 162, 178–80, 208, 213, 261 n.42
   and Austin 143
   autobiography 160–1
   Birds in Japan: A Field Guide 191
   Birds of Jehol 39, 114, 129–31
   "Bookplate of Alan Francis Brooke" 142
   on breeding poultry 135
   The Eating Habits of Manchurian Birds 137
   The Ecology and Protection of Japanese Birds 191
   "How to Breed Fancy Birds" 40
   Japanese Birds and Their Ecology 141–3
   A List of the Birds of Micronesia under Japanese Mandatory Rule 39
   and Manchurian birds 137–8
   Monograph of the Birds of Greater East Asia 140
   "A New Race of a Sparrow from Shansi (*sic*), China" 151
   research on water birds 40
   Shibuya laboratory 12, 51–2, 58, 161, 180, 188, 218
Yamashina Yoshimasa (b. 1927) 63
Yanagida Kunio (1875–1972) 61
Yankee elitism 37, 68, 165, 171, 179, 198
Yanosuke, Baron Iwasaki (1851–1908) 44–5, 47
Yarde-Buller, Olga Alice Muriel Walters (1906–92) 73
Yasar, Kerim 30
yashiki (residences/estates) 45
Yasuharu Matsudaira 73
Yasukuni Shrine 156
Y. Fukukita, *The Tea Cult of Japan* 143
Yi dynasty (1392–1910) 83, 176
Yokoi Eisuke 159–60
Yomiuri shimbun 155
Yoshida Shigeru (1878–1967) 46, 193
Yoshii Masashi 201–2, 204, 210–11
Yoshimi Yoshiaki 136
Yoshimura Junzô (1908–97) 206
Yoshi Nakada 223 n.12
Young, Louise 111
Yukio Mishima (1925–70), *Kamen no kokuhaku* 248 n.65
Yûkôsha 140

Zheng Xiaoxu (1860–1938) 125
Zimmer, John Todd (1889–1957) 181
Zoological Museum of Moscow University (ZMMU) 208
zoology 18, 109, 129, 198, 219
   as academic discipline in imperial Japan 27–32
   as imperial agent/s 37–40
   offshoot 17
   professionalization of 18, 27, 29
   zoologists 7, 11, 14, 21–2, 30–1, 38, 41, 43, 50, 135, 165, 167–8, 186 (*see also* ornithology, ornithologists)
zoonotic disease vectors 153

www.ingramcontent.com/pod-product-compliance
Lightning Source LLC
Chambersburg PA
CBHW052148300426
44115CB00011B/1570